Medicinal and Aromatic Plants in Agricultural Research When Considering Criteria of Multifunctionality and Sustainability

Medicinal and Aromatic Plants in Agricultural Research When Considering Criteria of Multifunctionality and Sustainability

Editors

Mario Licata
Antonella Maria Maggio
Salvatore La Bella
Teresa Tuttolomondo

MDPI • Basel • Beijing • Wuhan • Barcelona • Belgrade • Manchester • Tokyo • Cluj • Tianjin

Editors
Mario Licata
Università degli Studi di Palermo
Italy

Antonella Maria Maggio
Università degli Studi di Palermo
Italy

Salvatore La Bella
Università degli Studi di Palermo
Italy

Teresa Tuttolomondo
Università degli Studi di Palermo
Italy

Editorial Office
MDPI
St. Alban-Anlage 66
4052 Basel, Switzerland

This is a reprint of articles from the Special Issue published online in the open access journal *Agriculture* (ISSN 2077-0472) (available at: https://www.mdpi.com/journal/agriculture/special_issues/Medicinal_Aromatic_Plants_Agricultural_Research_Multifunctionality_Sustainability).

For citation purposes, cite each article independently as indicated on the article page online and as indicated below:

LastName, A.A.; LastName, B.B.; LastName, C.C. Article Title. *Journal Name* **Year**, *Volume Number*, Page Range.

ISBN 978-3-0365-4013-9 (Hbk)
ISBN 978-3-0365-4014-6 (PDF)

© 2022 by the authors. Articles in this book are Open Access and distributed under the Creative Commons Attribution (CC BY) license, which allows users to download, copy and build upon published articles, as long as the author and publisher are properly credited, which ensures maximum dissemination and a wider impact of our publications.

The book as a whole is distributed by MDPI under the terms and conditions of the Creative Commons license CC BY-NC-ND.

Contents

About the Editors ix

Mario Licata, Antonella Maria Maggio, Salvatore La Bella and Teresa Tuttolomondo
Medicinal and Aromatic Plants in Agricultural Research, When Considering Criteria of Multifunctionality and Sustainability
Reprinted from: *Agriculture* **2022**, *12*, 529, doi:10.3390/agriculture12040529 1

Pietro Catania, Raimondo Gaglio, Santo Orlando, Luca Settanni and Mariangela Vallone
Design and Implementation of a Smart System to Control Aromatic Herb Dehydration Process
Reprinted from: *Agriculture* **2020**, *10*, 332, doi:10.3390/agriculture10080332 5

Montserrat Fernández-Sestelo and José M. Carrillo
Environmental Effects on Yield and Composition of Essential Oil in Wild Populations of Spike Lavender (*Lavandula latifolia* Medik.)
Reprinted from: *Agriculture* **2020**, *10*, 626, doi:10.3390/agriculture10120626 25

Salvatore La Bella, Giuseppe Virga, Nicolò Iacuzzi, Mario Licata, Leo Sabatino, Beppe Benedetto Consentino, Claudio Leto and Teresa Tuttolomondo
Effects of Irrigation, Peat-Alternative Substrate and Plant Habitus on the Morphological and Production Characteristics of Sicilian Rosemary (*Rosmarinus officinalis* L.) Biotypes Grown in Pot
Reprinted from: *Agriculture* **2021**, *11*, 13, doi:10.3390/agriculture11010013 43

Vandana S. Singh, Shashikant C. Dhawale, Faiyaz Shakeel, Md. Faiyazuddin and Sultan Alshehri
Antiarthritic Potential of *Calotropis procera* Leaf Fractions in FCA-Induced Arthritic Rats: Involvement of Cellular Inflammatory Mediators and Other Biomarkers
Reprinted from: *Agriculture* **2021**, *11*, 68, doi:10.3390/agriculture11010068 59

Clarissa Clemente, Luciana G. Angelini, Roberta Ascrizzi and Silvia Tavarini
Stevia rebaudiana (Bertoni) as a Multifunctional and Sustainable Crop for the Mediterranean Climate
Reprinted from: *Agriculture* **2021**, *11*, 123, doi:10.3390/agriculture11020123 75

Salvatore Barreca, Salvatore La Bella, Antonella Maggio, Mario Licata, Silvestre Buscemi, Claudio Leto, Andrea Pace and Teresa Tuttolomondo
Flavouring Extra-Virgin Olive Oil with Aromatic and Medicinal Plants Essential Oils Stabilizes Oleic Acid Composition during Photo-Oxidative Stress
Reprinted from: *Agriculture* **2021**, *11*, 266, doi:10.3390/agriculture11030266 93

Salvatore La Bella, Francesco Rossini, Mario Licata, Giuseppe Virga, Roberto Ruggeri, Nicolò Iacuzzi, Claudio Leto and Teresa Tuttolomondo
Four-Year Study on the Bio-Agronomic Response of Biotypes of *Capparis spinosa* L. on the Island of Linosa (Italy)
Reprinted from: *Agriculture* **2021**, *11*, 327, doi:10.3390/agriculture11040327 107

Silvia Lazzara, Alessandra Carrubba and Edoardo Napoli
Cultivating for the Industry: Cropping Experiences with *Hypericum perforatum* L. in a Mediterranean Environment
Reprinted from: *Agriculture* **2021**, *11*, 446, doi:10.3390/agriculture11050446 125

Francesco Rossini, Giuseppe Virga, Paolo Loreti, Nicolò Iacuzzi, Roberto Ruggeri and Maria Elena Provenzano
Hops (*Humulus lupulus* L.) as a Novel Multipurpose Crop for the Mediterranean Region of Europe: Challenges and Opportunities of Their Cultivation
Reprinted from: *Agriculture* **2021**, *11*, 484, doi:10.3390/agriculture11060484 139

Luciana G. Angelini, Clarissa Clemente and Silvia Tavarin
Pre-Germination Treatments, Temperature, and Light Conditions Improved Seed Germination of *Passiflora incarnata* L.
Reprinted from: *Agriculture* **2021**, *11*, 937, doi:10.3390/agriculture11100937 161

Xue Yang, Yuzheng Li, Chunying Li, Qianqian Li, Bin Qiao, Sen Shi and Chunjian Zhao
Enhancement of Interplanting of *Ficus carica* L. with *Taxus cuspidata* Sieb. et Zucc. on Growth of Two Plants
Reprinted from: *Agriculture* **2021**, *11*, 1276, doi:10.3390/agriculture11121276 171

Fernando Pinto-Morales, Jorge Retamal-Salgado, María Dolores Lopéz, Nelson Zapata, Rosa Vergara-Retamales and Andrés Pinto-Poblete
The Use of Compost Increases Bioactive Compounds and Fruit Yield in Calafate Grown in the Central South of Chile
Reprinted from: *Agriculture* **2022**, *12*, 98, doi:10.3390/agriculture12010098 185

Zeinab Izadi, Abdolhossein Rezaei Nejad and Javier Abadía
Iron Chelate Improves Rooting in Indole-3-Butyric Acid-Treated Rosemary (*Rosmarinus officinalis*) Stem Cuttings
Reprinted from: *Agriculture* **2022**, *12*, 210, doi:10.3390/agriculture12020210 201

About the Editors

Mario Licata is a researcher in Agronomy and Crop Science at the Department of Agricultural, Food and Forest Sciences of the University of Palermo (Italy). He obtained an MSc in Agriculture and a PhD in Environmental Agronomy from the University of Palermo. His key topics of research include the production of medicinal and aromatic plants, the production of industrial crops for energy use, the turfgrass management of sports fields, the use of constructed wetlands for wastewater treatment, and the reuse of wastewater for irrigation purposes. He has published more than 70 scientific papers in national and international journals, peer-reviewed journals, and conference proceedings. He is involved in many research partnerships, and he serves as an Editorial Board Member and peer reviewer for many indexed journals.

Antonella Maria Maggio is an organic chemist. She is an Associate Professor at the STEBICEF Department of the University of Palermo. Her research activity is focused on the field of the chemistry of natural substances, and the history and teaching of chemistry. She is the author of more than 100 scientific papers published in international journals and the proceedings of national and international conferences. She has also published works about the history of chemistry, and is the co-author of a number of chemistry manuals. In 2019, she won the IUPAC Periodic Table Challenge Nobelium Contest with Prof. Palumbo Piccionello.

Salvatore La Bella is a Full Professor in Horticulture at the Department of Agriculture, Food and Forest Sciences of the University of Palermo (Italy). He obtained his PhD in the Productivity of Plant Cultivation at the University of Sassari (Italy). He has published more than 100 scientific papers in national and international journals and conference proceedings. His research activity focuses on the following topics: aromatic and medicinal plants, industrial crops, constructed wetlands for wastewater treatment and reuse for irrigation purposes, turfgrass, and green roofs.

Teresa Tuttolomondo is a Full Professor in Agronomy and Crop Science at the Department of Agriculture, Food and Forest Sciences of the University of Palermo (Italy). She obtained her PhD in Environmental Agronomy at the University of Palermo (Italy). She has published more than 100 scientific papers in national and international journals and conference proceedings. Her research activity focuses on the following topics: aromatic and medicinal plants, industrial crops, constructed wetlands for wastewater treatment and reuse for irrigation purposes, turfgrass, and green roofs.

Editorial

Medicinal and Aromatic Plants in Agricultural Research, When Considering Criteria of Multifunctionality and Sustainability

Mario Licata [1], Antonella Maria Maggio [2], Salvatore La Bella [1,*] and Teresa Tuttolomondo [1]

1. Department of Agricultural, Food and Forest Sciences, Università degli Studi di Palermo, Viale delle Scienze 13, Building 4, 90128 Palermo, Italy; mario.licata@unipa.it (M.L.); teresa.tuttolomondo@unipa.it (T.T.)
2. Department of Biological, Chemical and Pharmaceutical Sciences and Technologies, Università degli Studi di Palermo, Viale delle Scienze, Building 16, 90128 Palermo, Italy; antonella.maggio@unipa.it
* Correspondence: salvatore.labella@unipa.it

Over the last twenty years, agriculture has witnessed significant changes regarding energy requirements, advanced technologies and practices. This is in response to the impacts of crop production on the climate and environment and increasing awareness of the importance of agricultural sustainability through organic farming. Agriculture encompasses complex production systems, and certain aspects of multi-functionality and sustainability have become fundamental to these systems.

Agricultural activity can provide various functions in agro-ecosystems, such as producing food, managing natural resources, and conserving landscape and plant biodiversity, contributing to the cultural, historical and economic viability of rural areas. Agriculture must now adopt scientific innovations to produce food that consider not only human well-being and the environment but also the requirements of farmers. Aromatic and medicinal plants (MAPs), as open-field crops, can play an important role in multifunctional and sustainable agriculture, due to their low energy requirements for cultivation and their many uses, from the production of nutraceuticals, phytonutrients and phytotherapy to land valorization. Various MAPs are used in the food sector to flavor foods or prolong their shelf-life, while others are used in modern and traditional medicine in the production of phytocomplexes for human health and well-being. The cultivation of MAPs, when based on an integrated and sustainable approach, can contribute to the conservation of, and increase in, biodiversity in agro-ecosystems, as well as the recovery of degraded and marginal lands. One of the main aspects that highlights the quality of MAPs is the content and composition of essential oils, which are influenced by several factors, some of which depend on the plant (endogenous or genetic factors) and others on the environment (exogenous or environmental factors and biotic factors), while others concern the collection, preparation and conservation of the plant or the processed products.

On this basis, the main aim of the Special Issue "Medicinal and aromatic plants in agricultural research, when considering multifunctionality and sustainability criteria" is to illustrate the role of MAPs in agriculture under low-impact farming practices and the benefits they can generate in terms of functional products. A total of thirteen papers were published under this Special Issue, including twelve original research papers and one review article. Papers were submitted from six countries: Chile, China, India, Iran, Italy and Spain. The papers cover diverse scientific macro-areas related to MAPs, such as agronomy, chemistry and pharmacy, food and nutrition and ecology and provide new scientific data on natural products obtained from these species. In most papers (9), the authors investigated the effects of agronomic and environmental factors on the morphological, physiological and production characteristics of MAPs. In the remaining papers (4), the authors reported information on the biological activity of some MAPs and their metabolites, explained how MAPs could be used to enhance the quality of food products and provided technological data about the drying process of aromatic herbs.

Citation: Licata, M.; Maggio, A.M.; La Bella, S.; Tuttolomondo, T. Medicinal and Aromatic Plants in Agricultural Research, When Considering Criteria of Multifunctionality and Sustainability. *Agriculture* 2022, *12*, 529. https://doi.org/10.3390/agriculture12040529

Received: 4 April 2022
Accepted: 7 April 2022
Published: 8 April 2022

Publisher's Note: MDPI stays neutral with regard to jurisdictional claims in published maps and institutional affiliations.

Copyright: © 2022 by the authors. Licensee MDPI, Basel, Switzerland. This article is an open access article distributed under the terms and conditions of the Creative Commons Attribution (CC BY) license (https://creativecommons.org/licenses/by/4.0/).

In this Special Issue, La Bella et al. [1] highlighted the effects of irrigation and peat-alternative substrates on the morphological, aesthetic and production characteristics of potted Sicilian rosemary (*Rosmarinus officinalis* L.) biotypes with different habitus types. The authors used four types of substrates with varying percentages of peat and perlite and irrigated the plants, integrating 100% field capacity every four days and every two days. They concluded that the greatest percent content in essential oil was obtained when irrigation events were less frequent, and that the substrates with 20% and 30% compost led to excellent performance results.

In a study carried out in Chile, Pinto-Morales et al. [2] reported the effect of different doses of compost on productive and physiologic parameters, including the polyphenolic composition and antioxidant activity of the fruit of calafate (*Berberis microphylla* G. Forst) grown under an intensive agronomic management. The authors demonstrated that the use of increasing doses of compost was beneficial to the physiological, productive, and quality parameters of the species but, at the same time, generated an increase in organic matter in the soil and the nutritional content of the soil.

In Italy, Angelini et al. [3] investigated different chemical and physical treatments to overpass seed dormancy and enhance the seed germination rates of *Passiflora incarnata* L. Different pre-germination treatments (pre-chilling, gibberellic acid, leaching, and scarification) were examined under different light and temperature conditions. The authors showed that the pre-germination treatments stimulated a faster germination compared to the control, with the best results obtained in the dark and with high temperatures.

In a study conducted in Spain, Fernández-Sestelo and Carrillo [4] estimated the effect of variable climate and fixed factors, such as soil and geographic location, on the essential oil yield and quality of 34 Spanish populations of spike lavender (*Lavandula latifolia* Medik). They found that the composition of the soil influenced the essential oil yield and quality, as well as some climatic and geographical factors such as rain and altitude.

Lazzara at al. [5] assessed the yield and phytochemical composition of *three Hypericum perforatum* biotypes, obtained from different Italian geographical areas, with contrasting cultivation methods, pot and open-field cultivation. The authors highlighted that the cultivation of *Hypericum* required a properly tuned cropping technique, along with a sound choice of the genotype to be cultivated. Furthermore, they stated that pot cultivation did not reflect the performance obtained from open-field cultivation.

Clemente et al. [6] evaluated the agronomic and qualitative performances of nine *Stevia rebaudiana* (Bertoni) genotypes cultivated in open field conditions, under the Mediterranean climate of central Italy. The authors found high variability among genotypes and provided useful information on the influence of crop age and harvest time in defining quanti-qualitative traits in stevia.

In another study, La Bella et al. [7] investigated the agronomic and production behavior of some caper biotypes (*Capparis spinosa* L. subsp. *rupestris*), identified on the island of Linosa (Italy) for growing purposes. This article takes an underused species, such as caper, into consideration, and highlights its agronomic importance in the context of Linosa island, identifying accessions of interest for the introduction of innovation into the new caper field.

In Iran, Izadi et al. [8] studied the propagation of rosemary (*Rosmarinus officinalis* L.) by stem cuttings and found that iron chelate application promotes root emergence and improves root and shoot biomass, leaf photosynthetic pigment concentrations and survival percentage. In China, Yang et al. [9] demonstrated that interplanting *Ficus carica* L. with *Taxus cuspidata* Sieb. increased the plant growth biomass, photosynthesis, soil organic carbon, total nitrogen, and secondary metabolites, such as psoralen and paclitaxel, with respect to monocultures. The authors stated that these results could provide a feasible theoretical basis for the large-scale establishment of *Ficus carica* and *Taxus cuspidata* mixed forests and obtain high-quality medicine sources for extracting psoralen and paclitaxel. In a study carried out in India, Singh et al. [10] explored the antiarthritic potential of different fractions of Swallow wort (*Calotropis procera* Aiton) for the evaluation of antiarthritic potential using Freund's complete adjuvant model on wistar rats, as no such study has been carried out to date.

Regarding the impact of essential oil on the qualitative properties of food products, Barreca et al. [11] added different essential oils of Sicilian accessions of common sage (*Salvia officinalis* L.), oregano (*Origanum vulgare* L. ssp. *hirtum* (Link) Ietswaart), rosemary (*Rosmarinus officinalis* L.) and thyme (*Thymbra capitata* (L.) Cav.) to improve both the food shelf-life and aromatic flavour of extra-virgin olive oil. The results of this original study showed that no significant change in oleic acid percentage was detected in the mixture of extra-virgin olive oils with essential oil samples but seemed to highlight the presence of an antioxidant effect of essential oils of MAPs on extra-virgin olive oil.

Considering the technological aspects of MAPs, Catania et al. [12] designed a low-cost, real-time monitoring and control system for the drying process of sage (*Salvia officinalis* L.) and laurel (*Laurus nobilis* L.), and assessed drying efficacy in the microbial community associated with the studied MAPs. In particular, the authors found that the two species showed a different microbial stability with the adopted drying method and had a different shelf life.

In the only review included in this Special Issue, Rossini et al. [13] reported in-depth information on the cultivation, quality aspects, sustainable production and uses of hops (*Humulus lupulus* L) in the Mediterranean area.

The thirteen papers in this Special Issue of "Medicinal and aromatic plants in agricultural research, when considering multifunctionality and sustainability criteria" represent an excellent contribution to scientific research on MAPs. More than one author contributed several papers to this Special Issue, exploring various research fields regarding MAPs. We believe that the data provided by all published papers can greatly improve the knowledge of MAPs and prove useful for researchers, technicians and students.

Author Contributions: M.L., A.M.M., S.L.B. and T.T. made equal contributions to this article. All authors have read and agreed to the published version of the manuscript.

Acknowledgments: We would like to thank all the authors for submitting their manuscripts to this Special Issue and the editors and reviewers for their contribution. Furthermore, we are also grateful to the handling editors and staff of *Agriculture* for their support during the preparation and finalization of this Special Issue.

Conflicts of Interest: The authors declare no conflict of interest.

References

1. La Bella, S.; Virga, G.; Iacuzzi, N.; Licata, M.; Sabatino, L.; Consentino, B.B.; Leto, C.; Tuttolomondo, T. Effects of irrigation, peat-alternative substrate and plant habitus on the morphological and production characteristics of Sicilian rosemary (*Rosmarinus officinalis* L.) biotypes grown in pot. *Agriculture* **2021**, *11*, 13. [CrossRef]
2. Pinto-Morales, F.; Retamal-Salgado, J.; Lopéz, M.D.; Zapata, N.; Vergara-Retamales, R.; Pinto-Poblete, A. The use of compost increases bioactive compounds and fruit. *Agriculture* **2022**, *12*, 98. [CrossRef]
3. Angelini, L.G.; Clemente, C.; Tavarini, S. Pre-germination treatments, temperature, and light conditions improved seed germination of *Passiflora incarnata* L. *Agriculture* **2021**, *11*, 937. [CrossRef]
4. Fernández-Sestelo, M.; Carrillo, J.M. Environmental effects on yield and composition of essential oil in wild populations of spike lavender (*Lavandula latifolia* Medik.). *Agriculture* **2020**, *10*, 626. [CrossRef]
5. Lazzara, S.; Carrubba, A.; Napoli, E. Cultivating for the industry: Cropping experiences with *Hypericum perforatum* L. in a Mediterranean environment. *Agriculture* **2021**, *11*, 446. [CrossRef]
6. Clemente, C.; Angelini, L.G.; Ascrizzi, R.; Tavarini, S. *Stevia rebaudiana* (Bertoni) as a multifunctional and sustainable crop for the Mediterranean climate. *Agriculture* **2021**, *11*, 123. [CrossRef]
7. La Bella, S.; Rossini, F.; Licata, M.; Virga, G.; Ruggeri, R.; Iacuzzi, N.; Leto, C.; Tuttolomondo, T. Four-year study on the bio-agronomic response of biotypes of *Capparis spinosa* L. on the island of Linosa (Italy). *Agriculture* **2021**, *11*, 327. [CrossRef]
8. Izardi, Z.; Rezaei Nejad, A.; Abadía, J. Iron chelate improves rooting in indole-3-butyric acid-treated rosemary (*Rosmarinus officinalis*) stem cuttings. *Agriculture* **2022**, *12*, 210. [CrossRef]
9. Yang, X.; Li, Y.; Li, C.; Li, Q.; Qiao, B.; Shi, S.; Zhao, C. Enhancement of interplanting of *Ficus carica* L. with *Taxus cuspidata* Sieb. et Zucc. on growth of two plants. *Agriculture* **2021**, *11*, 1276. [CrossRef]

10. Singh, V.S.; Dhawale, S.C.; Shakeel, F.; Faiyazuddin, M.; Alshehri, A. Antiarthritic potential of *Calotropis procera* leaf fractions in FCA-induced arthritic rats: Involvement of cellular inflammatory mediators and other biomarkers yield in calafate grown in the central south of Chile. *Agriculture* **2021**, *11*, 68. [CrossRef]
11. Barreca, S.; La Bella, S.; Maggio, A.; Licata, M.; Buscemi, S.; Leto, C.; Pace, A.; Tuttolomondo, T. Flavouring extra-virgin olive oil with aromatic and medicinal plants essential oils stabilizes oleic acid composition during photo-oxidative stress. *Agriculture* **2021**, *11*, 266. [CrossRef]
12. Catania, P.; Gaglio, R.; Orlando, S.; Settanni, L.; Vallone, M. Design and implementation of a smart system to control aromatic herb dehydration process. *Agriculture* **2020**, *10*, 332. [CrossRef]
13. Rossini, F.; Virga, G.; Loreti, P.; Iacuzzi, N.; Ruggeri, R.; Provenzano, M.E. Hops (*Humulus lupulus* L.) as a novel multipurpose crop for the Mediterranean region of Europe: Challenges and opportunities of their cultivation. *Agriculture* **2020**, *11*, 484. [CrossRef]

Article

Design and Implementation of a Smart System to Control Aromatic Herb Dehydration Process

Pietro Catania, Raimondo Gaglio, Santo Orlando *, Luca Settanni and Mariangela Vallone

Department of Agricultural, Food and Forest Sciences (SAAF), University of Palermo, viale delle Scienze ed. 4, 90128 Palermo, Italy; pietro.catania@unipa.it (P.C.); raimondo.gaglio@unipa.it (R.G.); luca.settanni@unipa.it (L.S.); mariangela.vallone@unipa.it (M.V.)
* Correspondence: santo.orlando@unipa.it

Received: 18 June 2020; Accepted: 3 August 2020; Published: 5 August 2020

Abstract: Drying is a process aimed at reducing the water content in plant materials below a limit where the activity of microbes and decomposing enzymes deteriorate the quality of medicinal and aromatic plants. Today, the interest of consumers towards medicinal and aromatic herbs has registered a growing trend. This study aims at designing a low-cost real-time monitoring and control system for the drying process of aromatic herbs and evaluating drying efficacy on the microbial community associated with the studied herbs. Hot-air drying tests of sage and laurel leaves were carried out in a dryer desiccator cabinet at 40 °C and 25% relative humidity using three biomass densities (3, 4 and 5 kg/m^2). The prototype of the smart system is based on an Arduino Mega 2560 board, to which nine Siemens 7MH5102-1PD00 load cells and a DHT22 temperature and humidity sensor were added. The data acquired by the sensors were transmitted through Wi-Fi to a ThingSpeak account in order to monitor the drying process in real time. The variation in the moisture content of the product and the drying rate were obtained. The system provided a valid support decision during the drying process, allowing for the precise monitoring of the evolution of the biomass moisture loss and drying rate for laurel and sage. The three different biomass densities employed did not provide significant differences in the drying process for sage. Statistically significant differences among the three tests were found for laurel in the final part of the process. The microbial loads of the aromatic herbs after drying were influenced by the different leaf structures of the species; in particular, with laurel leaves, microbial survival increased with increasing biomass density. Finally, with the drying method adopted, the two species under consideration showed a different microbial stability and, consequently, had a different shelf life, longer for sage than laurel, as also confirmed by water activity (a_w) values.

Keywords: laurel; microbial load; oven drying; real time monitoring; sage; sensor

1. Introduction

In the last few years, the interest of consumers towards aromatic and medicinal herbs has registered a growing trend both in terms of product types and consumption [1]. In the past, these plants mainly concerned the derivatives and ingredients industries, while today we are witnessing a growing use in different sectors, such as functional food (nutraceutics) or infusion drinks (herbal teas) and bio-ecological cosmetics [2].

Plants from historical times are applied for fitness in addition to the supply of drug treatments; regarding the fact that 80–85% of the world-wide population depends on ancient medicines [3–5]. Furthermore, since ancient times, aromatic plants have been used in food preparation, not only to ameliorate the taste and flavor of the final products, but also as preservatives, due to their antimicrobial properties [6,7].

At least 2000 species of medicinal and aromatic plants are marketed in Europe. Italy has a contributes to 3% of overall European production. Italian imports are around 161 thousand tons with an outlay of around USD 1.14 billion. The Italian production of medicinal plants satisfies only 30% of the national needs, the remaining 70% of the herbs consumed in our country come from abroad; in particular, from countries of Eastern Europe and North Africa, but their cultivation is certainly expanding [8].

The cultivation of aromatic plants has many similarities with the cultivation of horticultural species in the open field and, therefore, a fairly intensive cultivation system, which provides for planting, frequent cultivation care and one or more harvests during the development cycle of the crop in the production season. The plants are harvested in the green field, in bloom. When the parts mature, except for a few and rare cases, they are a product that contain variable percentages of vegetation water and, therefore, cannot be conserved and transported, except in a short range. Harvesting can be done by hand or by machine, depending on the type of crop and the characteristics of the farm.

Once harvested, the biomass has a short life, from a few hours to a maximum of half a day. It needs immediate processing or stabilization. After harvesting, primary processes for the product to be marketed are required. These processes are specific to the type of farm and production chain—that is, each company definitively chooses its own basic production orientation: dried, essential oils, or fresh [9].

Drying is a process that consists in reducing the water content of the product below a value that limits the microbiological and enzymatic reactions responsible for its deterioration [10]. In addition to traditional methods, plants can be dried by automated methods with stoves or dryers.

Artificial drying is necessary in industrial production. It allows for drying a large quantity of material in a short time (24–48 h), always using low temperatures below 50 °C with an optimum of 40 °C [11–13]. It is a very expensive system both from an energy and an economic point of view, also affecting up to 50% of the production cost of the dry plant [14–17].

During the drying process, the aromatic herbs are subjected to chemical and physical changes that influence the quality of the finished product. The extent of these changes mainly depends on the drying conditions and the biological characteristics of the herbs [18].

The principle is to dry the free waters with a forced flow of dry air, passed through the biomass, spread over large surfaces on one or more layers. Hot-air drying, using convective ovens, is a fundamental technology for the postharvest preservation of aromatic and medicinal plants in non-tropical countries, since it allows a fully controlled operation, resulting in a desired end product. The most essential parameter influencing the end product quality of dried herbs is the temperature used [19].

In general, hot-air drying can lead to a drastic reduction in the quality of the end product compared to the original foodstuff [20]; principally high temperature causes the decomposition of bioactive ingredients and changes in colorful components [21,22]. On the other hand, drying at 35–50 °C allows for the preservation of heat-sensitive compounds [23,24]. In addition, high temperature has been used successfully to decrease the concentration of toxic compounds or reduce adverse taste characteristics in dried herbs. The main disadvantages of hot-air drying are the excessive shrinkage that is sometimes observed, color changes and overall product collapsing.

The consulted literature sources show that the optimal temperature to be used in the dehydration processes of aromatic herbs through the use of oven-drying is in the range of 40–45 °C. In fact, Diaz-Maroto et al. [11] confirm that by drying laurel leaves at 45 °C, the losses of volatile compounds are negligible compared to air drying, and Hadjibagher Kandi and Sefidkon [12] stated that, to obtain the highest yield value in essential oil from laurel leaves, the best drying method is oven-drying at 40 °C. Sadowska et al. [13] found that, by drying sage at 40 °C, polyphenols content and antioxidant activity did not show statistically significant differences compared to the values obtained with the naturally dry method.

From these studies, it appears that temperature control during the drying process is of fundamental importance for the quality of the final product. In fact, in recent years numerous studies have been carried out on the application of modern systems to control and monitor the main environmental parameters [25–27].

The aim of this study was the design of a low-cost real-time monitoring and control system for the drying process of aromatic herbs inside a dryer desiccator cabinet. During the process and at the end, microbiological analyses were carried out on biomass samples to evaluate the stability achieved in the dried product.

2. Materials and Methods

2.1. Plant Material

The texts were carried out in June 2019. The biomass was harvested from plants of the campus area of the Department of Agricultural, Food and Forest Sciences, University of Palermo, Italy (N 38°06′28″, E 13°21′00″, 31 m asl), where two experimental plots were selected (20 m × 20 m each). The climate of the area is Mediterranean with mild and humid winters and hot and dry summers; the climate is classified as Csa, according to Köppen and Geiger (hot-summer Mediterranean climate). The average temperature is 18.4 °C and the average annual rainfall is 605 mm. The soils have a sandy clay texture (Aric Regosol, 54% sand, 23% silt and 23% clay) with a pH of 7.6, 14 g kg^{-1} organic matter, 3.70% active limestone [28].

The sage (*Salvia officinalis* L.) was taken from 5 year-old plants, with a shrubby creeping habitus in a planting layout of 1 × 2 m; the apical leaves were collected using scissors.

The laurel (*Laurus nobilis* L.) plants had a shrub habitus, were 1 m high, 5 years-old, and had a planting lay out of 1 × 2 m. Using pruning shears, 1–2 year-old branches with leaves of about 0.30–0.40 m long, were taken.

The collected material was immediately transported to the laboratory, placed in polyethylene vented crates and put inside the drying chamber after about 1 h from harvest.

2.2. Drying

Hot-air drying tests of sage and laurel leaves were carried out in a dryer desiccator cabinet (KW Apparecchi Scientifici s.r.l, Italy). The cabinet is entirely made of AISI 304 stainless steel, it has 95 × 60 × 150 cm internal dimensions, equipped with two split opening doors with sealing. It is 1250 W powered and the temperature ranges from +5 °C above ambient temperature to 130 °C, controlled by means of a thermostat. The dryer desiccator cabinet is also equipped with a ventilation system to control relative humidity. Heating is achieved with specific air heaters with ascending flow to facilitate the evacuation of vapors. The steam discharge inside the chamber occurs by means of a centrifugal electro-aspirator with 40 m^3/h flow rate, placed in the upper part of the cabinet.

Three biomass densities (kg/m^2) were used for each species. The tests, named 1, 2 and 3, respectively, with biomass densities of 3, 4 and 5 kg/m^2, were placed inside the shelves of the cabinet using polypropylene vented crates (dimensions 40 × 60 × 22 cm, 52 L capacity, 62% hollow surface). Three replicates for each test, named A, B and C, were randomly arranged inside the drying chamber (Figure 1). This was possible because the drying chamber was equipped with an automatic heating and ventilation system that guarantees to maintain both constant temperature and relative humidity inside, and because the contact surface between the biomass and the air inside the chamber was the same in the different tests.

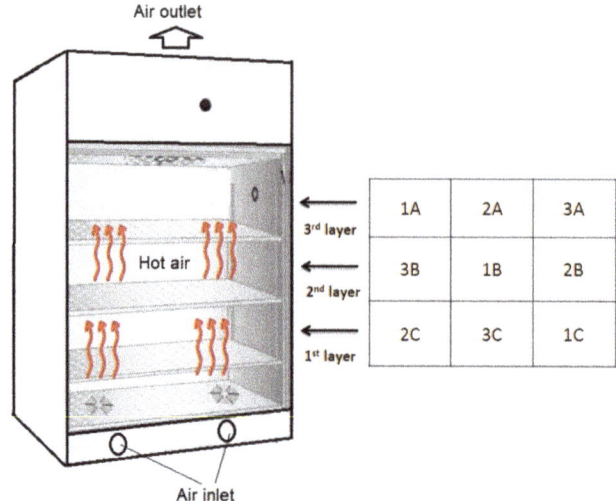

Figure 1. Scheme of the drying chamber and test repetitions.

After preheating the dryer desiccator cabinet, the drying temperature was maintained constant at 40 °C since a higher temperature may result in a darkening of the samples due to non-enzymatic browning and leading to a loss of quality. Relative humidity was kept at 25% for the entire duration of the process [18].

Overall, about 30 kg of leaves were taken to carry out the tests, 15 kg of sage and 15 kg of laurel. A sample of leaves was brought to the laboratory for the determination of the dry weight of each species. Dry weight was determined by weighing 10 g of herb samples and drying until a constant weight at 105 °C in the drying oven (Termaks TS 800, Norway). The initial moisture content of sage was found to be 73.2 ± 0.6% (wet basis); for laurel it was found to be 55.3 ± 0.5% (wet basis). Water activity (a_w) was measured to evaluate microbiological stability using a Rotronic Hygropalm HC2-AW (Rotronic AG, Bassersdorf, Switzerland), both on fresh and dried herbs (i.e., before and after dehydration).

The variation in the moisture content of the product during drying was experimentally determined through the evaluation of the following parameters.

The moisture content at time t MC_t [%] was calculated as:

$$MC_t = \frac{W_{bt} - W_{red\, H_2Ot}}{MC_i} \tag{1}$$

where: W_{bt} is the normalized biomass weight [g] at time t [h]; $W_{red\, H_2Ot}$ is the reduced water weight [g] at time t [h]; MC_i is the initial biomass moisture content [g].

The drying rate at time t, DR_t [g/g h^{-1}] was then calculated as:

$$DR_t = \frac{W_{red\, H_2O\, t}}{W_{b\, t}} t^{-1} \tag{2}$$

The experimentally obtained curves for MC_t and DR_t are the variation during drying of the moisture content of the product and the drying rate as a function of time.

The end of drying was defined when the biomass weight remained stable for more than 10 min and the moisture content was lower than 14% for sage and 32% for laurel.

2.3. Smart Sensors System Structure

The system is based on sensors capable of real time monitoring the biomass weight of the individual samples and temperature and relative humidity inside the drying chamber.

The choice of the components to be used in the realization of the prototype of the biomass drying measuring system was based on low-cost, reliable and small-sized components, which can be assembled inside the cabinet with the plastic boxes used to hold the sage and laurel samples. The system is based on an Arduino Mega 2560 board which, thanks to the large availability of inputs/outputs, allowed for the connection of (Figure 2):

- Five HX711 modules that power, amplify and convert to digital signal of the nine load cells;
- A DS3231 Real Time Clock (RTC) module that provides the system with the year, month, day, hours, minutes and seconds time reference;
- An ESP8266-05 module to transmit, through Wi-Fi, the data acquired by the various sensors to a ThingSpeak account every 15 min;
- A DHT22 sensor that allows for the detection of relative humidity and ambient temperature and to digitally transmit it to a micro-controller;
- Nine Siemens 7MH5102-1PD00 load cells capable of measuring static loads with medium precision with a maximum capacity of 5 kg;
- A Serial OpenLog module that allows the acquired data to be stored on a microSD card in the form of appropriately structured strings, so as to contain the data of the individual sensors in the various fields (e.g., Data/Time; Sensor 1; Sensor 2; … … ; Sensor N), and transmitted from Arduino Mega serial output 2. This represents a security system that, even in the absence of a Wi-Fi connection, allows data recording from the sensors.

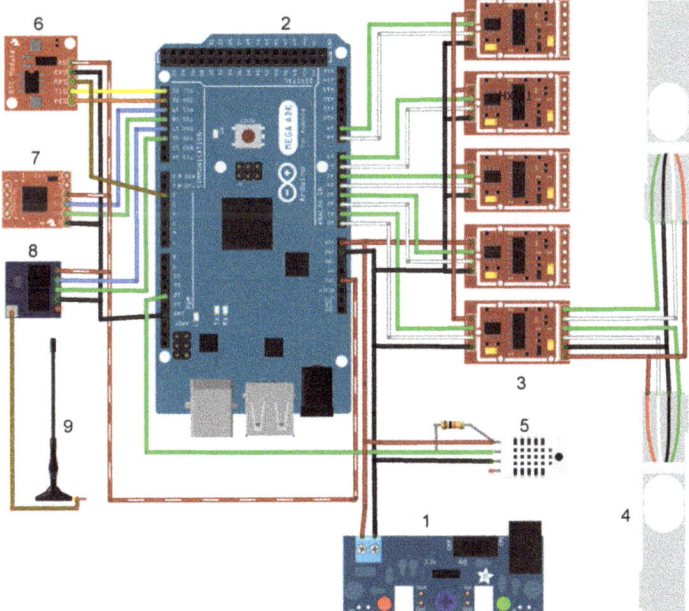

Figure 2. Smart sensors system structure. **1**: Power supply; **2**: Arduino Mega 2560 R3; **3**: Load Cell Amplifier HX711; **4**: Load Cell; **5**: Digital temperature and humidity sensor DHT22; **6**: Real-time clock DS3231; **7**: OpenLog Data Logger; **8**: Transceiver Wireless Module ESP8266-05; **9**: External antenna.

Arduino Mega 2560 R3 is a module based on the ATmega2560 microcontroller, which can be programmed using the Arduino IDE through the USB interface, which also acts as a power connection.

This card provides developers of simple embedded systems with a fast prototyping tool, thanks to the discrete calculation power and to the possibility of being interfaced with other cards or sensors through the different inputs and outputs available (54 I/O Digital Pin and 16 Analog Inputs Pin), allowing for the rapid creation of relatively complex systems. In fact, the Arduino Mega choice as a micro-controller was dictated by the need to use the numerous I/O pins available to simultaneously acquire the weight of the different medicinal herb samples placed in the cabinet, as well as temperature and relative humidity.

The specifications of the components used are:

- A/D converter module: Based on the HX711 chip, it is a 24-bit Analog-to-Digital Converter (ADC) for weighing scales specifically designed to interface the load cells, which exploit the Wheatstone bridge, to a microcontroller, such as Arduino or another that is compatible. It is also intended for process control applications. The module has two differentiated input channels, selectable through a multiplexer: a programmable and low noise gain amplifier allowing for the selection of the gain at 32, 64 and 128; a load cell on-chip power regulator; an analog ADC power supply; an on-chip oscillator that requires no external component intervention with external crystal. Since the output signal from a load cell is an analog signal proportional to the deformation of the cell, and is therefore proportional to the applied force, the signal is acquired through the analog to digital converter which reads the voltage difference in the Wheatstone bridge over the cells and then converts it into a 24-bit digital string, which can be acquired by a microcontroller (Arduino) via serial communication. The data are sampled at a frequency of 80 Hz.
- 2C real time clock module (RTC): Based on the DS3231 chip, it is a low consumption clock/calendar that allows the Arduino to have a time reference to assign to the various analog samples acquired. Communication with the Arduino board takes place on the I2C serial bus, which is accessed in SLAVE mode. The time count is based on a 32 kHz quartz oscillator, which ensures good accuracy.
- SP8266-05 module with external antenna: This is a Wi-Fi integrated SoC (system-on-a-chip) with a 32 bit LX106 Micro Controller Unit (MCU). It is a system suitable for providing Wi-Fi connectivity to an Arduino-like card but it is possible to use it to directly create IoT (Internet of Things) projects. As a simple Wi-Fi module, it is possible to manage it via AT commands and it is programmable with the Arduino IDE. It supports 802.11 b/g/n transmission protocols and Wi-Fi Direct (P2P) and soft-AP modes. It integrates the TR switch, the RF amplifier and the antenna, a PLL (power regulator).
- The DHT22 is a basic digital temperature and humidity sensor. It uses a capacitive humidity sensor and a thermistor to measure the surrounding air, and gives out a digital signal on the data pin, while no analog input pins are needed and the specifications are: 3 to 5 V power and I/O, 2.5 mA max current use during conversion (while requesting data), 0–100% relative humidity range with 2–5% accuracy, −40 to 80 °C temperature range ±0.5 °C accuracy, 0.5 Hz maximum sampling rate (once every 2 s), dimensions 15.1 mm × 25 mm × 7.7 mm, 4 pins with 0.1" spacing.
- The single point, medium precision (class III) load cell—Siemens 7MH5102-1PD00—is designed for use in platform scales equipped with a single load cell. It is very easy to use and apply in a wide variety of applications, where the center of force acting is 62.5 mm from the vertical axis of the load cell. It offers good performance and very small dimensions (25 × 40 × 150 mm). The cell is made of aluminum and the recommended excitation voltage is between 5 and 12 V DC, while the output at nominal load is 2 mV/V. It has an IP65 protection degree and allows 4 or 6 wire connections.
- Openlog module: A simple serial datalogger, based on an ATmega328 microprocessor clocked at 16 MHz, capable of handling large capacity microSD cards (up to 64 GB) with FAT16 and FAT32. The main connection interface is FTDI type, with a configurable baud rate (up to 115,200 bps); 4 pins are also available for the SPI interface.

All the electronic components are assembled inside a box from which only the power cables, connecting to the load cells and the external antenna, come out. Each load cell is fixed on one side on a stable steel support base, on the other side of a metal weighing plate. The whole system takes on a Z shape. Nine of them were realized and placed under each polypropylene vented crate containing the biomass samples inside the drying chamber.

2.4. System Calibration and Data Processing

For each individual cell, connected to its HX711 module, calibration was performed through 4 load conditions (0, 1.5, 3 and 5 kg), which allowed us to determine the offset and compensation values of each individual sensor. Considering that the system had to work inside the dryer desiccator cabinet ($\approx$40 °C), the response of the sensor to three different temperature conditions (10, 25, 40 °C) was also verified to assess the possible need for thermal compensation.

Using the Arduino IDE, a first calibration sketch was developed, which allowed us to determine the calibration factors subsequently inserted in the operating sketch. The acquisition of the values of weight, temperature and relative humidity was expected every 15 min, as the respective average of a sample of 20 measurements, as well as the login parameters to the Wi-Fi network and the API keys for writing the data in the various ThingSpeak channels, were set up.

ThingSpeak is a cloud platform provided by Mathworks intended for IoT applications, capable of collecting, displaying and analyzing data from sensors in real time. In addition to allowing instantaneous displays of data from IoT devices, the platform can run MATLAB code on these data, for advanced "live" analysis and possibly for sending alerts.

After the complete assembly and the loading of the operating sketch, some tests were carried out to verify the correct functioning of the sensors, the transmission system (Wi-Fi and ThingSpeak) and data storage under operating conditions. To this end, the system and sensors were placed in the dryer desiccator cabinet and measurement tests were started. Metal loads were applied to the weighing plates and heating was started until the temperature reached 45 °C was reached, keeping it constant for 48 h. From the analysis of the data collected during these preliminary tests, no significant errors in the measurements emerged, both during the heating transient (22–45 °C) and during the steady state.

2.5. Microbiological Analysis

The microbial loads of sage and laurel were determined before and after each drying process. Ten grams of each herb sample were suspended in 90 mL of Ringer's solution (Sigma-Aldrich, Milan, Italy) and subjected to homogenization by means of the stomacher BagMixer® 400 (Interscience, Saint Nom, France) for 2 min at the highest speed. The decimal serial dilutions of the cell suspensions were plated and incubated as follows: total mesophilic count (TMC) were spread plated on plate count agar (PCA), incubated aerobically at 30 °C for 72 h; members of the *Enterobacteriaceae* family were pour plated on double-layer violet red bile glucose agar (VRBGA), incubated aerobically at 37 °C for 24 h; total yeast spread plated on yeast extract peptone dextrose (YPD) nutrient agar incubated at 28 °C for 48 h; molds spread plated on malt agar (MA), incubated aerobically at 25 °C for 7 days. To inhibit the growth of bacteria, chloramphenicol (0.05 mg/mL) was added to YPD and MA. Microbiological counts were performed in triplicate.

2.6. Statistical Analysis

The simple regression and the polynomial regression procedures were used to construct statistical models describing the impact of a single quantitative factor—that is time (t) in our study—on a dependent variable as the parameters moisture content (MC) and drying rate (DR) expressed through (1) and (2). A linear model involving t or a polynomial model involving t and powers of t were considered appropriate to fit to the data. In addition, the analysis of variance (ANOVA) was considered in order to test if there were differences in the three biomass densities (3, 4 and 5 kg/m^2) in MC during time (a value every 6 h throughout the process was considered, named t_0, t_6, etc.). To test if there were statistical differences, Tukey's test was performed. Differences were considered significant at 5% level of significance.

Microbiological data were subjected to the one-way analysis of variance (ANOVA), too. The comparison of treatment means was achieved by Tukey's test; differences were considered significant at 5% level of significance.

All the analyses were performed with the statistical software package Statgraphics centurion, version XV (Statpoint Inc., The Plains, VA, USA, 2005).

3. Results and Discussion

3.1. Dehydration Process

The data continuously acquired by the drying process monitoring system are displayed through the ThingSpeak platform, which provides diagrams of the type shown in Figure 3, referring to a 24-h time span for three tests performed on the sage samples, in which the progressive weight decrease during drying is observed.

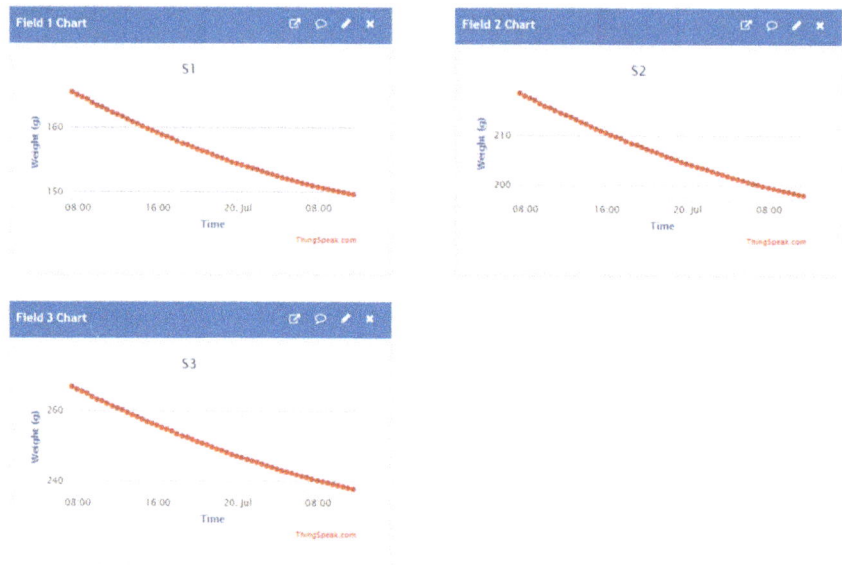

Figure 3. Example of real time data acquisition through ThingSpeak during the sage dehydration process for the three tests conditions (S1 = 3 kg/m^2; S2 = 4 kg/m^2 and S3 = 5 kg/m^2).

The a_w values of fresh and dried herbs are reported in Table 1. The dehydration process consistently reduced the availability of water for the microbial development in laurel and, especially, sage. Fresh herbs were characterized by a_w 0.993 and 0.968 (sage and laurel, respectively), which reduced during the drying process inversely to the biomass density with the lowest value (0.335) shown by the trial S1 involving sage. The levels of a_w after the drying of laurel were significantly higher (0.766 on average) than those showed by the corresponding biomass densities of sage, and this is undoubtedly imputable to the different leaf structure of the two aromatic plants.

Table 1. Water activity (a_w) values before and after dehydration.

Sage		Laurel	
Test	a_w	Test	a_w
Fresh	0.993 ± 0.001	Fresh	0.968 ± 0.001
S1	0.335 ± 0.002	L1	0.679 ± 0.001
S2	0.339 ± 0.001	L2	0.787 ± 0.003
S3	0.377 ± 0.003	L3	0.831 ± 0.002

The curves of the three biomass densities show a very similar trend (Figure 4A). The whole process has a total duration of 65 h. The starting MC mean value is 73% and the final mean value is 13%. In the first 14 h of the process, the curves relating to the three tests are overlapping. After 32 h, halfway through the process, MC decreases about 59% of the total. These results are in agreement with those obtained in [29].

Figure 4 shows the progress of the sage dehydration process for the three biomass densities (S1 = 3 kg/m²; S2 = 4 kg/m² and S3 = 5 kg/m²).

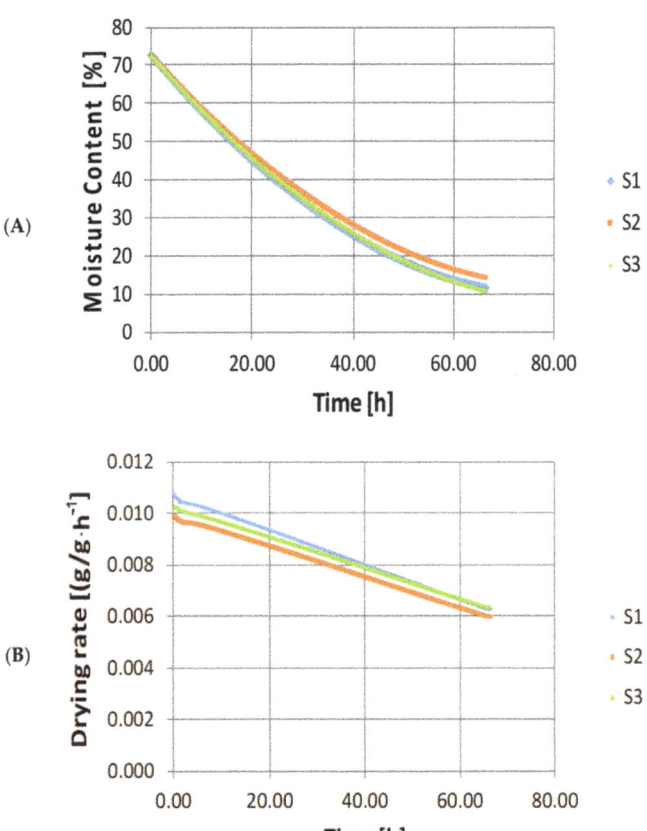

Figure 4. Curves of moisture content MC_t (**A**) and drying rate DR_t (**B**) as a function of time for sage dehydration process. Data are means of three replications (S1 = 3 kg/m²; S2 = 4 kg/m² and S3 = 5 kg/m²).

The MC value of dried sage at the end of the process represents the bound water and is reflected in the water activity value obtained after dehydration in the three tests.

As for the DR values obtained (Figure 4B), a falling drying rate trend can be observed, due to the difficulty of the capillary forces in transferring water from inside the tissues to the surface of the plant. DR starts from a value of 0.01 g/g h^{-1}; then it constantly decreases, reaching the value of 0.006 g/g h^{-1} at the end of the process.

Regarding the laurel dehydration process, MC continuously decreases with the drying time (Figure 5A). The drying of laurel leaves takes more than 40 h. Test L3 ends the dehydration process with a moisture content below 30%, while tests L1 and L2 show an MC value higher than 30%. In the first 10 h of the process, the curves related to the three tests are nearly coincident. After 20 h, in the middle of the process, MC decreases by about 60% of the total value. The results obtained from this study are in agreement with those obtained in [30]. DR (Figure 5B) shows a sudden decrease in the first 3 h of the process, going from 0.02 to 0.01 g/g h^{-1}. Then, the curves take on a horizontal trend up to half of the drying process. Starting from the 20th hour, in fact, there is a DR reduction up to the final value of 0.006 g/g h^{-1}.

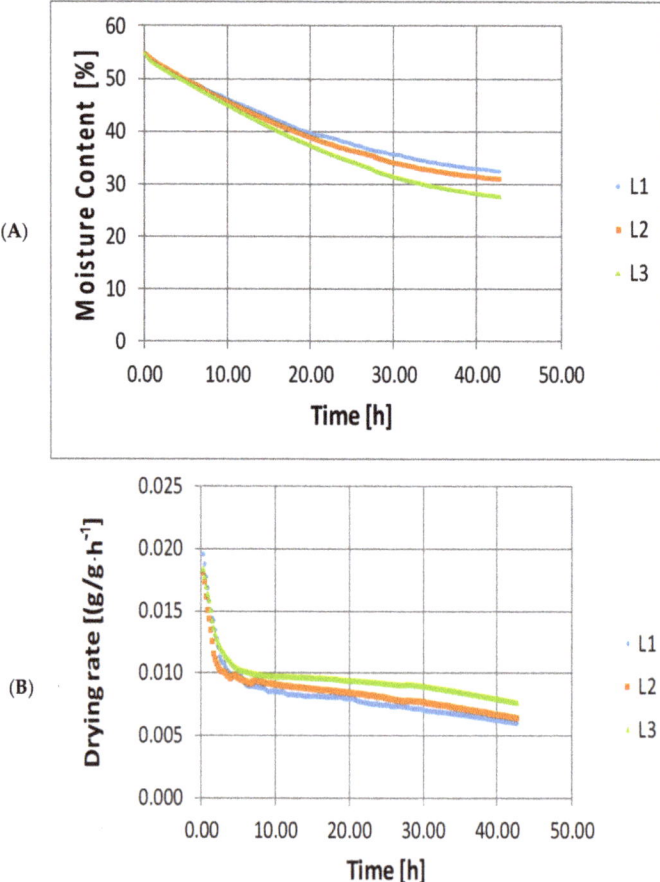

Figure 5. Curves of moisture content MC_t (**A**) and drying rate DR_t (**B**) as a function of time for laurel dehydration process. Data are means of three replications (L1 = 3 kg/m^2, L2 = 4 kg/m^2 and L3 = 5 kg/m^2).

The MC value of present dried laurel at the end of the process is reflected in the water activity obtained after dehydration, which was equal to 0.766 on average.

The DR curves both for sage and laurel are representative of the drying kinetics of the two species under the test conditions applied. The difference in the drying kinetics found in the two species is due to the different structural characteristics of the biomass processed. Sage leaves have a lower vegetative thickness than laurel leaves, that are also leatherier.

The laurel drying process was stopped after 42 h from the start, since the smart system detected, for the three tests, MC values on average equal to 30%. This is in accordance with the mean water activity value of the laurel obtained after dehydration (0.766). Considering that, in the literature, there are no sources regarding laurel water activity, the only explanation that can be provided is related to the particular physical–mechanical characteristics of the leaf of this species compared to those of other aromatic herbs. The laurel leaf, in fact, is very leathery and the internal cells offer great resistance to the release of water. Therefore, it would have been necessary to increase the process temperature to reach MC values similar to those obtained in sage. This choice was not made, since it would have been in contrast with many studies present in the literature. Sellami et al. [31], in fact, showed that increasing drying temperature resulted in a significant decrease in the concentration of most volatiles. Additionally, Hadjibagher Kandi et al. [12] state the importance of keeping temperature below 45 °C to limit the essential oil content losses.

As for sage, on the other hand, the dehydration process ended after 66 h, therefore requiring a drying time higher than 30% compared to laurel. Furthermore, also in this case, the process was stopped by the smart system when the detected MC variations were negligible. This made it possible to reach an average final MC value of 13%. This corresponds to the water activity value obtained after dehydration (0.350). The process time and temperature used for sage are in agreement with [13]; they dehydrated sage using temperatures of 40 °C for 77 h and the drying process was conducted until achieving 10% of humidity of drying material. As a consequence, the two species have a different "expiration dates", laurel's being shorter than that of sage.

Regarding the polynomial regression analysis applied to the MC values for sage in the three tests, since the p-value in Table 2 is less than 0.05, there is a statistically significant relationship between MC and t at the 95% confidence level. The R-squared statistic indicates that the model, as fitted, explains the reported percentage of the variability in MC. The order of the polynomial is appropriate as the p-value on the highest order term of the polynomial equals 0.0.

Table 2. Polynomial regression analysis results for moisture content MC as a function of time for sage dehydration process in the three tests (S1 = 3 kg/m^2; S2 = 4 kg/m^2 and S3 = 5 kg/m^2).

Parameter	Estimate	Standard Error	T Statistic	p-Value
constant	73.0722	0.01008	7246.57	0.0000
t	−1.57941	0.00070	−2254.84	0.0000
t^2	0.00992658	0.00001	973.36	0.0000

Source	Sum of Squares	Df	Mean Square	F-Ratio	p-Value
Model	107,623	2	53,811.3	14,211,543.51	0.0000
Residual	1.24953	330	0.00378645		
Total	107,624	332			
R-squared	99.9988%				
MC equation for S1:		MC_S1 = 73.0722 − 1.57941·t + 0.00992658·t^2			

Parameter	Estimate	Standard Error	T Statistic	p-Value
constant	73.0656	0.01029	7101.54	0.0000
t	−1.4671	0.00071	−2052.77	0.0000
t^2	0.00885007	0.00001	850.511	0.0000

Table 2. Cont.

Source	Sum of Squares	Df	Mean Square	F-Ratio	p-Value
Model	97,873.4	2	48,936.7	12,414,254.04	0.0000
Residual	1.30085	330	0.00394198		
Total	97,874.7	332			
R-squared	99.9987%				
MC equation for S2:			MC_S2 = 73.0656 − 1.4671·t + 0.00885007·t^2		

Parameter	Estimate	Standard Error	T Statistic	p-Value
constant	73.0712	0.01040	7022.69	0.0000
t	−1.51967	0.00072	−2102.55	0.0000
t^2	0.0088421	0.00001	840.245	0.0000

Source	Sum of Squares	Df	Mean Square	F-Ratio	p-Value
Model	109,701	2	54,850.3	13,605,034.19	0.0000
Residual	1.33043	330	0.00403162		
Total	109,702	332			
R-squared	99.9988%				
MC equation for S3:			MC_S3 = 73.0712 − 1.51967·t + 0.0088421·t^2		

With reference to the simple regression analysis applied to sage DR in the three tests, since the p-value in Table 3 is less than 0.05, there is a statistically significant relationship between DR and t at the 95.0% confidence level. The R-squared statistic indicates that the model, as fitted, explains the reported percentage of the variability in DR.

Table 3. Simple regression analysis results for drying rate DR as a function of time for sage dehydration process in the three tests (S1 = 3 kg/m^2; S2 = 4 kg/m^2 and S3 = 5 kg/m^2).

Parameter	Least Squares Estimate	Standard Error	T Statistic	p-Value
intercept	0.0107142	0.000003	3922.41	0.0000
slope	−0.0000667179	7.10 × 10^{-8}	−939.53	0.0000

Source	Sum of Squares	Df	Mean Square	F-Ratio	p-Value
Model	0.000542987	1	0.000542987	882,716.62	0.0000
Residual	2.02993 × 10^{-7}	330	6.15 × 10^{-10}		
Total	0.00054319	331			
R-squared	99.9626%				
DR equation for S1:			DR_S1 = 0.0107142 − 0.0000667179·t		

Parameter	Least Squares Estimate	Standard Error	T Statistic	p-Value
intercept	0.00994843	0.000003	3386.19	0.0000
slope	−0.0000593194	7.64 × 10^{-8}	−776.65	0.0000

Source	Sum of Squares	Df	Mean Square	F-Ratio	p-Value
Model	0.000429237	1	0.000429	603,193.33	0.0000
Residual	2.34831 × 10^{-7}	330	7.12 × 10^{-10}		
Total	0.000429472	331			
R-squared	99.9453%				
DR equation for S2:			DR_S2 = 0.00994843 − 0.0000593194·t		

Parameter	Least Squares Estimate	Standard Error	T Statistic	p-Value
intercept	0.0103152	0.000002	5467.19	0.0000
slope	−0.0000594371	4.90 × 10^{-8}	−1211.77	0.0000

Source	Sum of Squares	Df	Mean Square	F-Ratio	p-Value
Model	0.000430944	1	0.000431	1,468,382.14	0.0000
Residual	9.6849 × 10^{-8}	330	2.93 × 10^{-10}		
Total	0.00043104	331			
R-squared	99.9775%				
DR equation for S3:			DR_S3 = 0.0103152 − 0.0000594371·t		

Tables 4 and 5 show the polynomial regression analysis results respectively for MC and DR as a function time for laurel dehydration process.

Table 4. Polynomial regression analysis results for moisture content MC as a function of time for laurel dehydration process in the three tests (L1 = 3 kg/m^2; L2 = 4 kg/m^2 and L3 = 5 kg/m^2).

Parameter	Estimate	Standard Error	T Statistic	p-Value
constant	54.1725	0.02741	1976.49	0.0000
t	−0.875258	0.00297	−294.44	0.0000
t^2	0.00879715	0.00007	130.24	0.0000

Source	Sum of Squares	Df	Mean Square	F-Ratio	p-Value
Model	8491.8	2	4245.9	233,311.76	0.0000
Residual	3.83986	211	0.0181984		
Total	8495.64	213			
R-squared	99.9548%				
MC equation for L1:	MC_L1 = 54.1725 − 0.875258·t + 0.00879715·t^2				

Parameter	Estimate	Standard Error	T Statistic	p-Value
constant	54.6339	0.01557	3509.28	0.0000
t	−0.981658	0.00169	−581.39	0.0000
t^2	0.0101928	0.00004	265.68	0.0000

Source	Sum of Squares	Df	Mean Square	F-Ratio	p-Value
Model	10,204.7	2	5102.33	868,994.45	0.0000
Residual	1.23889	211	0.00587153		
Total	10,205.9	213			
R-squared	99.9879%				
MC equation for L2:	MC_L2 = 54.6339 − 0.981658·t + 0.0101928·t^2				

Parameter	Estimate	Standard Error	T Statistic	p-Value
constant	54.5935	0.02813	1940.74	0.0000
t	−1.05426	0.00305	−345.56	0.0000
t^2	0.00990105	0.00007	142.83	0.0000

Source	Sum of Squares	Df	Mean Square	F-Ratio	p-Value
Model	13,458.6	2	6729.28	351,040.99	0.0000
Residual	4.04477	211	0.0191695		
Total	13,462.6	213			
R-squared	99.97%				
MC equation for L3:	MC_L3 = 54.5935 − 1.05426·t + 0.00990105·t^2				

Table 5. Polynomial regression analysis results for drying rate DR as a function of time for laurel dehydration process in the three tests (L1 = 3 kg/m^2; L2 = 4 kg/m^2 and L3 = 5 kg/m^2).

Parameter	Estimate	Standard Error	T Statistic	p-Value
constant	0.016539	0.00021	77.44	0.0000
t	−0.00178689	0.00007	−25.97	0.0000
t^2	0.000128877	0.00001	19.78	0.0000
t^3	−0.00000381636	2.28 × 10^{-7}	−16.70	0.0000
t^4	3.87457 × 10^{-8}	2.65 × 10^{-9}	14.63	0.0000

Source	Sum of Squares	Df	Mean Square	F-Ratio	p-Value
Model	0.000861156	4	0.000215289	586.36	0.0000
Residual	0.0000763696	208	3.67161 × 10^{-7}		
Total	0.000937526	212			
R-squared	91.8541%				
DR equation for L1:	DR_L1 = 0.016539 − 0.00178689·t + 000128877·t^2 − 0.0000038164·t^3 + 3.87457 × 10^{-8}·t^4				

Table 5. Cont.

Parameter	Estimate	Standard Error	T Statistic	p-Value
constant	0.0143628	0.00022	66.06	0.0000
t	−0.0012179	0.00007	−17.39	0.0000
t^2	0.0000893886	0.00001	13.48	0.0000
t^3	−0.00000270083	2.33×10^{-7}	−11.61	0.0000
t^4	2.76817×10^{-8}	2.70×10^{-9}	10.27	0.0000

Source	Sum of Squares	Df	Mean Square	F-Ratio	p-Value
Model	0.00049453	4	0.000123633	324.94	0.0000
Residual	0.0000791405	208	$3.80483 \cdot 10^{-7}$		
Total	0.000573671	212			
R-squared	86.2045%				
DR equation for L2:	$DR\_L2 = 0.0143628 - 0.0012179 \cdot t + 0.000089389 \cdot t^2 - 0.0000027008 \cdot t^3 + 2.76817 \times 10^{-8} \cdot t^4$				

Parameter	Estimate	Standard Error	T Statistic	p-Value
constant	0.0160073	0.00017	92.84	0.0000
t	−0.00143272	0.00005	−25.79	0.0000
t^2	0.000105325	0.000005	20.02	0.0000
t^3	−0.00000311549	1.84×10^{-7}	−16.89	0.0000
t^4	3.12868×10^{-8}	2.14×10^{-9}	14.63	0.0000

Source	Sum of Squares	Df	Mean Square	F-Ratio	p-Value
Model	0.000491812	4	0.000122953	513.82	0.0000
Residual	0.0000497726	208	2.39292×10^{-7}		
Total	0.000541584	212			
R-squared	90.8098%				
DR equation for L3:	$DR\_L3 = 0.0160073 - 0.00143272 \cdot t + 0.00010532 \cdot t^2 - 0.0000031155 \cdot t^3 + 3.12868 \times 10^{-8} \cdot t^4$				

ANOVA results on MC values for sage taken every 6 h during the dehydration process do not show statistically significant differences between the three tests under study (Table 6).

Table 6. Analysis of variance results on moisture content values every six hours (MC) for sage dehydration process in the three tests (S1 = 3 kg/m^2; S2 = 4 kg/m^2 and S3 = 5 kg/m^2). Data are means ± SD (n = 3).

Time	S1	S2	S3	p-Value	Significance [1]
t_0	72.7 ± 1.9	72.7 ± 1.3	72.7 ± 2.1	1.0000	n.s.
t_6	63.9 ± 2.4	64.5 ± 1.9	64.2 ± 1.9	0.9379	n.s.
t_{12}	55.5 ± 1.9	56.7 ± 1.4	56.0 ± 1.7	0.6860	n.s.
t_{18}	47.8 ± 2.2	49.5 ± 2.5	48.5 ± 1.3	0.6206	n.s.
t_{24}	40.8 ± 2.0	42.9 ± 2.4	41.6 ± 1.7	0.4942	n.s.
t_{30}	36.4 ± 2.3	37.0 ± 2.0	35.4 ± 1.6	0.3542	n.s.
t_{36}	29.0 ± 2.0	31.7 ± 2.0	29.8 ± 2.2	0.3285	n.s.
t_{42}	24.2 ± 2.2	27.0 ± 2.0	24.8 ± 2.5	0.3232	n.s.
t_{48}	20.1 ± 1.9	23.0 ± 1.0	20.5 ± 1.9	0.1388	n.s.
t_{54}	16.7 ± 1.5	19.6 ± 1.8	16.8 ± 2.1	0.1582	n.s.
t_{60}	14.0 ± 2.0	16.9 ± 2.2	13.7 ± 2.2	0.1986	n.s.
t_{60}	12.1 ± 2.2	14.8 ± 2.2	11.3 ± 1.5	0.1572	n.s.

[1] n.s. = not significant.

ANOVA results on MC values for laurel taken every 6 h during the dehydration process (Table 7) show statistically significant differences between the three biomass densities in the last part of the process; in particular at t_{30}, t_{36} and t_{42}, (corresponding to the MC value registered after 30, 36 and 42 h from the beginning of the process).

Table 7. Analysis of variance results on moisture content values every six hours (MC) for laurel dehydration process in the three tests (L1 = 3 kg/m^2; L2 = 4 kg/m^2 and L3 = 5 kg/m^2). Data are means ± SD (n = 3). Values in each row with different letters are significantly different from one another at $p < 0.05$ (Tukey's test).

Time	L1	L2	L3	p-Value	Significance [1]
t_0	55.0 ± 1.4	55.0 ± 1.3	55.0 ± 2.0	0.9986	n.s.
t_6	49.3 ± 1.7	49.1 ± 1.2	48.7 ± 1.7	0.8789	n.s.
t_{12}	45.1 ± 1.1	44.3 ± 1.6	43.5 ± 1.6	0.4040	n.s.
t_{18}	41.1 ± 1.2	40.2 ± 0.8	38.8 ± 1.5	0.1123	n.s.
t_{24}	38.4 ± 1.9	37.0 ± 1.0	35.0 ± 1.2	0.0683	n.s.
t_{30}	35.9 ± 1.8 a	34.3 ± 1.4 ab	31.6 ± 1.4 b	0.0334	
t_{36}	34.1 ± 1.9 a	32.5 ± 1.7 ab	29.4 ± 1.9 b	0.0479	
t_{42}	32.8 ± 1.4 a	31.4 ± 1.6 ab	28.0 ± 1.7 b	0.0207	

[1] n.s. = not significant.

3.2. Microbiological Aspects

The results of the microbiological analyses are reported in Figure 6. The drying processes completely eliminated members of *Enterobacteriacea* family, yeasts and molds from sage, since their levels were below the detection limit (reported as zero in the graphic), independently on the biomass density applied (Figure 6A). TMM were still detected after drying, but the thermal process consistently decreased their levels (until 3.47–3.74 CFU/g). Even though no statistical differences were registered among the biomass densities during drying, the trend observed is that microbial mortality decreased when the biomass density increased inside the drying chamber.

The same drying process applied to laurel indicated how the different leaf structure influenced microbial survival. In this case, all four microbial groups object of investigation showed a higher resistance than those searched in sage, because their levels were clearly detectable (Figure 6B). The drying process did not significantly reduce the levels of TMM and molds when the biomass density was 5 kg/m^2. With laurel leaves, it is undoubtedly shown that the biomass drying density influenced microbial death kinetics, since the viability of all populations registered at 3 kg/m^2 was at the lowest levels and increased with biomass drying density.

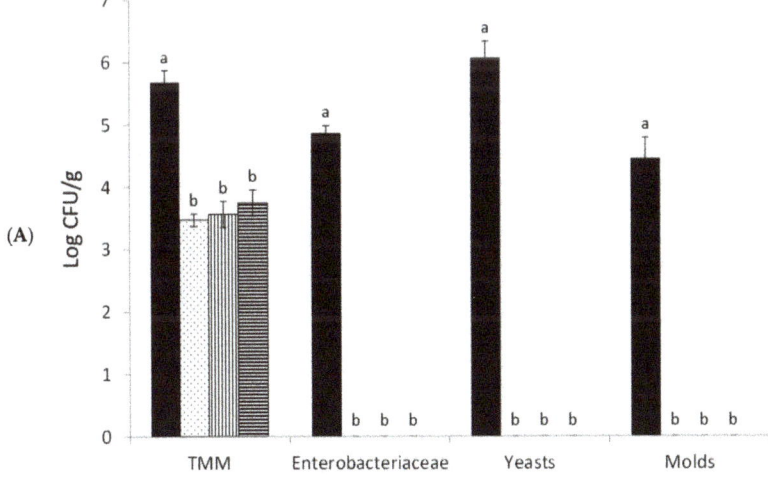

Figure 6. *Cont.*

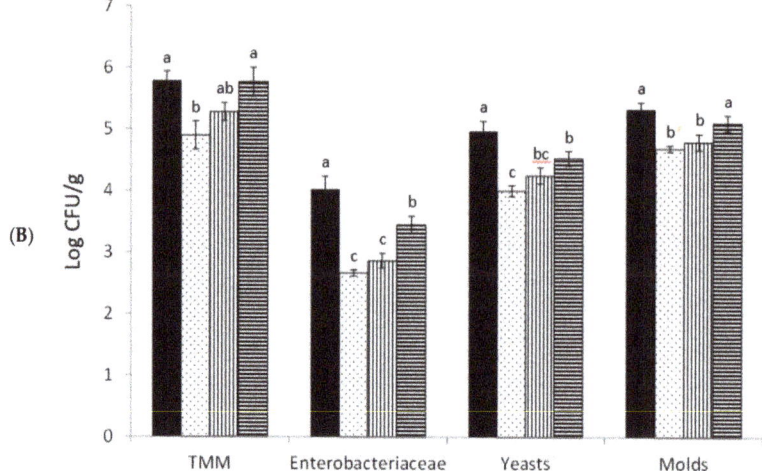

Figure 6. Microbial loads of aromatic herbs subjected to drying. (**A**), sage; (**B**), laurel.

The results of TMM in our study are little below the levels of total bacterial counts at 22 and 37 °C detected by [32], where the authors analyzed different dried herbs, including sage, cultivated in family-managed fields. The differences could not be explained in terms of drying process, since the drying process reported in [32] consisted of a ventilated drying at 40 °C for 24 h, but the information about the load density of the leaves was not reported. The same authors also reported for sage total coliforms, which are part of *Enterobacteriaceae* family, values below 104 CFU/g. In [33], several potential pathogenic species were found, including *Yersinia intermedia*, *Shigella* spp. and *Enterobacter* spp. in several commercialized dried spices and herbs, highlighting the importance of acquiring information on the efficacy of the drying process. Considering that the microbiological quality of commercialized herbs can influence the final safety of different formulated foods [34], the drying process assumes a relevant role during the processing of fresh herbs, even though the conditions applied during conservation are of paramount importance [35].

To our knowledge, this is the first work performed on the microbiological characterization of laurel leaves. Our data clearly show how these leaves exert a protective effect on bacteria, yeasts and molds, highlighting the importance of acquiring information on the microbiological characteristics of herbs used in food formulation and preparation. The disappearance of all microbial groups from sage is easily explained by the very low water activity values registered for the species after the dehydration process, which was in the range 0.335–0.377 for the different biomass densities. Water activity represents an intrinsic parameter of paramount importance for the development of microorganisms in food. This factor greatly influences microbial thermal resistance; thus, low moisture foods are particularly hostile for microbial survival [36]. On the contrary, laurel structure did not allow a consistent removal of water, determining the higher survival of bacteria and fungi in comparison to sage. However, unlike sage, laurel is not commonly consumed raw and generally undergoes a thermal treatment before consumption, such as boiling (e.g., for the preparation of chestnuts boiled in water flavored with fennel seeds or laurel leaves) [37] or roasting (when added to meat) [38].

4. Conclusions

The smart system tested has provided valid decision support during the entire drying process, allowing the precise monitoring of the evolution of the biomass moisture loss and drying rate. It allowed us to stop the drying process at time "t" in which the biomass gave negligible weight variations—that is, at the temperature of 40 °C it was no longer possible to lower MC for both species. This made it possible to limit the loss of volatile and phenolic compounds of the dried product. In fact, in the literature it is well known that very long drying times and temperatures above 40 °C lead to a reduction in the quality of the final product. The two species examined, after drying at the same temperature, achieved a different microbial stability, higher in sage than in laurel. This difference can be attributed to the different structural characteristics of the leaves of the two species, which influence the quantity of bound water present in the plant tissues. The three different biomass densities employed (3, 4 and 5 kg/m^2) did not provide significant differences in the drying process for sage. Statistically significant differences among the three tests were found for laurel in the final part of the process. It follows that biomass density up to 5 kg/m^2 is not a limiting factor in the duration of the process. Further tests with higher biomass densities will be necessary in order to identify the biomass density beyond which the duration of the drying process is higher than that obtained for the two species. Monitoring and controlling the process also allows us to identify any critical points during drying, including the malfunction of the drying chamber, with temperature variations inside, such as to compromise the product in the event of temperatures above 40 °C, or to be carried out for very long process times in the event of temperatures below 40 °C. The microbial loads of the aromatic herbs after drying was influenced by the different leaf structures of the species; in particular, with laurel leaves, microbial survival increased with increasing biomass density. Finally, with the drying method adopted, the two species under consideration showed a different microbial stability and, consequently, will have a different shelf life, longer for sage than for laurel.

Author Contributions: Conceptualization, P.C.; methodology, P.C.; software, S.O.; validation, S.O. and L.S.; formal analysis, M.V. and L.S.; investigation, P.C. and S.O.; resources, P.C. and L.S.; data curation, S.O. and R.G.; writing—original draft preparation, M.V. and R.G.; writing—review and editing, M.V.; visualization, R.G.; supervision, M.V. and L.S.; project administration, P.C. All authors have read and agreed to the published version of the manuscript.

Funding: This research received no external funding.

Conflicts of Interest: The authors declare no conflict of interest.

References

1. Gunjan, M.; Naing, T.W.; Saini, S.R.; Ahmad, A.; Naidu, J.R.; Kumar, I. Marketing trends and future prospects of herbal medicine in the treatment of various disease. *World J. Pharm. Res.* **2015**, *4*, 132–155.
2. Chandrasekara, A.; Shahidi, F. Herbal beverages: Bioactive compounds and their role in disease risk reduction—A review. *J. Tradit. Complement Med.* **2018**, *8*, 451–458. [CrossRef] [PubMed]
3. Serraino, M.; Thompson, L.U. The effect of flaxseed supplementation on the initiation and promotional stages of mammary tumorigenesis. *Nutr. Cancer* **1992**, *17*, 153–159. [CrossRef]
4. Ignacimuthu, S.; Ayyanar, M.; Silverman, S.K. Ethnobotanical investigations among tribes in Madurai district of Tamil Nadu (India). *J. Ethnobiol. Ethnomed.* **2006**, *2*, 25. [CrossRef] [PubMed]
5. Elujoba, A.A.; Odeleye, O.M.; Ogunyemi, C.M. Traditional medicine development for medical and dental primary health care delivery system in Africa. *Afr. J. Tradit. Complement. Altern. Med.* **2005**, *2*, 46–61. [CrossRef]
6. El-Sayed, S.M.; Youssef, A.M. Potential application of herbs and spices and their effects in functional dairy products. *Heliyon* **2019**, *5*, e01989. [CrossRef]
7. Gottardi, D.; Bukvicki, D.; Prasad, S.; Tyagi, A.K. Beneficial effects of spices in food preservation and safety. *Front. Microbiol.* **2016**, *7*, 1394. [CrossRef]
8. ISMEA. Piante Officinali in Italia: Un'istantanea Della Filiera e dei Rapporti tra i Diversi Attori. 2013. Available online: https://www.ismea.it (accessed on 17 June 2020).
9. Basso, F. *Piante Officinali, Aromatiche e Medicinali. Aspetti Bioagronomici Aromatici e Fitoterapeutici*; Editor Pitagora: Bologna, Italy, 2009; ISBN 9788837117320.
10. Bimbenet, J.J.; Duquenoy, A.; Trystram, G. *Genie des Procedes Alimentaires: Des Bases Aux Applications*, 2nd ed.; Dunod: Paris, France, 2002; ISBN 13-978-2100763696.
11. Diaz-Maroto, M.C.; Perez-Coello, M.S.; Cabezudo, M.D. Effect of drying method on the volatiles in bay leaf (*Laurus nobilis* L.). *J. Agric. Food Chem.* **2002**, *50*, 4520–4524. [CrossRef]

12. Hadjibagher Kandi, M.N.; Sefidkon, F. The influense of drying methods on essential oil content and composition of *Laurus nobilis* L. *J. Essent. Oil Bear. Plant* **2011**, *14*, 302–308. [CrossRef]
13. Sadowska, U.; Kopeć, A.; Kourimska, L.; Zarubova, L.; Kloucek, P. The effect of drying methods on the concentration of compounds in sage and thyme. *J. Food Process. Pres.* **2017**, *41*, e13286. [CrossRef]
14. El-Mesery, H.S.; Mwithiga, G. Mathematical modelling of thin layer drying kinetics of onion slices hot-air convection, infrared radiation and combined infrared-convection drying. *Adv. Environ. Biol.* **2014**, *8*, 1–19.
15. Kudra, T. Energy aspects in drying. *Dry. Technol.* **2004**, *22*, 917–932. [CrossRef]
16. Motevali, A.; Minaei, S.; Banakar, A.; Ghobadian, B.; Khoshtaghaza, M.H. Comparison of energy parameters in various dryers. *Energ. Convers. Manag.* **2014**, *87*, 711–725. [CrossRef]
17. Sarsavadia, P.N. Development of a solar-assisted dryer and evaluation of energy requirement for the drying of onion. *Renew. Energy* **2007**, *32*, 2529–2547. [CrossRef]
18. Chakraborty, R.; Tilottama, D. Drying protocols for traditional medicinal herbs: A critical review. *Int. J. Eng. Technol.* **2016**, *4*, 1–8.
19. Rodriguez, J.; Melo, E.C.; Mulet, A.; Bon, J. Optimization of the antioxidant capacity of thyme (*Thymus vulgaris* L.) extracts: Management of the convective drying process assisted by power ultrasound. *J. Food Eng.* **2013**, *119*, 793–799. [CrossRef]
20. Ratti, C. Hot air and freeze-drying of high-value foods: A review. *J. Food Eng.* **2001**, *49*, 311–319. [CrossRef]
21. Fennell, C.W.; Light, M.E.; Sparg, S.G.; Stafford, G.I.; Van Staden, J. Assessing African medicinal plants for efficacy and safety: Agricultural and storage practices. *J. Ethnopharmacol.* **2004**, *95*, 113–121. [CrossRef]
22. Tunde-Akintunde, T.Y.; Ogunlakin, G.O. Influence of drying conditions on the effective moisture diffusivity and energy requirements during the drying of pretreated and untreated pumpkin. *Energ. Convers. Manag.* **2011**, *52*, 1107–1113. [CrossRef]
23. Antal, T.; Figiel, A.; Kerekes, B.; Sikolya, L. Effect of drying methods on the quality of the essential oil of spearmint leaves (*Mentha spicata* L.). *Dry. Technol.* **2011**, *29*, 1836–1844. [CrossRef]
24. Müller, J.; Reisinger, G.; Kisgeci, J.; Kotta, E.; Tesic, M.; Mühlbauer, W. Development of a greenhouse-type solar dryer for medicinal plants and herbs. *Solar Wind Technol.* **1989**, *6*, 523–530. [CrossRef]
25. Aiello, G.; Giovino, I.; Vallone, M.; Catania, P.; Argento, A. A decision support system based on multisensor data fusion for sustainable greenhouse management. *J. Clean. Prod.* **2018**, *172*, 4057–4065. [CrossRef]
26. Vallone, M.; Aiello, G.; Sciortino, R.; Catania, P. First results of iButton loggers and infrared camera application inside a greenhouse. *Acta Hortic.* **2017**, *1170*, 283–291. [CrossRef]
27. Catania, P.; Vallone, M.; Farid, A.; De Pasquale, C. Effect of O_2 control and monitoring on the nutraceutical properties of extra virgin olive oils. *J. Food Eng.* **2016**, *169*, 179–188. [CrossRef]
28. Tuttolomondo, T.; La Bella, S.; Leto, C.; Gennaro, M.; Calvo, R.; D'Asaro, F. Biotechnical characteristics of root systems in erect and prostrate habit rosmarinus officinalis L. accessions grown in a Mediterranean climate. *Chem. Eng. Trans.* **2017**, *58*, 769–774. [CrossRef]
29. Doymaz, I.; Karasu, S. Effect of air temperature on drying kinetics, colour changes and total phenolic content of sage leaves (*Salvia officinalis*). *Qual. Assur. Saf. Crop.* **2018**, *10*, 269–276. [CrossRef]
30. Demir, V.; Gunhan, T.; Yagcioglu, A.K.; Degirmencioglu, A. mathematical modelling and the determination of some quality parameters of air-dried bay leaves. *Biosyst. Eng.* **2004**, *88*, 325–335. [CrossRef]
31. Sellami, I.H.; Wannes, W.A.; Bettaieb, I.; Berrima, S.; Chahed, T.; Marzouk, B.; Limam, F. Qualitative and quantitative changes in the essential oil of *Laurus nobilis* L. leaves as affected by different drying methods. *Food Chem.* **2011**, *126*, 691–697. [CrossRef]
32. Vitullo, M.; Ripabelli, G.; Fanelli, I.; Tamburro, M.; Delfine, S.; Sammarco, M.L. Microbiological and toxicological quality of dried herbs. *Lett. Appl. Microbiol.* **2011**, *52*, 573–580. [CrossRef]
33. Sospedra, I.; Soriano, J.M.; Mañes, J. Assessment of the microbiological safety of dried spices and herbs commercialized in Spain. *Plant Food. Hum. Nutr.* **2010**, *65*, 364–368. [CrossRef]
34. Witkowska, A.M.; Hickey, D.K.; Alonso-Gomez, M.; Wilkinson, M.G. The microbiological quality of commercial herb and spice preparations used in the formulation of a chicken supreme ready meal and microbial survival following a simulated industrial heating process. *Food Control* **2011**, *22*, 616–625. [CrossRef]
35. Santos, J.; Herrero, M.; Mendiola, J.A.; Oliva-Teles, M.T.; Ibáñez, E.; Delerue-Matos, C.; Oliveira, M.B.P.P. Fresh-cut aromatic herbs: Nutritional quality stability during shelf-life. *LWT-Food Sci. Technol.* **2014**, *59*, 101–107. [CrossRef]

36. Syamaladevi, R.M.; Tang, J.; Villa-Rojas, R.; Sablani, S.; Carter, B.; Campbell, G. Influence of water activity on thermal resistance of microorganisms in low-moisture foods: A review. *Compr. Rev. Food Sci. Food Saf.* **2016**, *15*, 353–370. [CrossRef]
37. Bellini, E. The chestnut and its resources: Images and considerations. *Acta Hortic.* **2005**, *693*, 85–96. [CrossRef]
38. Alejo-Armijo, A.; Altarejos, J.; Salido, S. Phytochemicals and biological activities of laurel tree (*Laurus nobilis*). *Nat. Prod. Commun.* **2017**, *12*, 743–757. [CrossRef] [PubMed]

© 2020 by the authors. Licensee MDPI, Basel, Switzerland. This article is an open access article distributed under the terms and conditions of the Creative Commons Attribution (CC BY) license (http://creativecommons.org/licenses/by/4.0/).

Article

Environmental Effects on Yield and Composition of Essential Oil in Wild Populations of Spike Lavender (*Lavandula latifolia* Medik.)

Montserrat Fernández-Sestelo [1,2,*] and José M. Carrillo [3]

[1] Plant Genetic Resources Center, Instituto Nacional Investigación y Tecnología Agraria y Alimentaria, Autovía de Aragón km 36.200, Finca La Canaleja INIA, 28800 Alcalá de Henares, Spain
[2] Biotechnology and Genetic Resources of Plants and Associated Microorganisms, School of Agricultural Engineering, Universidad Politécnica de Madrid, Avenida Puerta de Hierro 2, 28040 Madrid, Spain
[3] Department of Biotechnology and Plant Biology, School of Agricultural Engineering, Universidad Politécnica de Madrid, Avenida Puerta de Hierro 2, 28040 Madrid, Spain; josem.carrillo@upm.es
* Correspondence: montserrat.fernandez@inia.es

Received: 9 November 2020; Accepted: 10 December 2020; Published: 12 December 2020

Abstract: Spike lavender, *Lavandula latifolia* Medik., is a species of economic importance for its essential oil (EO). The purpose of this study was to estimate the effect of the variable climate and fixed factors such as soil and geographic location on EO yield and quality. The study material was collected in 34 populations from four different Spanish bioregions for three years. The EO extraction from spike lavender leaves and flowers was done with simple hydrodistillation, in Clevenger. Soil samples were also collected. Climate data were provided by the State Meteorological Agency. The EO average yield was obtained for the bioregion mean and in each bioregion. The higher EO yield is related clearly to the climate condition. A greater amount of annual rainfall produced a higher EO yield in the four bioregions and of better quality. Soils richer in organic matter and minerals produced higher EO yield but with less quality. The altitude had little effect on EO yield. Higher altitude favored obtaining higher EO quality. At lower latitude, further south, the populations obtained a higher EO yield. The evaluation of the environmental effect on the EO yield and quality could allow better natural conservation and more accurate selection of the best populations for breeding and spike lavender cultivation protocols.

Keywords: spike lavender; essential oil; 1,8-cineole; linalool; camphor; edaphic characteristics; altitude; latitude; longitude

1. Introduction

The genus Lavandula of the Lamiaceae family comprises about 39 species [1]. This genus is made up of small perennial green shrubs, with aromatic flowers and forage from which essential oil (EO) can be obtained. The main species of the genus from which commercial EO is obtained are Lavandula latifolia Medik. with an estimated average annual production of 200 t, *Lavandula angustifolia* Mill. with another 200 t, and the hybrid of the two previous species, called Lavandin (L x intermedia Emerik ex Loisel) with about 1000 t [2]. In France, in 2018, 4662 ha of *L. angustifolia* Mill were cultivated with a production of 116.62 t of EO and 20,770 ha of Lavandin with a production of 1646.13 t of EO [3], which represents slightly more than 3 times kg/ha yield than lavender.

The spike lavender (*L. latifolia* Medik.) is native to the Mediterranean region, growing wild mainly in the former Yugoslavia, Italy, France, Spain, and Portugal [1]. It grows in forest clearings, especially in limestone rocky or dry pastures on sunny slopes, in basic substrates and alluvial sands [1]. It prefers areas between 600–1000 m of altitude [4]. It is collected in the field and cultivated for its EO, to which antibacterial and antifungal, sedative, and antidepressant properties have been attributed, highly appreciated in aromatherapy and phytotherapy [5–7], and is a source of natural antioxidants [8,9]. The spike lavender has been in the past in Spanish regions the species with the highest incidence in its spontaneous collection for the perfume industry [4].

Due to its economic importance, the composition of the spike lavender EO has been widely studied (bibliographic reviews by Boelens [10], García-Vallejo [11], Lis-Balchin [12]). Spike lavender EO consists mainly of monoterpenes and is produced and stored in the glandular trichomes that cover the surface of the aerial parts of the plant, although its production and composition are different in the flower than in other parts of the plant [13]. The main monoterpenes are 1,8-cineole, linalool, and camphor, which determine the olfactory body of EO and comprise about 80% of EO [10,12]. The commercial value of an aromatic plant is determined by the EO yield and composition. Higher levels of linalool and lower amounts of 1,8-cineole and camphor in *L. latifolia* Medik. are positive factors for the pharmaceutical and cosmetic industry and are considered higher quality EO [10,14]. EO with high proportions of camphor is used in the phytosanitary industry [7,15].

The EO of spike lavender, like that of other EOs, is the end product of a complex biological process and its production and composition can vary considerably at the intraspecific level depending on the genotype, on the part of the plant that is used for extraction [10,13], on environmental factors such as climatic conditions and soil composition, geographic location [16], and date of collection [17]. Despite the commercial importance of spike lavender EO, the influence of some environmental factors on its production and quality has been poorly characterized.

Spain is the largest producer of spike lavender EO with 150–200 t per year [2] and its EO is the most important of the essential oils commercially produced in Spain [12]. The Spanish Ministry of Agriculture publishes joint data on lavender and lavandin on acreage and biomass production. In 2018, there were 4725 ha and 15,844 t of biomass [18]. Most studies on the production and chemical composition of EO have been carried out in wild populations in different Spanish regions. The most extensive study of populations has been carried out by Herraiz-Peñalver et al. [16], analyzing 194 samples from 6 different biogeographic Spanish regions. Muñoz-Bertomeu et al. [13] analyzed the differences in production and quality in 7 populations of the Valencian region. Salido et al. [17] analyzed the seasonal behavior in samples from 3 different localities in Andalucía. Its behavior has also been analyzed under cultivation conditions in Spanish regions adapted to the species [4,7].

The quality of an aromatic plant is determined by its secondary metabolite content and its biomass yield. The prospecting and chemical characterization of wild populations and the analysis of the factors that influence their quantity and quality provide essential starting information for the conservation programs and for the selection of parental lines that allow obtaining in future adapted cultivars. Productions from wild collections carry products that are difficult to trace and can pose environmental problems due to overexploitation of natural populations. It is more convenient to collect the species plants in culture.

The objective of this study is to estimate the environmental effects on the production and composition of spike lavender EO in wild populations of various Spanish regions of the North, Center, and Southeast with the dual purpose of conserving natural populations and selecting populations in order to obtain suitable genotypes for spike lavender cultivation and as progenitors of new Lavandin varieties in crosses with *L. angustifolia*.

2. Materials and Methods

2.1. Vegetable Samples

Plant samples were collected in 34 wild populations of *L. latifolia* Medik. for three consecutive years in 2011, 2012, and 2013 in three Biogeographic Regions: Atlantic European 4, Cévenno-Pyrenean 7, Mediterranean Central Iberian 18, defined by Rivas-Martínez and Rivas-Saenz [19]. The populations were distributed in four bioregions dividing the Mediterranean Biogeographic region in two due to its extension. The Cantabroatlantic bioregion 4a (C-4a) with 9 populations, the Prepyrenean 7a (P-7a) with 9 populations, the Mediterranean Castillian 18a (MC-18a) with 7 populations, and the Mediterranean Oroiberian 18b (MO-18b) with 9 populations were the four bioregions (Figure 1). Biogeographic regions layer source Ministerio para la Transición Ecológica (MITECO) (10 February 2018). Software Diva-GIS is available at https://www.diva-gis.org/download. Table S1 shows the latitude and longitude coordinates and collection date for three years of the 34 populations studied.

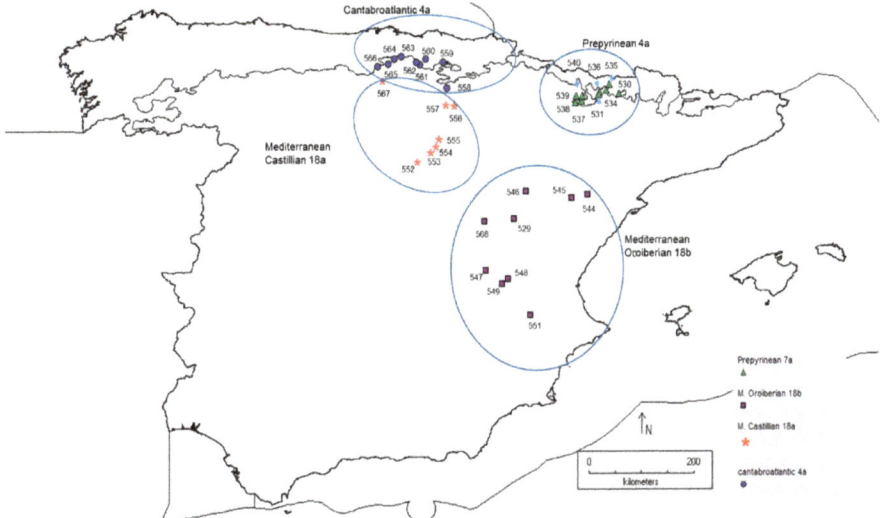

Figure 1. Map with location of the 34 wild populations of *Lavandula latifolia* Medik. by bioregions.

The plant samples collected consisted of flowers and leaves of about 25 plants per population until reaching an approximate weight of 500 g. The collections were made at the time of full flowering, with more than 50% of the plants with open flowers [20] at the end of August–September, depending on the geographical location. The samples from each population were dried at room temperature in shade, and they reached a constant weight in a week.

2.2. Soils

To establish the edaphic characteristics, soil samples were collected from each population in which the plant material was collected. By removing the first centimeters of soil to eliminate the vegetation cover, a kg of soil was collected from the first 20 cm or until reaching bedrock. The characteristics that were analyzed are: pH using a CRISON BASIC 20 (CRISON INSTRUMENTS, S.A., Alella, Spain) [21] model pH meter, % fines, percentage of soil less than 2 mm in diameter, electrical conductivity (EC) using a model conductivity meter micro CM 2200 CRISON (CRISON INSTRUMENTS, S.A., Alella, Spain) [22], the oxidizable carbon content in the soil (SOC) according to Walkley and Black [23], the total nitrogen content of the soil (SN) according to the Kjeldahl method, the assimilable phosphorus (P) content [24] using a spectrophotometer Genesys 10S UV-VIS (Thermo Scientific, Madison, WI, USA), and the bioavailable potassium (K) content [25] using an inductively coupled plasma atomic emission spectrometer ICP-AES Óptima 5300 DV (Perkin Elmer, Wellesley, MA, USA).

2.3. Climatic Data

The climatic variables were obtained from the information provided by the State Meteorological Agency (AEMET) Ministry of Agriculture, Food, and Environment.

For the meteorological data, the three campaigns from August 2010 to August 2013 were considered, the end of the agricultural year coinciding with the collection of the material to be distilled. From the data of extreme temperatures by months and monthly precipitation, the quarterly values were calculated taking the agricultural years: Quarter 1: from September to November; Quarter 2: from December to February; Quarter 3: from March to May; Quarter 4: from June to August.

A Gaussen climogram was constructed [26] to represent the climatic variability to which the sampled populations were subjected.

2.4. Analysis and Quantification of Chemical Parameters

The EO was obtained following the methodology proposed by the European Pharmacopoeia with simple hydrodistillation, in Clevenger [27]. Each sample (180 g of leaves and flowers of dry material) is introduced so as to form a fixed bed in an alembic with water (2L) and boiled for 150 min. The identification of the volatile active principles that make up the EO is analyzed with Gas–Liquid chromatography. The analyses were performed on a Hewlett-Packard Agilent HP 6890N GC system equipped with a quadrupole mass spectrometer Agilent 5973N (Agilent Technologies, S.L., Palo Alto, CA, USA) and DB-5 capillary column with stationary phase phenyl methyl silicone (non-polar) of 30 m long (0.25 mm in diameter and 0.25 µm in film thickness), applying a temperature gradient from 343.15 K to 513.15 K, with an increase of 276.15 K per minute, maintaining the final temperature for two minutes. Additionally, an Agilent 5975 B (Agilent Technologies, S.L., Santa Clara, CA, USA) model gas chromatograph (GC/MS) was used, coupled to an electronic impact mass spectrometer (70 eV) with a column equal to that used in the FID gas chromatograph, to check the active principles.

For the identification of the EO components, n-alkanes standards from C6 (hexane) to C25 (pentacosane) have been injected into the GC/MS column under the same conditions as the samples, the relative retention times of pure substances (standards) and the corresponding Kovats retention indices (RI) were used. The quantification of the percentages of the components is performed according to the areas of the chromatographic peaks. With this methodology, there are several samples in which some active principles do not separate well, so they are considered together. These active principles are: sabinene + β pinene and limonene + 1,8-cineole.

The EO yield and the active principles that were detected with a percentage higher than 1% that appear in the ISO-4719 [28] were EO yield, α pinene, camphene, sabinene + β pinene, limonene + 1,8-cineole, linalool, camphor, borneol, and α-terpineol. Given the low percentage of limonene, around a mean value of 1% [13,16,17,29], the result tables will show only 1-8-cineole instead of limonene + 1,8-cineole, so the percentage of 1,8-cineole will be somewhat overrated.

2.5. Statistical Analysis

In order to compare the different variables, they were standardized. The means are presented with their standard differences (±S.D). One-way analysis of variance (ANOVA) was used to test the effect of experimental years and different geographical origins on the variables EO yield, 1,8-cineole, linalool, camphor, and rainfall. Tukey's test was performed to determine differences between treatment means. The Pearson correlation coefficients between EO yield, 1,8-cineole, linalool, and camphor were also determined. Linear regressions of the dependent variables, EO yield, 1,8-cineole, linalool, and camphor, were made on edaphic and location characteristics of the populations. The level of statistical significance was set at $p < 0.05$ and $p < 0.01$ (**). All statistical analyses were performed with the InfoStat statistical package (InfoStat v2016; Grupo InfoStat, FCA, Universidad Nacional de Córdoba, Argentina, 2016).

3. Results

3.1. EO Yield and Composition

The EO yield mean value in the 34 populations of the four bioregions for three years was 2.95%, with values between 1.75–4.58% (Table 1). Of the 8 components that represent around 90% of the EO, three components account for 80% of the EO: 1,8-cineole represents 36.62%, linalool 26.74%, and camphor 17.23%.

Table 1. Essential oil (EO) yield (% of dry weight) and major components (in EO %) in 34 wild populations of *L. latifolia* Medik. Year mean and by bioregions.

Bioregion	Component	n[a]	Min.	Mean	S.D.	Max.
All populations	essential oil yield	34	1.75	2.95	0.71	4.58
	α-pinene	34	0.85	1.73	0.53	3.13
	camphene	34	0.37	0.81	0.27	1.45
	sabinene+β-pinene	34	1.52	2.85	0.87	4.75
	1,8-cineole	34	20.96	36.62	9.89	54.30
	linalool	34	11.42	26.74	7.74	45.36
	camphor	34	4.23	17.23	7.97	31.39
	borneol	34	1.05	2.36	1.00	5.97
	α-terpineol	34	0.38	0.88	0.21	1.39
Cantabroatlantic (C-4a)	essential oil yield	9	1.75	2.52 b	0.63	3.51
	1,8-cineole	9	21.51	38.20 a	8.63	46.91
	linalool	9	25.29	30.69 a	6.92	45.36
	camphor	9	7.23	12.34 b	4.46	20.73
Prepyrenean (P-7a)	essential oil yield	9	2.37	2.86 ab	0.47	3.70
	1,8-cineole	9	20.96	24.46 b	2.18	26.82
	linalool	9	22.43	29.39 ab	4.73	38.81
	camphor	9	23.17	27.04 a	2.58	31.39
Mediterranean Castillian (MC-18a)	essential oil yield	7	2.30	2.77 b	0.46	3.55
	1,8-cineole	7	35.45	41.84 a	6.19	51.41
	linalool	7	20.01	25.60 ab	5.58	36.28
	camphor	7	7.76	13.43 b	7.04	24.83
Mediterranean Oroiberian (MO-18b)	essential oil yield	9	2.28	3.63 a	0.74	4.58
	1,8-cineole	9	30.40	43.13 a	7.41	54.30
	linalool	9	11.42	21.03 b	9.49	36.60
	camphor	9	4.23	15.28 b	6.80	25.61

n[a] Number of populations in which the component appears. Different letters indicate significant differences among bioregions.

There are significant differences among the bioregions in the EO yield and the proportion of its components (Table 1). The MO-18b region is with the highest significant EO mean yield, 3.63%. The other three bioregions obtain a similar yield, P-7a with 2.86%, MC-18a with 2.77%, and the C-4a region having the lowest value, 2.52%.

The EO composition is different in the four bioregions (Table 1). The major component is 1,8-cineole, with values around 40% in the two Mediterranean regions and in C-4a, while in the P-7a region its proportion falls to less than 25%. Linalool is the first EO component in P-7a with 29.39% and is the second in the other bioregions, 30.69% in C-4a and somewhat lower in the two Mediterranean regions, 25.60% in MC-18a and 21.03% in MO-18b. The proportion of camphor is 27.04% in P-7a, about double the values of the other bioregions.

There are significant differences in EO yield among years (Table 2). The average value of the 34 populations in 2013 is the highest EO yield (3.99%), followed by 2011 (2.77%), and with the lowest yield is 2012 (2.06%). The same order of years in EO yield is maintained in the bioregions C-4a (Table S2),

P-7a (Table S3), MC-18a (Table S4), and MO-18b (Table S5). In MC-18a, the difference is not significant in EO yield between 2011 and 2012.

Table 2. EO yield (% of dry weight) and major components (in EO %) in 34 wild populations of L. latifolia Medik. for each of the years analyzed.

Year	Component	n[a]	Min.	Mean	S.D.	Max.
2011	essential oil yield	33	0.83	2.77 b	1.08	5.78
	1,8-cineole	34	13.05	32.89 b	8.99	50.51
	linalool	31	0.23	21.17 b	15.95	48.94
	camphor	31	4.17	20.07 a	10.15	36.67
2012	essential oil yield	34	0.97	2.06 c	0.64	3.18
	1,8-cineole	34	14.08	42.04 a	13.55	66.13
	linalool	34	3.03	24.36 b	11.13	49.94
	camphor	34	4.94	17.11 ab	7.83	31.97
2013	essential oil yield	34	1.69	3.99 a	1.30	6.80
	1,8-cineole	34	17.09	34.92 b	10.67	55.03
	linalool	34	5.31	34.00 a	10.18	54.08
	camphor	34	3.60	15.49 b	7.65	31.82

n[a] Number of populations in which the component appears. Different letters indicate significant differences among years.

In each of the three years, the order from highest to lowest proportion of 1,8-cineole, linalool, and camphor is maintained in the EO composition (Table 2). However, their percentages differ among the years. In 2012, the percentage of 1-8 cineol is 42.04%, significantly higher than in the other two years. In 2013, linalool presents a significantly higher proportion, 34.00%, and camphor, a significantly lower value, 15.49%, than in 2012 and 2011. In each bioregion (Tables S2–S5), there are the same orders in the percentages of the EO components in the years 2012 and 2013 that occur in the 34 population means.

3.2. Climate

Comparing rainfall among years (Table 3), the agricultural year 2013 was the year of significantly higher rainfall in all bioregions, followed by 2011, and 2012 with the worst record; however, the differences between these two years are not statistically significant, except in P-7a where, in 2012, precipitation is significantly lower.

Table 3. Values of precipitations by agricultural quarters (TRI1: September to November; TRI2: December to February; TRI3: March to May; TRI4: June to August) by bioregions in each year.

Year	Bioregion	TRI1	TRI2	TRI3	TRI4	Annual	*
2011	C-4a	239.34 a	197.07 a	150.51 a	119.63 b	706.54 a	B
	P-7a	238.67 a	157.18 ab	162.07 a	177.56 a	735.47 a	B
	MC-18a	134.01 b	145.69 ab	194.06 a	89.48 bc	563.24 ab	AB
	MO-18b	128.92 b	98.8 b	178.94 a	49.17 c	455.84 b	B
2012	C-4a	180.36 a	196.79 a	200.73 a	71.35 bc	649.23 a	B
	P-7a	188.87 a	40.33 b	240.69 a	131.22 a	601.11 a	C
	MC-18a	108.37 b	66.15 b	165.11 b	76.91 bc	416.55 b	B
	MO-18b	110.99 b	36.92 b	107.67 c	51.34 c	306.92 b	B
2013	C-4a	218.28 b	430.94 a	281.3 a	119.72 b	1,050.25 a	A
	P-7a	296.47 a	232.69 b	288.2 a	204.16 a	1,021.51 a	A
	MC-18a	162.02 bc	173.29 bc	257.01 a	97.52 b	689.85 b	A
	MO-18b	197.07 c	95.89 c	206.58 a	124.34 b	623.87 b	A

Lowercase letters indicate the difference in precipitation among bioregions within each year. * Capital letters in the last column indicate the difference among years for each bioregion for annual precipitation. Source of meteorological data: AEMET.

Comparing rainfall among bioregions (Table 3), MO-18b has the lowest annual rainfall in the three years, followed by MC-18a with slightly higher values. The two northern bioregions, C-4a and P-7a, have significantly higher annual rainfall than the two Mediterranean bioregions for the three years. They have very similar annual values, but taking into account the rainfall of the fourth quarter of the plant cycle (June, July, and August), P-7a has a significantly higher value than C-4a in the three years. In this fourth quarter, the two Mediterranean bioregions have rainfall values significantly lower than P-7a, and with respect to C-4a, the difference is significant only in 2011. Figure S1 shows a Gaussen climogram with the maximum and minimum temperatures in °C and quarterly precipitation (mm) during the three years and in the four bioregions.

3.3. Edaphic Characteristics

The soils of the populations of the four bioregions (Table 4) present a similar pH value, between 7.5 and 8, these are basic soils. The P-7a bioregion shows lower mean values than the other bioregions in all the edaphic variables and with more homogeneous values in its 9 populations than the other bioregions. It has a lower percentage of fines, that is, more stony soil and a lower electrical conductivity (EC), which may indicate a lower presence of salts in its soil and also lower values with respect to the other bioregions in organic soil (SOC) in soil nitrogen content (SN), potassium (K), and phosphorus (P).

The soil values of MO-18b populations present a greater dispersion, less in the physical texture of the soil, with a high value of fines %, somewhat lower than the other Mediterranean bioregion. It also presents higher mean values than the other bioregions in EC, SOC, SN, P, and a high K value.

The EO yield % does not show a relationship with the variability of the values of the different edaphic variables of the 34 populations. In the EO composition, the 1-8 cineol increases significantly with higher values of fine grain % (Figure 2A), EC (Figure 2B), SOC (Figure 2C), and SN (Figure 2D), and the proportion of linalool decreases significantly with higher values of fine grain % (Figure 2E), EC (Figure 2F), SOC (Figure 2G), and P (Figure 2H). The different pH values in the populations show no relationship with changes in the EO component proportions, although there is a trend of more EO yield in more basic soil.

3.4. Altitude, Latitude, and Longitude of the Populations

The altitude mean of the populations (Table 5) is similar in the Mediterranean bioregions and higher than in the Nord bioregions. The altitude is 997 in MC-18a and is 937 m in MO-18b. The C-4a bioregion has a lower mean height of 700 m. The P-7a is the most homogeneous in altitude, almost all the localities with a height greater than 800 m and an average value of 900 m.

To analyze whether the population altitude was related to the EO yield and components, regressions were performed of the 34 populations' mean values in EO yield, linalool, 1,8-cineole, and camphor on their altitude. None regression gave significant values, and there seems to be a trend towards a higher EO yield as the altitude of the population increases.

Analyzing the regression by bioregions, none have shown any relationship of greater EO yield as altitude increases. Regarding the relationship with the EO components (Figure 3), there is a significant negative relationship in P-7a with 1,8-cineole (Figure 3A); with linalool, there is a positive relationship in the two Mediterranean, MO-18b (Figure 3B); and with camphor, there is a positive relationship in C-4a (Figure 3C) and negative in the two Mediterranean, M-18b (Figure 3D).

Table 4. Edaphic characteristics of the 34 *L. latifolia* Medik. populations by bioregions.

CAT	pH	% Fine-Grain	EC	SOC	SN	K	P	Altitude
C-4a								
558	8.12	46.89	157.7	3.46	0.460	158.85	1.51	530
559	7.56	69.92	163.4	37.09	2.360	418.97	14.95	566
560	7.45	41.21	158.7	16.42	1.321	108.96	6.41	676
561	7.45	60.19	143.0	13.12	1.271	391.81	5.13	584
562	8.04	54.91	173.7	30.56	2.615	321.50	11.32	710
563	7.95	43.55	183.0	26.65	2.041	378.44	8.83	769
564	8.25	58.90	143.1	7.09	0.535	86.92	10.11	676
565	8.04	57.62	163.4	13.94	0.958	235.93	4.99	848
566	7.99	59.75	134.0	27.15	2.105	301.09	6.23	939
Mean	7.87 a	54.77 ab	157.78 a	19.50 a	1.52 a	266.94 a	7.72 a	699.78 b
S.D.	0.30	9.22	16.62	11.37	0.79	125.15	4.01	134.79
P-7a								
530	7.81	38.51	154.3	6.34	0.555	68.38	5.55	977
531	7.60	24.85	118.3	1.36	0.584	28.54	3.48	838
534	7.63	43.23	174.6	22.20	1.788	102.24	9.69	781
535	7.46	65.38	159.9	5.83	0.550	90.91	1.57	802
536	8.33	55.55	133.3	6.76	0.370	139.15	18.43	973
537	7.52	54.56	155.2	12.96	1.066	61.12	6.83	1048
538	7.88	56.46	147.3	5.38	0.510	64.53	3.84	993
539	8.19	62.46	151.3	4.21	0.412	83.45	3.71	793
540	7.59	54.27	146.7	8.58	0.773	115.77	5.42	892
Mean	7.78 a	50.59 b	149.00 a	8.18 a	0.734 a	83.79 b	6.50 a	899.67 ab
S.D.	0.30	12.80	16.00	6.13	0.447	32.89	5.04	100.58
MC-18a								
552	7.52	63.76	194.0	34.54	2.830	45.63	22.66	1133
553	7.75	77.35	174.9	17.72	0.884	36.30	6.83	940
554	7.87	74.80	157.5	15.27	1.335	40.96	5.40	1008
555	7.75	66.51	158.9	10.69	0.794	24.69	4.76	1130
556	7.85	58.16	152.2	1.42	0.314	35.04	1.58	835
557	7.59	61.46	199.6	28.04	2.245	186.34	11.90	879
567	8.10	60.21	168.4	37.26	3.215	323.31	11.11	1056
Mean	7.78 a	66.04 a	172.21 a	20.70 a	1.66 a	98.90 ab	9.18 a	997.29 a
S.D.	0.19	7.38	16.44	13.10	1.11	113.80	6.95	117.76
MO-18b								
529	8.00	57.63	149.9	3.37	0.292	126.13	2.86	1255
544	8.42	62.58	151.0	5.70	0.417	40.87	3.42	604
545	8.17	57.91	162.9	5.45	0.384	153.03	3.42	968
546	7.20	60.46	234.0	17.14	1.384	154.64	8.53	941
547	7.65	66.78	208.0	77.98	4.164	687.71	23.66	1012
548	8.08	50.20	181.2	13.57	0.766	144.84	7.98	777
549	7.82	59.54	143.2	3.32	0.298	129.83	3.71	757
551	7.73	66.85	357.5	87.35	5.004	353.52	75.01	790
568	7.30	45.90	631.5	28.60	2.711	417.79	13.95	1333
Mean	7.82 a	58.65 ab	246.58 a	26.94 a	1.71 a	245.37 ab	15.84 a	937.44 a
S.D.	0.40	6.97	159.12	32.71	1.81	204.13	23.20	238.49

CAT: Population code. Edaphic variables and units: pH; % fine-grain < 2 mm; EC: Electric Conductivity (µs/cm); SOC: Soil Organic Carbon. (g/kg); SN: Soil Nitrogen (g/kg); K: Potassium assimilable (mg/kg); P: Phosphorus Olsen (mg/kg). Different letters indicate significant differences among bioregions.

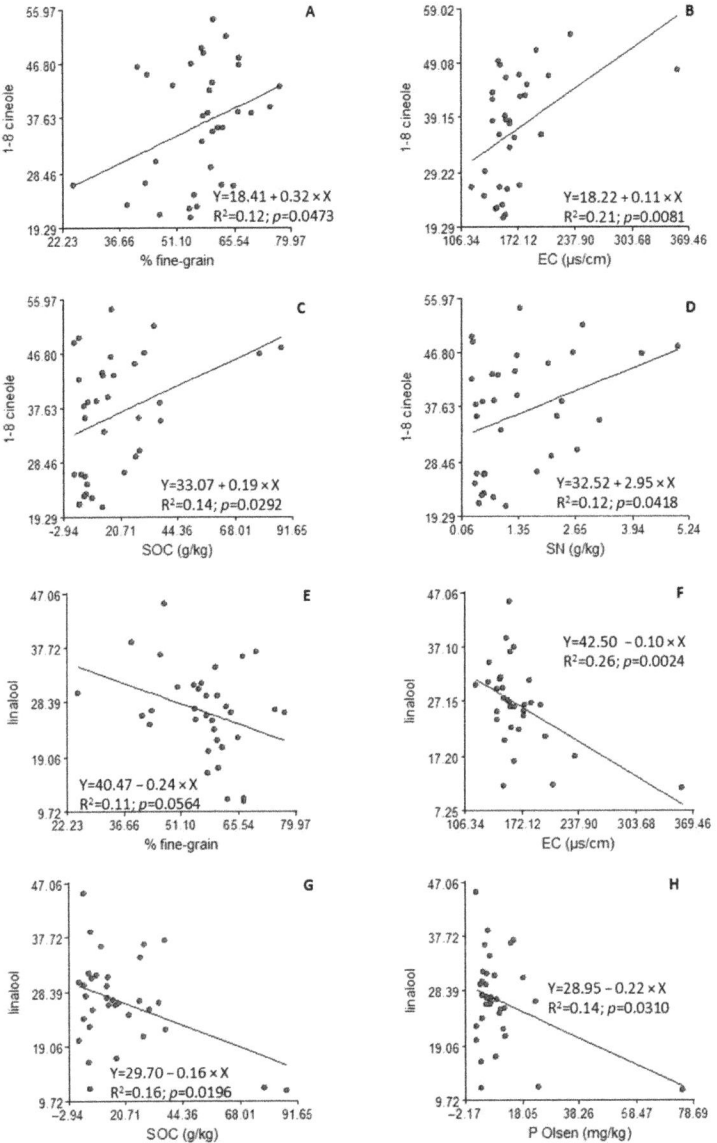

Figure 2. Regressions of 1,8-cineole on % fine-grain < 2 mm (**A**); EC: Electric Conductivity (μm/cm) (**B**); SOC: Soil Organic Carbon (g/kg) (**C**); and SN: Soil Nitrogen (g/kg) (**D**) and regressions of linalool on % fine-grain < 2 mm (**E**); EC: Electric Conductivity (μm/cm) (**F**); SOC: Soil Organic Carbon (g/kg) (**G**); and P Olsen: P assimilable (g/kg) (**H**).

Table 5. Altitude (m), EO yield (% on dry weight), and major components (EO %) of the 34 wild populations of *L. latifolia* Medik. by bioregion.

CAT	Bioregion	Altitude	Essential Oil Yield	1,8-Cineol	Linalool	Camphor
558		530	3.51	21.51	45.36	14.52
559		566	1.75	38.53	37.15	7.23
560		676	2.01	46.36	26.19	11.60
561		584	2.35	43.69	29.43	7.85
562	C-4a	710	1.90	46.91	25.42	9.24
563		769	2.38	45.09	26.89	10.37
564		676	2.87	38.57	25.29	12.33
565		848	2.52	33.69	26.17	20.73
566		939	3.41	29.41	34.31	17.18
	Mean	699.78 b	2.52 b	38.20 a	30.69 a	12.34 b
	S.D.	134.79	0.63	8.63	6.92	4.46
530		977	2.37	23.16	38.81	23.17
531		838	2.48	26.49	30.05	28.25
534		781	3.38	26.82	24.54	28.61
535		802	2.61	26.20	22.43	29.25
536	P-7a	973	3.01	24.81	30.75	26.51
537		1048	2.47	20.96	27.35	25.08
538		993	3.70	22.74	31.65	26.58
539		793	3.13	26.51	27.70	24.54
540		892	2.55	22.47	31.27	31.39
	Mean	899.67 ab	2.86 b	24.46 b	29.39 ab	27.04 a
	S.D.	100.58	0.47	2.18	4.73	2.58
552		1133	3.26	51.41	26.59	7.76
553		940	2.30	43.07	26.56	8.64
554		1008	2.55	39.51	27.07	10.87
555	MC-18a	1130	2.50	38.78	36.28	8.76
556		835	3.55	48.70	20.01	10.93
557		879	2.74	35.98	20.72	24.83
567		1056	2.47	35.45	21.94	22.25
	Mean	997.29 a	2.77 b	41.84 a	25.60 ab	13.43 b
	S.D.	117.76	0.46	6.19	5.58	7.04
529		1255	3.94	49.46	29.53	4.23
544		604	3.68	35.99	11.69	25.61
545		968	3.55	37.98	16.32	21.04
546		941	2.28	54.30	17.10	9.39
547	MO-18b	1012	2.80	46.72	11.89	13.23
548		777	3.35	43.12	30.95	10.22
549		757	4.58	42.41	23.76	19.19
551		790	4.33	47.82	11.42	20.44
568		1333	4.11	30.40	36.60	14.15
	Mean	937.44 a	3.63 a	43.13 a	21.03 b	15.28 b
	S.D.	238.49	0.74	7.41	9.49	6.80

CAT: population code. Different letters indicate significant differences between bioregions.

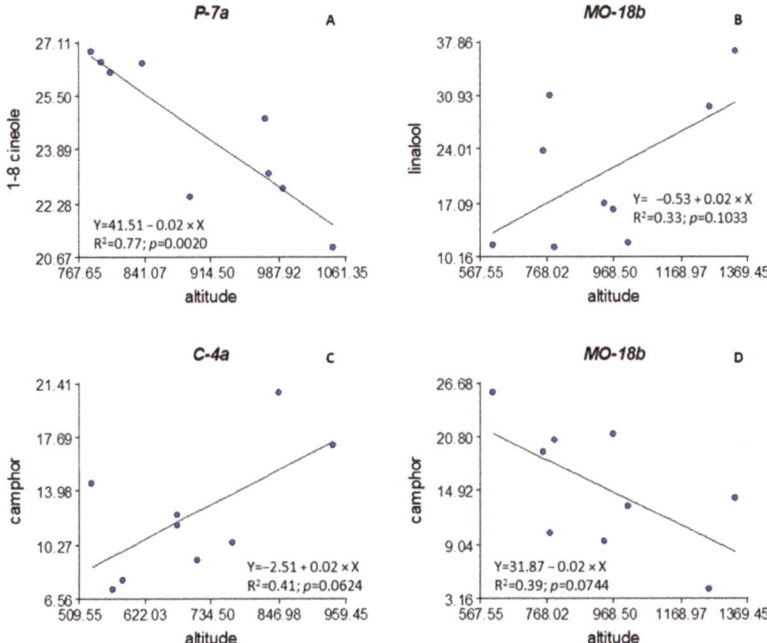

Figure 3. Regressions on altitude: (**A**) 1,8-cineole in P-7a; (**B**) linalool in MO-18b; (**C**) camphor in C-4a; (**D**) camphor in MO-18b.

Populations with lower latitude have significantly higher EO yield (Figure 4A), a higher proportion of 1,8-cineole (Figure 4B), and a lower proportion of linalool in EO composition (Figure 4C). Therefore, populations with higher latitude produce less EO, a lower proportion of 1,8-cineole, and a significantly higher proportion of linalool.

Regarding the longitude location of the populations, the EO yield and the linalool proportion do not have a significant relationship with the longitude. There are significant values of 1,8-cineole and camphor with a greater longitude, negative and positive, respectively (Figure 4D,E).

3.5. Relationship between EO Yield Main Components

Considering the mean values of the 34 populations for the three years, the higher or lower EO yield does not show a correlation with the variability of the values of the three main EO components (Table 6). For the mean values in each of the four bioregions, there is only a significant and negative correlation in C-4a between EO yield and 1,8-cineole (−0.82 **) (Table S6).

Table 6. Relationship between EO yield main components means of 34 populations of *L. latifolia* Medik. for three years based on the Pearson correlation matrix with standardized variables.

	Essential Oil Yield	1,8-Cineole	Linalool	Camphor
essential oil yield	1.00			
1,8-cineole	−0.02	1.00		
linalool	−0.15	−0.48 **	1.00	
camphor	0.15	−0.76 **	−0.13	1.00

** Correlation is significant at the 0.01 level.

In the relationship between the three main components (Table 6), 1-8 cineol shows a significant and negative correlation with camphor (−0.76 **) and with linalool (−0.48 **). In the bioregions (Table S6), the correlations between 1-8 cineol and camphor and linalool show negative non-significant values; only in C-4a, it is significant with linalool (−0.78 *).

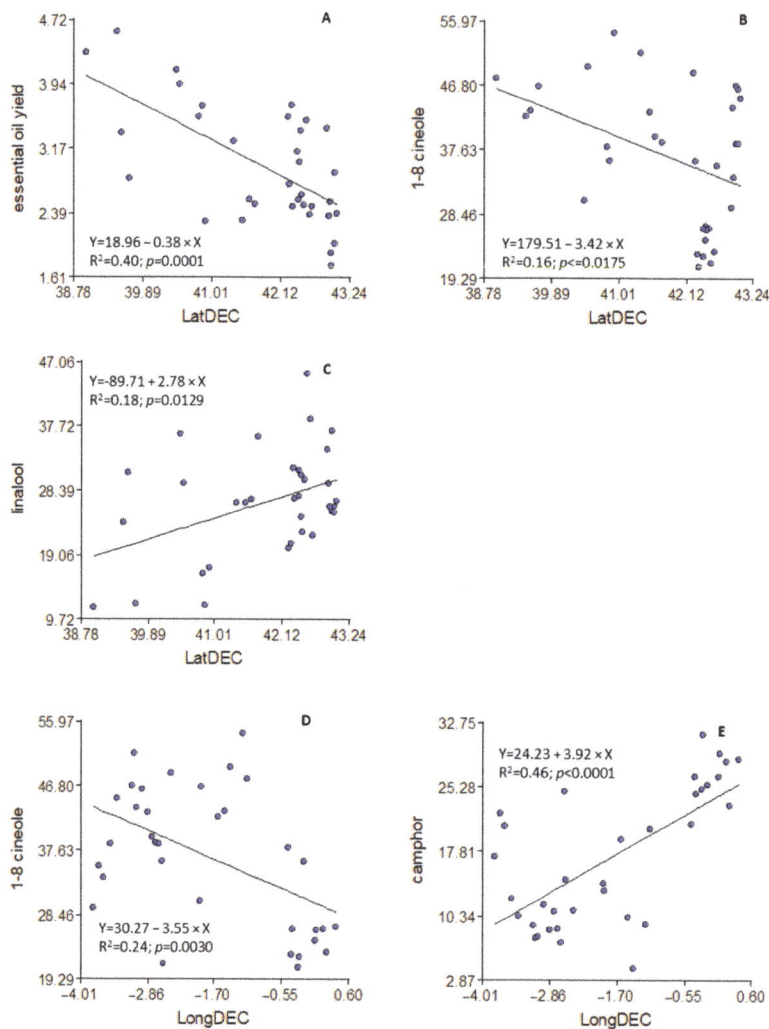

Figure 4. (**A**) Regression of EO yield on decimal latitude (LatDEC); (**B**) Regression of 1,8-cineole on LatDEC; (**C**) Regression of linalool on LatDEC; (**D**) Regression of 1,8-cineole on decimal longitude (LongDEC); and (**E**) Regression of camphor on LongDEC.

4. Discussion

4.1. EO Yield and Composition

The EO yield mean value of 2.95% with respect to the dry weight of flowers and leaves of the plant could be a fairly real estimate of the average EO yield of *L. latifolia* in its natural environment and on a national scale (Table 1). That mean value has been obtained as an average of three years of different climatology and of four different bioclimatic and edaphological regions.

There are almost no estimates of EO yield in *L. latifolia* from a high number of wild populations and several years. The study by Herraiz-Peñalver et al. [16] stands out, with an average EO yield of 4.2%, in a two-year analysis and with a wide sample of populations. The higher value obtained in their study could be due to the weather being more favorable or to the fact that only flowers were used for extraction [13]. To be able to compare EO yield per unit area extracting from flowers and leaves or only from flowers, it would be necessary to know the percentage of the flower weight with respect to flowers and leaves weight in the plant.

The estimate of the average EO yield in three years in the four bioregions is also a good approximation to the production capacity of each of the bioregions. The MO-18b bioregion stands out in a significant way with respect to the other bioregions, with EO yield of 3.63%. The C-4a bioregion has the lowest EO yield with 2.52%, but it does not differ significantly from P-7a and MC-18a, with 2.86% and 2.77%, respectively. Herraiz-Peñalver et al. [16] did not find significant yield differences due to biogeographic origin. There is little information on EO yield in wild populations of spike lavender. Burillo [4] found significant differences in yield trials between different localities in the region of Aragón. Salido et al. [17] found values of 1.5%–2.2% of EO yield in 6 populations from southern Spain.

The EO composition is relevant because of the implication in its quality and commercial value. There is usually a high variability between populations and between plants within the same population [13]. The EO composition may vary depending on the part of the plant, flowers, or leaves, which is used for its extraction [13], the geographical situation, and the climatic differences [10]. García-Vallejo [11] in a bibliographic review on the proportion of the main EO constituents gives ranges of 20.5–44.8% for 1,8-cineole, 11.0–53.9% for linalool, and of 5.3–35.3% for camphor. Linalool is usually the main component in most studies, so in wild populations [16], and in cultivated plots [7,29,30]. The linalool content is higher if the EO is extracted only from flowers instead of all parts of the plant because the glandular trichomes of the leaves do not secrete linalool, only 1,8-cineole and camphor [13]. In our study, the main EO component is 1,8-cineole, 36.62%, followed by linalool, 26.74%, and camphor, 17.13%, in the mean values of the 34 populations (Table 1). The average values do not give high-quality oil for perfumery, even though the % of camphor is low.

The EO composition in the bioregions (Table 1) gives a more realistic idea of the EO quality extracted in the natural populations of spike lavender. The bioregion P-7a is that diverges the most in the EO composition from the others. Linalool (29.39%) is its first EO component, while in the other three bioregions, 1,8-cineole is the main component with values around 40%. P-7a differs significantly in a higher % of camphor (27.04%), twice of the others, and in a lower % of 1,8-cineole (24.46%). The two northern bioregions have a significantly higher value in linalool % with respect to the two Mediterranean ones. The highest camphor % in the Prepyrenean region, P-7a, coincides with Herraiz-Peñalver et al. [16] and differs in which the Mediterranean regions have lower, not higher, linalool mean.

4.2. Climate

The weather had a great impact on EO yield since the differences between years are significant (Table 3), with 2013 having the highest yield with 3.99%, 2011 with 2.77%, and 2012 with the lowest production, 2.06%. The yield difference between years also occurs in all bioregions (Tables S2–S5). Herraiz-Peñalver et al. [16] found yield differences in two years, 4.6% and 3.9%, attributing the difference to the high importance of annual climatic conditions.

In the present work, it is verified that the EO yield difference in spike lavender between years has clear correspondence with the different rainfall between years. There are significant differences in the annual rainfall quantity between years (Table 3). The agricultural year 2013 is the rainiest in all the bioregions, followed by 2011 and 2012. The greater or lesser rainfall coincides with the greater or lesser EO yield. The year 2013 differs significantly from the other two years both in rainfall and in

EO yield. Moreover, 2012 is the year with the lowest rainfall and EO yield (Tables 2 and 3). The same relationship between rainfall and EO yield occurs in the 4 bioregions for the three years (Tables S2–S5). The EO yield is estimated at a fixed weight of dry matter; therefore, the higher EO yield may be due to a greater number of glandular trichomes or to the fact that they store more EO. In a situation of more rain, the production per unit area would probably be higher than that shown in the tables since it has been found that the EO yield per ha in spike lavender in cultivation follows a similar pattern to the yield in vegetable matter, which is usually superior in rainy conditions [4].

The rainfall influences the yield and also the EO composition (Tables 2 and 3). In the wettest year 2013, the higher EO yield is accompanied by a higher proportion of linalool, 34%, significantly higher than the proportions in 2012 and 2011, and at the same time a significantly lower proportion of camphor with respect to the other two years. The year 2012, the one with the lowest yield and rainfall, significantly increased 1,8-cineole, 42%, with respect to the other two years. The same pattern is observed in the four bioregions (Tables S2–S5). In *L. angustifolia*, the water deficit increases the proportion of 1,8-cineole in the essential oil [31]. These results partially contradict what was obtained in experimental plots of spike lavender by Burillo [4] who affirms that the scarcity of water influences the biosynthesis of linalool, or by Usano-Alemany [32] that the high percentage of linalool obtained in lavender and lavandin could be due to the drought during the quarter before harvest. In our experiment, the fourth agronomic quarter is always the driest and could influence the biosynthesis of more linalool, but the 4th quarter of 2013 is the least dry of the three years and is the year that shows the highest proportion of linalool. With more annual rainfall, a higher EO yield with higher quality was obtained.

However, rainfall was not the only determining factor in the EO yield in bioregions. MO-18b was the bioregion with the highest EO yield in the three years and was the one with the lowest rainfall during the three years. The other Mediterranean region does not differ significantly from the other two northern bioregions in EO yield, although it differs significantly in lower annual rainfall over the three years.

4.3. Edaphic Characteristics

The components of the soil of the different bioregions could influence the EO yield and characteristics. The soil of MO-18b populations, with the highest significant EO yield (Table 1), presents a set of characteristics conducive to greater production, such as a less stony soil due to its high percentage of fines, a lower presence of salts due to their higher significant EC, and higher values than in the other bioregions in SOC, in N, K, and P (Table 4). The edaphic variables could also influence the EO composition. Considering the means of 34 populations, the 1,8-cineole proportion increases significantly with higher values of the edaphic variables (Figure 2A–D) and the linalool proportion decreases significantly (Figure 2E–H), and the camphor proportion show a tendency to lower values. The two Mediterranean regions fulfill this composition on their oil. The inverse situation could be applied to the P-7a bioregion, which has a higher significant value of camphor and presents lower mean values than the other bioregions in all edaphic variables. We have not found bibliographic references about the relationship between edaphic variants and the EO yield and composition of the spike lavender. In *L. angustifolia*, the EO yield was remained unaffected under N and P different levels, and 1,8-cineole and camphor of leaves EO were affected [33]. On the other hand, different levels of K application affected both the EO yield and constituents [34].

4.4. Altitude, Latitude, and Longitude of the Populations

Other factors that could influence the EO yield and composition of the spike lavender are the fixed and different factors of altitude, latitude, and longitude of the populations. The bioregions differ in the mean altitude values, and the C-4a bioregion shows a significantly lower mean altitude than the two Mediterranean bioregions (Table 5) and presents the lowest EO yield (Table 1) but was significant only with respect to MO-18b. It could be one of the causes of lower C-4a EO yield comparing to the other bioregions. However, there is a considerable disparity between altitude and production data in the bioregions (Table 5), and the regression of the altitude of all populations on the EO yield does not give significant values, although it seems that there is a trend of greater EO yield at a greater altitude.

The EO composition in bioregions could be affected by their altitude. In P-7a, there is a significant negative relationship between 1,8-cineole proportion and altitude (Figure 3A). In MO-18b, there is a positive relationship between higher altitude and higher linalool proportion (Figure 3B). These data may coincide with Muñoz-Bertomeu et al. [13] who, in spike lavender populations in the Valencia region, described that the higher altitude favored populations with EO rich in linalool, and the lower altitude favored the accumulation of 1,8-cineole. The camphor proportion showed a tendency in the bioregion populations toward a lower content with higher altitude (Figure 3D), but not in C-4a where it was positive (Figure 3C). Herraiz-Peñalver et al. [16] found a significant and negative correlation between altitude and camphor concentration in their spike lavender population study. Our results show that higher altitude favors obtaining plants with higher EO quality, more content in linalool and less in 1,8-cineole and camphor.

The geographical location in latitude and longitude could be a factor influencing the EO biosynthesis. There are differences between the populations in EO yield according to their latitudinal situation. The populations with a lower latitude, further south, show significantly higher EO yield, such as the MO-18b populations, and in the populations with higher latitude, further north, such as the C-4a and P-7a populations, their EO yield is lower (Figure 4A). The longitudinal, west-east situation of the populations shows no relationship with EO yield. The effect of the geographical location on the EO yield and composition of *L. latifolia* has not been directly addressed by other authors. Herraiz-Peñalver et al. [16] in their study in populations of different peninsular geographic regions did not find differences in EO yield. They find differences in the EO composition, the northern and eastern regions have higher camphor content, and the central and southern regions are characterized by higher linalool content. In our study, it is agreed that the populations further north and east have significantly higher camphor content (Figure 4E) as Prepyrenean bioregion, but it differs with respect to linalool content. The populations located at less latitude, further south, have less linalool and more 1,8-cineole (Figure 4B,C).

Between the EO yield and its main components (Table 6), there is no significant correlation in the mean values of the 34 populations and in the bioregions. There is a negative and significant correlation between 1,8-cineole and linalool and camphor. There is a weak, non-significant negative correlation between camphor and linalool, also in bioregions (Table S6), as it was described by Herraiz-Peñalver et al. [16].

From the data provided on the EO yield and composition of the populations, valuable genotypes could be selected and propagated to transfer them to culture (Table 5). Selecting for EO yield and better quality in relation to linalool/camphor, in the C-4a bioregion, population 558 could be selected with three-year mean data of 3.5% EO yield, and a 45.4%/14.1% ratio; in MC-18a, the population 552 with 3.3% EO yield and a ratio of 26.6%/7.8%; and in MO-18b, the population 568 with 4.1% EO yield and a ratio of 36.6%/14.1%. These last two populations are located in localities above 1000 m of altitude, and the genotypes obtained for cultivation could be competitive in that altitude compared to the hybrid Lavandin adapted to more thermophilic conditions. In the P-7a region, there are genotypes with high camphor % and could be selected to obtain EO for phytosanitary uses. The selected genotypes could serve also as genitors in crosses with *L. angustifolia* to obtain new Lavandin lines since the current lines in culture are of French origin [32] and hybrid lines with wider adaptation would be useful.

5. Conclusions

The new information provided in this work about EO yield and composition and the climatic, edaphic, and geographic influence could serve as a reference for better conservation and possible restoration of spike lavender populations. The study shows a fairly real value of the production and quality of spike lavender EO obtained from wild populations, both on a national scale and in different

Spanish regions. The influence of falling rain on the EO yield and quality is verified. Higher annual rainfall results in higher EO yield in all bioregions and also better quality. There are other factors that also influence EO yield and quality since the bioregion with the highest yield is the one with the least rainfall. It is found that the composition of the soil influences the EO yield and quality. Soils richer in organic matter and minerals result in more production and poorer quality. The altitude at which the populations are found has little effect on their EO yield. It seems that the higher altitude influences the quality of oils. The geographical location of the population could influence their production and quality. At lower latitude, further south, the populations obtain a higher EO yield. The populations located to the northeast have a higher content of camphor, and further south, less linalool. The detailed provision of data from all populations, both in their location and in their behavior in EO yield and composition, allows the selection of the best genotypes for different breeding plans and useful information to improve culture protocols for *L. latifolia* Medik.

Supplementary Materials: The following are available online at http://www.mdpi.com/2077-0472/10/12/626/s1, Table S1: Decimal coordinates of the 34 populations of bioregions and collection date in the three years, Table S2: EO yield (% on dry weight) and major components (in EO %) in 34 wild populations of *L. latifolia* Medik. for years and the mean for each year. For the Cantabroatlantic (C-4a) bioregion, Table S3: EO yield (% on dry weight) and major components (in EO %) in 34 wild populations of *L. latifolia* Medik. for the Prepyrinean (P-7a) bioregion, Table S4: EO yield (% on dry weight) and major components (in EO %) in 34 wild populations of *L. latifolia* Medik. for the Mediterranean Castillian (MC-18a) bioregion, Table S5: EO yield (% on dry weight) and major components (in EO %) in 34 wild populations of *L. latifolia* Medik. for the Mediterranean Oroiberian (MO-18b) bioregion, Table S6: Relationship between EO yield main components means of 34 populations of *L. latifolia* Medik. for three years based on the Pearson correlation matrix with standardized variables by bioregions. With significance level $p < 0.05$ (*) and $p < 0.01$ (**). Figure S1: Gaussen's climatic diagrams by year quarters in each bioregion.

Author Contributions: M.F.-S. and J.M.C. conceived and designed the study; M.F.-S. performed the experiment; M.F.-S. and J.M.C. analyzed the data; M.F.-S. wrote the manuscript; M.F.-S. and J.M.C. reviewed and edited the manuscript. All authors have read and agreed to the published version of the manuscript.

Funding: Spanish Ministerio de Economía y Competitividad RTA2012-00057-C03-02 supported the research work.

Acknowledgments: The authors would like to thank Federico Varela for helping in collecting plants and Mª Brigida Fernández de Simón for helping in the interpretation of chromatography.

Conflicts of Interest: The authors declare no conflict of interest.

References

1. Morales, R. Lavandula. In *Flora Iberica Volume 12*; Morales, R., Quintanar, A., Cabezas, F., Pujadas, A.J., Cirujano, S., Eds.; Real Jardín Botánico, CSIC: Madrid, Spain, 2010; pp. 484–496.
2. Lesage-Meessen, L.; Bou, M.; Sigoillot, J.C.; Faulds, C.B.; Lomascolo, A. Essential oils and distilled straws of lavender and lavandin: A review of current use and potential application in white biotechnology. *Appl. Microbiol. Biotechnol.* **2015**, *99*, 3375–3385. [CrossRef] [PubMed]
3. Agreste. La statistique, l'évaluation et la prospective du ministère de l'Agriculture et de l'Alimentation. Available online: https://agreste.agriculture.gouv.fr/agreste-web/disaron/SAANR_DEVELOPPE_2/detail/ (accessed on 8 December 2020).
4. Burillo, J. *Investigación y Experimentación de Plantas Aromáticas y Medicinales En Aragón. Cultivo, Transformación y Analítica*; Burillo, J., Ed.; Gobierno de Aragón, Departamento de Agricultura, Dirección General de Tecnología Agraria: Zaragoza, Spain, 2003.
5. Lis-Balchin, M.; Hart, S. Studies on the mode of action of the essential oil of Lavender (Lavandula angustifolia P. Miller). *Phyther Res.* **1999**, *13*, 540–542. [CrossRef]
6. Haig, T.J.; Haig, T.J.; Seal, A.N.; Pratley, J.E.; An, M.; Wu, H. Lavender as a source of novel plant compounds for the development of a natural herbicide. *J. Chem. Ecol.* **2009**, *35*, 1129–1136. [CrossRef] [PubMed]
7. Santana, O.; Cabrera, R.; Giménez, C.; Sánchez-Vioque, R.; de los Mozos-Pascual, M.; Rodríguez-Conde, M.F.; Laserna-Ruiz, I.; Usano-Alemany, J. Perfil químico y biológico de aceites esenciales de plantas aromáticas de interés agro-industrial en Castilla-La Mancha (España). *Grasas y Aceites* **2012**, *63*, 214–222. [CrossRef]
8. Méndez-Tovar, I.; Herrero, B.; Pérez-Magariño, S.; Pereira, J.A.; Asensio-S-Manzanera, M.C. By-product of Lavandula latifolia essential oil distillation as source of antioxidants. *J. Food Drug Anal.* **2015**, *23*, 225–233. [CrossRef] [PubMed]

9. Quílez, M.; Ferreres, F.; López-Miranda, S.; Salazar, E.; Jordán, M.J. Seed oil from mediterranean aromatic and medicinal plants of the lamiaceae family as a source of bioactive components with nutritional. *Antioxidants* **2020**, *9*, 510. [CrossRef] [PubMed]
10. Boelens, M.H. The essential oil of spike Lavender Lavandula latifolia Vill. (L. spica DC). *Perfum. Flavorist* **1986**, *11*, 43–63.
11. García Vallejo, M.I. *Aceites Esenciales De Las Lavandulas Ibericas Ensayo Ide La Quimiotaxonomi*; Universidad Complutense de Madrid: Madrid, Spain, 1992; Available online: http://bibdigital.rjb.csic.es/PDF/García-Vallejo_Aceites_esenciales_Lavandulas_ibéricas_Tesis_2002.pdf (accessed on 20 March 2020).
12. Lis-Balchin, M. Chemical composition of essential oils from different species, hybrids and cultivars of Lavandula. In *Lavender*; Taylor & Francis Inc.: New York, NY, USA, 2002; pp. 251–262.
13. Muñoz-Bertomeu, J.; Arrillaga, I.; Segura, J. Essential oil variation within and among natural populations of Lavandula latifolia and its relation to their ecological areas. *Biochem. Syst. Ecol.* **2007**, *35*, 479–488. [CrossRef]
14. Cavanagh, H.M.A.; Wilkinson, J.M. Biological activities of lavender essential oil. *Phyther. Res.* **2002**, *16*, 301–308. [CrossRef]
15. Kaloustian, J.; Pauli, A.M.; Pastor, J. Evolution of camphor and others components in the essential oils of two labiate species during the biological cycle. *Analusis* **2000**, *28*, 308–315. [CrossRef]
16. Herraiz-Peñalver, D.; Cases, M.Á.; Varela, F.; Navarrete, P.; Sánchez-Vioque, R.; Usano-Alemany, J. Chemical characterization of Lavandula latifolia Medik. essential oil from Spanish wild populations. *Biochem. Syst. Ecol.* **2013**, *46*, 59–68. [CrossRef]
17. Salido, S.; Altarejos, J.; Nogueras, M.; Sánchez, A. Chemical composition and seasonal variations of spike lavander oil from southern Spain. *J. Essent Oil Res.* **2004**, *16*, 206–210. [CrossRef]
18. Mapa. Ministerio de Agricultura, Pesca y Alimentación. Available online: https://www.mapa.gob.es/es/ (accessed on 8 December 2020).
19. Rivas-Martínez, S.; Rivas-Saenz, S. *Worldwide Bioclimatic Classification System*; Phytosociological Research Center: Madrid, Spain, 2011.
20. Cases, M.A.; Navarrete, P.; Calvo, R.; López-Ceper, P.; Pérez-Mao, D.; Varela, F. Recolección y caracterización química de espliego (Lavandula latifolia Medik.) en Las Alcarrias (Guadalajara y Cuenca). Primer paso de un programa de selección y mejora. *IV Congr Mejor Genética Córdoba* **2008**, *51*, 355–356.
21. Thomas, G.W. Soil pH and soil acidity. In *Methods of Soil Analysis. Part. 3. Chemical Methods*; Sparks, D.L., Ed.; Soil Science Society of America, Inc.: Madison, WI, USA, 1996; pp. 475–490.
22. Rhoades, J.D. Salinity: Electrical conductivity and total dissolved solid. In *Methods of Soil Analysis. Part. 3. Chemical Methods*; Sparks, D.L., Ed.; Soil Science Society of America, Inc.: Madison, WI, USA, 1996; pp. 417–435.
23. Walkley, A.; Black, I.A. An examination of Degtjareff method for determining soil organic matter and a proposed modification of the chromic acid titration method. *Soil Sci.* **1934**, *37*, 29–38. [CrossRef]
24. Watanabe, F.S.; Olsen, S.R. Test of an ascorbic acid method for determining phosphorous in water and NaHCO$_3$ extracts from soil. *Soil Sci. Soc. Am. J.* **1965**, 677–678. [CrossRef]
25. Pratt, P.F. Availability indexes. Potassium. In *Methods of Soil Analysis. Part. 2. American Society of Agronomy*; Soil Science Society of America, Inc.: Madison, WI, USA, 1965; pp. 1027–1028.
26. Bagnouls, F.; Gaussen, H. Saison seche el régime xerothermique, [Dry season and xerothermic regime]. Documents pour les Cartes des Production Vegetates, Sér. Géneralités. *Toulouse* **1953**, *3*, 193–239.
27. Council of Europe. *Eurpean Pharmacopoeia*, 3rd ed.; Council of Europe: Strasbourg, France, 1996.
28. International Organization for Standardization. *ISO 4719: Oil of Spike Lavender [Lavandula Latifolia (L.f.) Medikus], Spanish Type*; International Organization for Standardization: Vernier, Switzerland, 1999.
29. Carrasco, A.; Martinez-Gutierrez, R.; Tomas, V.; Tudela, J. Lavandula angustifolia and Lavandula latifolia Essential Oils from Spain: Aromatic Profile and Bioactivities. *Planta Med.* **2015**, *82*, 163–170. [CrossRef]
30. Guillen, M.D.; Cabo, N.; Burillo, J. Characterisation of the essential oils of some cultivated aromatic plants of industrial interest. *J. Sci. Food Agric.* **1996**, *70*, 359–363. [CrossRef]
31. Chrysargyris, A.; Laoutari, S.; Litskas, V.D.; Stavrinides, M.C.; Tzortzakis, N. Effects of water stress on lavender and sage biomass production, essential oil composition and biocidal properties against Tetranychus urticae (Koch). *Sci. Hortic.* **2016**, *213*, 96–103. [CrossRef]
32. Usano-Alemany, J.; Herraiz-Peñalver, D.; Cuadrado Ortiz, J.; De López, B.B.; Ruiz, O.S.; Palá-Paúl, J. Ecological production of lavenders in Cuenca province (Spain). A study of yield production and quality of the essential oils. *Bot. Complut.* **2011**, *35*, 147–152. [CrossRef]

33. Chrysargyris, A.; Panayiotou, C.; Tzortzakis, N. Nitrogen and phosphorus levels affected plant growth, essential oil composition and antioxidant status of lavender plant (Lavandula angustifolia Mill.). *Ind. Crops Prod.* **2016**, *83*, 577–586. [CrossRef]
34. Chrysargyris, A.; Drouza, C.; Tzortzakis, N. Optimization of potassium fertilization/nutrition for growth, physiological development, essential oil composition and antioxidant activity of Lavandula angustifolia Mill. *J. Soil Sci. Plant. Nutr.* **2017**, *17*, 291–306. [CrossRef]

Publisher's Note: MDPI stays neutral with regard to jurisdictional claims in published maps and institutional affiliations.

© 2020 by the authors. Licensee MDPI, Basel, Switzerland. This article is an open access article distributed under the terms and conditions of the Creative Commons Attribution (CC BY) license (http://creativecommons.org/licenses/by/4.0/).

Article

Effects of Irrigation, Peat-Alternative Substrate and Plant Habitus on the Morphological and Production Characteristics of Sicilian Rosemary (*Rosmarinus officinalis* L.) Biotypes Grown in Pot

Salvatore La Bella [1,†], **Giuseppe Virga** [2,†], **Nicolò Iacuzzi** [1], **Mario Licata** [1,*], **Leo Sabatino** [1,*], **Beppe Benedetto Consentino** [1], **Claudio Leto** [1,2] and **Teresa Tuttolomondo** [1]

[1] Department of Agricultural, Food and Forest Sciences, Università Degli Studi di Palermo, Viale delle Scienze 13, Building 4, 90128 Palermo, Italy; salvatore.labella@unipa.it (S.L.B.); nicolo.iacuzzi@unipa.it (N.I.); beppebenedetto.consentino@unipa.it (B.B.C.); claudio.leto@unipa.it (C.L.); teresa.tuttolomondo@unipa.it (T.T.)

[2] Research Consortium for the Development of Innovative Agro-Environmental Systems (Corissia), Via della Libertà 203, 90143 Palermo, Italy; giuseppe.virga@corissia.it

* Correspondence: mario.licata@unipa.it (M.L.); leo.sabatino@unipa.it (L.S.)

† These authors are equally contributed.

Abstract: Irrigation and growing substrate are considered as essential cultivation practices in order to obtain good productive and qualitative performance of potted rosemary plants. In pot growing, the chemical, physical and biological characteristics of the substrate must be stable over time in order to allow regular plant growth. However, the effects of cultivation techniques on the characteristics of potted rosemary are little known. Peat is traditionally used as the organic growing medium; however, despite numerous advantages, its use has determined a degradation of peatlands in the northern hemisphere and an increase in greenhouse gases in the atmosphere. The aim of the present study was to assess the effects of irrigation and peat-alternative substrates on the morphological, aesthetic and production characteristics of potted Sicilian rosemary biotypes with different habitus types. Two years, two different irrigation levels, three peat-alternative substrates and three types of rosemary plant habitus were tested in a split-split-split-plot design for a four-factor experiment. The results highlight that irrigation and substrate determined significant differences for all tested parameters. Rosemary plants demonstrated the best performances when irrigation was more frequent; vice versa, the greatest percent content in essential oil was obtained when irrigation events were less frequent. The chemical–physical characteristics of peat-alternative substrates changed with decreases in the peat content and increases in the compost content. The erect habitus biotype showed the best adaptation capacity to the various treatments. Our results suggest that irrigation and peat-alternative substrates significantly affect the growth of rosemary plants and should, therefore, be taken into consideration in order to improve the cultivation of this species in pots for ornamental purposes.

Keywords: aromatic species; alternative substrates; irrigation; plant habitus; sustainable cultivation

1. Introduction

Rosemary (*Rosmarinus officinalis* L.) is a xerophytic, evergreen shrub widely used for food and ornamental purposes, and is a long-time favourite for pot plants and private gardens [1–4]. Due to its biotechnical characteristics and hardiness in times of environmental stress [5,6], it is also used to protect against soil erosion and as a pioneer species during reforestation in fire-damaged areas [5,7].

Its wealth of bioactive compounds is also considered to be highly effective, also reported in the Pharmacopoeia [8,9].

Numerous phytochemical studies have demonstrated the presence of polyphenolic derivatives in the essential oil (EO) that provides the species with a number of pharmacological and medicinal properties, including antibacterial, anti-inflammatory, antioxidant, antitumor and antidiabetic actions [1,10–12]. Recent studies have demonstrated that the accumulation and composition of secondary metabolites in rosemary and, in general, in medicinal and aromatic plants, are highly influenced by genetic [8,13–16], environmental [17–22] and cultivation factors [23–27]. With regard to cultivation aspects, a number of studies have focused on the effects of some agronomic practices on growth, productivity and essential oil constituents of rosemary, in open-field conditions. Some authors [28] reported that the growth, quality and quantity of rosemary essential oil varied with organic and inorganic fertilizers. It was demonstrated [29] that combined application of vermicompost and chemical fertilizers helped to increase crop productivity and sustained the soil fertility. It was found [30] that the composition of rosemary oil could be altered by fertilization programs in regions of poor soil. It was observed [22] that the application of deficit irrigation affected the morphological and physiological characteristics of rosemary, while humidity influenced parameters related with plant–water relations. Singh [31] found that plant spacing, fertilizer and irrigation regimes affected herbage and oil yield but did not influence oil-content percentage of the species. Other authors [32] highlighted that growth media and regulators influenced significantly the vegetative propagation of rosemary. However, the effects of cultivation techniques on the morphological and production characteristics in pot-grown rosemary are, as yet, little known. Most studies focus on the influence of fertilization [33,34] and type of substrate [34–38] on plant growth and essential-oil production of rosemary. In particular, the choice of the growing medium seems to be crucial for cultivation of rosemary and, in general, of aromatic and medicinal plants due to significant effects on their vegetative and productive characteristics.

Peat is traditionally used as the organic growing medium in pot cultivation [38,39]. Peat is partially decayed organic matter, the result of the degradation of bog plants and bryophyte moss [40,41]. Although its use has many advantages—such as a reduction in pH and salinity, good hydraulic retention capacity, a decrease in pathogen load and weeds, and greater ease of handling and mixing—over time the continuous harvesting of peat for agricultural purposes has led to the degradation of peatlands in the northern hemisphere and an inevitable increase in greenhouse gases in the atmosphere [42]. As a consequence, many countries have begun to impose restrictions on the use of this material. Peatlands are home to a wide range of natural habitats that guarantee biological diversity and the survival of a number of species currently considered at risk. This important ecosystem not only plays a fundamental role in carbon-fixing and in storing natural water resources but also safeguards the historical and geochemical memory of our planet [43–45].

In recent times, peat has become increasingly expensive and difficult to obtain [46], leading to the search for alternative substrates [47]. The use of alternative substrates in pot plants presupposes, however, that a number of chemical–physical and hydraulic properties, such as bulk density, pH, electrical conductivity, cation-exchange capacity, hydraulic-retention capacity, organic-matter content and porosity, can be guaranteed [48–51]. A number of studies [37,47,52–56] have demonstrated that organic residues, such as municipal solid waste, sewage sludge and pruning residues (following an adequate composting process) can be used as alternative growing substrates to peat, with optimal results. Some authors [34] studied the effect of growing substrates on the EO content of rosemary, concluding that all growing substrates are suitable when fertilization and irrigation practices are well-controlled. It was demonstrated [35] that peat can be replaced with a mixture of compost, chicken manure and biochar in rosemary cultivation, while other authors [37] achieved better rooting and length/weight of rosemary roots when using vermicompost.

The main aim of this study was to assess the effects of irrigation, peat-alternative substrates and plant habitus on the morphological, aesthetic and production characteristics of Sicilian rosemary biotypes grown in pots.

2. Materials and Methods

2.1. Rosemary Experimental Field

Tests were carried out in the two years 2016 and 2017 in Sciacca (Italy) (37°30′33″ N–13°05′20″ E, 60 m a.s.l.), in an experimental field belonging to the Department of Agriculture, Food and Forest Sciences of the University of Palermo. Three Sicilian rosemary biotypes (code RSM) were gathered from a collection field of 10-year-old rosemary mother plants. Plant material was characterized taxonomically using analytical keys and by comparing it with exsiccata that had been previously prepared. The voucher specimen codes of the exsiccata (SAAF-S/0357) were deposited at the Department of Agricultural, Food and Forest Sciences Herbarium of University of Palermo (Italy). The plants were selected according to different growth habitus types: RSM_2 (SAAF-S/0357_a) presented an erect habitus, RSM_6 (SAAF-S/0357_b) a semi-erect habitus and RSM_5 (SAAF-S/0357_c) a prostrate habitus (Figure 1).

(a)   (b)   (c)

Figure 1. Types of plant habitus of rosemary biotypes in the study. (**a**) refers to erect habitus, (**b**) refers to semi-erect habitus, (**c**) refers to prostate habitus.

Each accession was propagated using stem cuttings from the apical plant parts 6 cm in length. The cuttings were treated with 1-Naphthaleneacetic acid (0.50%) and then rooted in an open cold frame. The rooted cuttings were transplanted into 18-cm pots on 20 May 2016 and 15 May 2017. Four types of substrate were used with varying percentages of peat (70, 50, 40 and 20%), compost (20, 30 and 50%) and perlite (constant at 30%). The substrate was fortified with a slow-release N–P–K fertilizer + microelements (15–19−15 + 2 + 5) at a rate of 3 gr L^{-1} prior to transplanting.

A drip irrigation system was used for water delivery. Irrigation management consisted of integrating 100% field capacity every 4 days or 2 days. A split-split-split plot experimental design was used for a 4-factor experiment with 3 replicates. The main plot was year (Y) with two treatment levels: Y_1 (2007) and Y_2 (2008). The sub-plot factor was irrigation (I) with two treatment levels: I_1 (integration 100% field capacity every 4 days) and I_2 (integration 100% field capacity every 2 days). The sub-sub-plot factor was the substrate (S) with four treatment levels: S_1 (50% peat, 20% compost, 30% perlite); S_2 (40% peat, 30% compost, 30% perlite); S_3 (30% peat, 40% compost, 30% perlite) and S_4 (70% peat, 30% perlite) as the control. The sub-sub-sub-plot factor was plant habitus (H) with three treatment levels: H_1 (erect habitus); H_2 (semi-erect habitus) and H_3 (prostrate habitus).

2.2. Morphological, Aesthetic and Production Characteristics of the Plants

For all of the rosemary treatments in the experiment, the main morphological and aesthetic characteristics were determined approx. 90 days from transplanting in order to attribute an ornamental value to the plant: plant height, plant diameter, height-to-diameter ratio, number of primary and secondary branches per plant, branch length and width, number of leaves per cm of branch and general appearance of the plant. Flowering stage was assessed only when 50% of the plant had flowers and was determined using a visible value scale between 1 (few flowers) and 3 (abundant flowers). The fresh-matter weight of plant parts was also determined. The plant material was subsequently dried in an oven at 65 °C for 48 h until it reached a constant weight; the plant dry-matter weight was then calculated. Essential oil content was obtained by hydrodistillation of air-dried plant material (50–100 g) for 3 h in accordance with international guidelines [57].

2.3. Chemical-Physical Properties of Compost

At the beginning of plant growth, the main chemical–physical analyses were carried out on all of the substrates used: pH, electrical conductivity, bulk density, total porosity, air capacity at pF1 and available water capacity. In particular, the compost used was supplied by the company SIRTEC Sistemi Ambientali SRL (Alcamo, Italy). The composting process used by the company included treatment of the preselected organic waste (organic matter from recycled municipal waste or organic residues from agro-industrial processing).

2.4. Weather Data

Data on rainfall and temperature were collected from a meteorological station belonging to Agro-Meteorological Information Service of the Sicilian Government [58]: the station was situated close to the experimental field. The station was equipped with an MTX datalogger (model WST1800) and various climate sensors. More specifically, a temperature sensor MTX (model TAM platinum PT100 thermo-resistance with anti-radiation screen) and a rainfall sensor MTX (model PPR with a tipping bucket rain gauge) provided data on average daily air temperatures (°C), total daily rainfall frequency (d mm > 1) (%) and rainy days per year (d mm > 1) (%).

2.5. Statistical Analysis

Statistical analyses were carried out using MINITAB 19 for Windows. The data were compared using analysis of variance. The difference between means of values was carried out using the Tukey test.

3. Results

3.1. Temperature Trends

Maximum and minimum temperature trends for the experimental site over the two years 2016/2017 are shown in Figure 2.

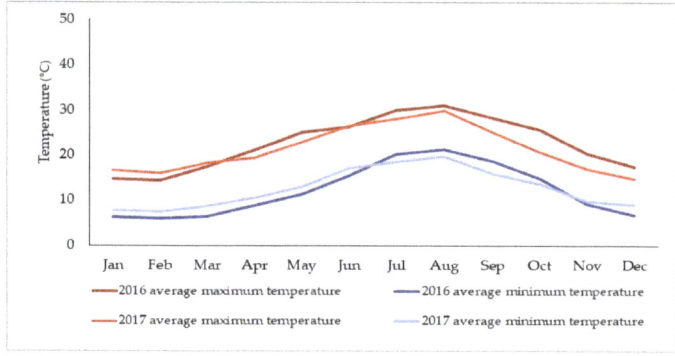

Figure 2. Maximum and minimum air temperature trends during the test period.

Average air temperatures over the approx. 90 days of testing in both years increased between May and August. Maximum air temperatures were on average higher in 2016 than 2017 in May, July and August. In June, average maximum temperatures were similar in both years. Regarding minimum air temperatures, average values were highest in July and August in 2016, whilst in May and June in 2017.

3.2. Chemical–Physical Properties of Substrates

Average values for electrical conductivity and pH (Figure 3) of the peat-alternative substrates increased as the percentage content of the compost in the mix increased in the 4 test substrates.

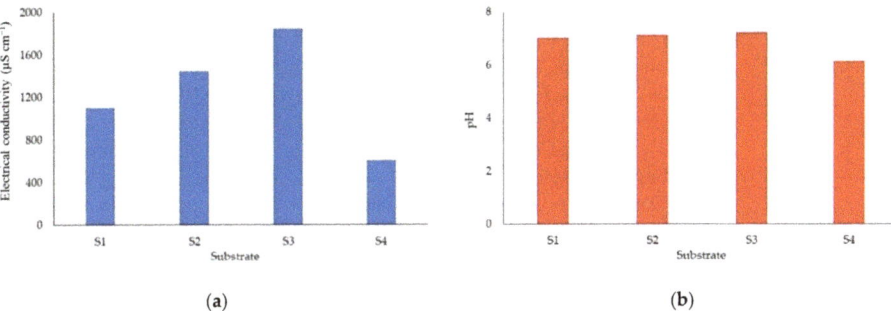

Figure 3. Chemical properties of substrates. Graph (**a**) refers to electrical conductivity values, while graph (**b**) refers to pH values of substrates.

Furthermore, an increase in average electrical conductivity and pH was recorded following a decrease in the peat content in the substrates.

From a physical point of view, it was noted that the gradual substitution of the peat in the mix also determined an increase in bulk density and a slight decrease in the air capacity at pF1 and in porosity. Regarding available water content, the substrates with a greater compost content produced higher average values compared to the control substrate (Table 1).

Table 1. Physical properties of substrates. Average values are shown.

Parameter	Growing Substrate			
	S_1	S_2	S_3	S_4
Bulk density (g cm^{-3})	0.23	0.25	0.33	0.17
Total porosity (%)	89.59	88.88	86.07	91.86
Air capacity at pF1 (%)	41.16	43.42	32.22	46.44
Available water (%)	20.65	19.81	24.33	18.32

3.3. Effects of Year, Substrate, Irrigation and Plant Habitus on Rosemary Plants

Data on the morphological, aesthetic and production characteristics of the rosemary plants, under the influence of the main factors, in years 2016 and 2017, are shown in Tables 2 and 3 and in Figures 4–6.

No significant differences were found for the factor year regarding all of the parameters in the study except for branch width. However, the factor irrigation produced significant effects for all of the parameters tested. Looking more closely at the effect of the two levels of irrigation on the morphological and production parameters, seemingly contrasting results were found. The highest average values for primary branching, flowering stage, number of leaves per cm/branch and percent content of EO were found under irrigation level I_1, whereas highest average values for height, diameter, height-to-diameter ratio, general appearance of the plants, fresh and dry weight, number of secondary branches, and length and width of branches were found under irrigation level I_2.

The plants demonstrated the best morphological, aesthetic and production performances when irrigation was more frequent. Vice versa, the greatest percent content in EO was obtained when irrigation events were less frequent (Table 3, Figure 4).

The factor substrate determined significant differences for all of the morphological and production parameters, with the exception of plant height-to-diameter ratio and flowering. More specifically, greatest average plant heights (29.46 cm) were recorded when the plants were cultivated using the control substrate, whereas the lowest average values were observed in the other substrates. Likewise, the highest average plant diameters (48.71 cm) and general appearance scores (6.10) were recorded for the control substrate. The greatest number of primary (15.29) and secondary (19.41) branches was obtained in rosemary plants grown in the control substrate; the smallest number of primary and secondary branches was found with substrates S_1 and S_3, respectively. The greatest average values relating to branch length and width were found for substrate S_3, whilst the lowest average values were obtained with the control substrate. Regarding production parameters, the greatest average fresh weights and dry weights were obtained using the control substrate, confirming trends observed for most of the morphological parameters. The percent content of EO ranged between 0.68% (S_1 and S_3) and 0.66% (S_2 and S_4); the difference between the different substrates in terms of average percentages in EO was, therefore, minimal (Table 3, Figure 5).

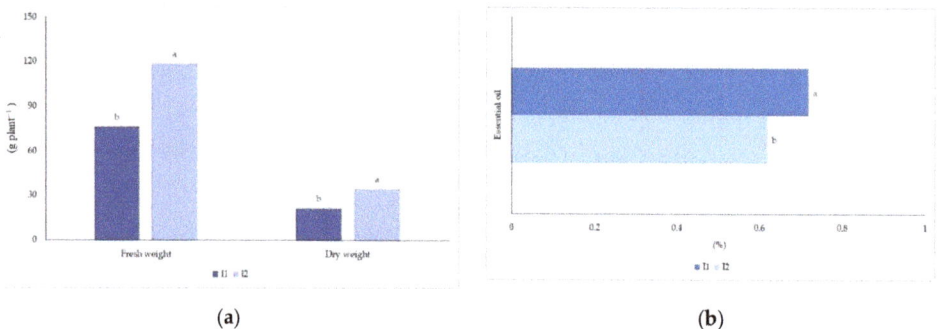

Figure 4. Effect of irrigation on production characteristics of rosemary biotypes. Graph (**a**) refers to effect of irrigation on fresh and dry weight, while graph (**b**) refers to effect of irrigation on essential oil. Means followed by the same letter are not significantly different for $p \leq 0.05$ according to Tukey's test.

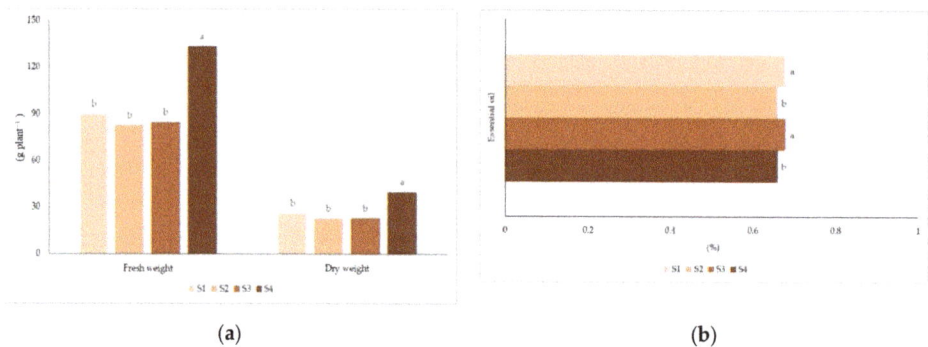

Figure 5. Effect of substrate on production characteristics of rosemary biotypes. Graph (**a**) refers to effect of substrate on fresh and dry weight, while graph (**b**) refers to effect of substrate on essential oil. Means followed by the same letter are not significantly different for $p \leq 0.05$ according to Tukey's test.

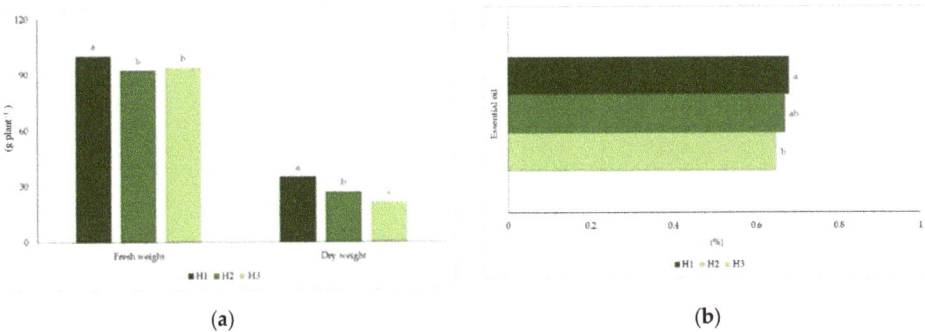

(a) (b)

Figure 6. Effect of plant habitus on production characteristics of rosemary biotypes. Graph (a) refers to effect of plant habitus on fresh and dry weight, while graph (b) refers to effect of plant habitus on essential oil. Means followed by the same letter are not significantly different for $p \leq 0.05$ according to Tukey's test.

The factor plant habitus had a significant effect on all of the parameters in the study. More specifically, biotype RSM_2, with erect growth habitus, differed from the others in terms of greater height (39.29 cm), greater height-to-diameter ratio (1.29), greater number of primary branches, better general appearance, more abundant flowering, greater fresh (106.08 g) and dry (35.05 g) weight and greater EO percent content (0.68%). The biotype RSM 6 with semi-erect growth habitus, showed greater diameter (50.73 cm) and greater branch length (33.12 cm) and width (3.80 cm) than other biotypes. The EO percent content was intermediate between the other two biotypes. Biotype RSM_5 with prostrate habitus showed the highest number of secondary branches but also obtained the lowest EO percentage (0.65%) of the three rosemary biotypes (Table 3, Figure 6).

Considering the interaction between the main factors (Tables 2 and 3), analysis of the variance showed that the interaction of the factor year with the other factors did not determine significant effects on any of the morphological and production parameters. The irrigation-by-plant habitus interaction determined significant differences for plant diameter and the number of secondary branches; more specifically, the highest average diameter was obtained with the I_2-by-H_2 interaction, whilst the lowest average value was found with I_1-by-H_1 interaction, as showed in Supplementary Table S1.

The substrate-by-plant habitus interaction had significant effects on a series of characteristics, such as plant diameter, fresh and dry weight, the development of secondary branching, branch length and width, and EO content. With specific reference to the treatments, the greatest plant diameter was recorded with the interaction between the control substrate and the semi-erect habitus biotype. The lowest value was recorded with the interaction between S_3 (which contained the highest compost content) and H_3. The greatest number of secondary branches was observed with the interaction between the control substrate and the prostrate habitus biotype, whilst the smallest number was found with the interaction between the control substrate and the erect habitus biotype. The highest average values for branch length and width were obtained with the S_3-by-H_2 interaction. Regarding production parameters, the interaction between the control substrate and the prostrate habitus biotypes produced the highest average value for fresh weight.

Considering dry weight, highest averages were found when the control substrate interacted with the erect habitus biotype, whereas the lowest average values were produced with the interaction between substrate S_3 and the biotype with prostrate habitus. A closer look at EO content showed that the S_1-by-H_1 interaction produced the highest EO percentage (0.73%). The lowest average EO percentage (0.64%) was recorded with the S_4-by-H_1 and S_2-by-H_3 interactions, as showed in Supplementary Materials (Table S1).

Table 2. Morphological and aesthetic characteristics of rosemary plants in response to year, irrigation, peat-alternative substrate and plant habitus.

Treatment	Plant Height (cm)	Plant Diameter (cm)	Height-to-Diameter Ratio	No. Primary Branches (per Plant)	No. Secondary Branches (per Plant)	Plant Branch Length (cm)	Plant Branch Width (cm)	No. Leaves cm Branch^{-1}	General Appearance of Plant	Flowering
Year (Y)										
Y1	26.18 a	39.60 a	0.73 a	14.51 a	14.22 a	29.01 a	3.43 a	8.55 a	5.40 a	0.95 a
Y2	25.27 b	38.66 b	0.74 a	14.25 a	14.01 a	28.53 a	3.32 b	8.67 b	5.26 a	0.94 a
Irrigation (I)										
IW$_1$	24.10 b	36.68 b	0.71 b	15.39 a	10.61 b	27.96 b	3.29 b	8.89 a	5.17 b	1.58 a
IW$_2$	27.83 a	41.58 a	0.76 a	15.37 b	17.63 a	29.58 a	3.46 a	8.33 b	5.50 a	0.31 b
Substrate (S)										
S$_1$	25.35 b	37.92 b	0.72 a	13.36 b	12.92 b	28.70 b	3.32 b	9.07 a	5.21 b	0.83 a
S$_2$	24.59 b	34.78 b	0.77 a	15.51 a	12.43 b	29.67 b	3.47 ab	8.81 ab	4.99 b	0.91 a
S$_3$	24.46 b	35.09 b	0.74 a	13.37 b	11.71 b	31.18 a	3.63 a	8.54 ab	5.05 b	1.11 a
S$_4$	29.46 a	48.71 a	0.70 a	15.29 a	19.41 a	26.54 c	3.08 c	8.02 b	6.10 a	0.94 a
Plant habitus (H)										
H$_1$	39.29 a	30.29 c	1.29 a	18.69 a	6.45 b	27.96 b	3.58 b	8.90 a	5.77 a	1.20 a
H$_2$	22.47 b	50.73 a	0.44 b	10.83 c	5.37 b	33.12 a	3.80 a	8.54 a	5.38 b	0.79 b
H$_3$	16.15 c	36.37 b	0.46 b	13.62 b	30.18 a	25.23 c	2.74 c	8.39 a	4.85 c	0.85 b
Interactions (significance)										
Y × I	n.s.	n.s.	n.s.	n.s.	n.s.	n.s.	n.s.	n.s.	n.s.	n.s.
Y × S	n.s.	n.s.	n.s.	n.s.	n.s.	n.s.	n.s.	n.s.	n.s.	n.s.
Y × H	n.s.	n.s.	n.s.	n.s.	n.s.	n.s.	n.s.	n.s.	n.s.	n.s.
I × S	n.s.	n.s.	n.s.	n.s.	*	n.s.	n.s.	n.s.	n.s.	n.s.
I × H	n.s.	*	n.s.	n.s.	*	*	n.s.	n.s.	n.s.	n.s.
S × H	n.s.	*	n.s.	n.s.	n.s.	*	*	n.s.	n.s.	n.s.
Y × I × S	n.s.	n.s.	n.s.	n.s.	n.s.	n.s.	n.s.	n.s.	n.s.	n.s.
Y × I × H	n.s.	n.s.	n.s.	n.s.	n.s.	n.s.	n.s.	n.s.	n.s.	n.s.
Y × S × H	n.s.	n.s.	n.s.	n.s.	*	n.s.	n.s.	n.s.	n.s.	n.s.
I × S × H	*	n.s.	*	n.s.	n.s.	n.s.	n.s.	n.s.	n.s.	n.s.
Y × I × S × H	n.s.	n.s.	n.s.	n.s.	n.s.	n.s.	n.s.	n.s.	n.s.	n.s.

Means followed by the same letter are not significantly different for $p \leq 0.05$ according to Tukey's test; n.s. = not significant; * = significant at $p \leq 0.05$.

Table 3. Production characteristics of rosemary plants in response to year, irrigation, peat-alternative substrate and plant habitus.

Treatment	Fresh Weight (g plant^{-1})	Dry Weight (g plant^{-1})	EO Content (%)
Year			
Y_1	98.35 a	28.25 a	0.67 a
Y_2	96.66 a	27.61 a	0.67 a
Irrigation			
I_1	76.31 b	21.38 b	0.72 a
I_2	118.76 a	34.49 a	0.62 b
Substrate			
S_1	89.78 b	26.01 b	0.68 a
S_2	82.83 b	22.99 b	0.66 b
S_3	84.63 b	22.89 b	0.68 a
S_4	133.28 a	39.74 a	0.66 b
Plant habitus			
H_1	100.08 a	35.05 a	0.68 a
H_2	92.43 b	27.01 b	0.67 ab
H_3	94.01 b	21.75 c	0.65 b
Interactions (significance)			
Y × I	n.s.	n.s.	n.s.
Y × S	n.s.	n.s.	n.s.
Y × H	n.s.	n.s.	n.s.
I × S	n.s.	n.s.	n.s.
I × H	n.s.	n.s.	n.s.
S × H	*	*	*
Y × I × S	n.s.	n.s.	n.s.
Y × I × H	n.s.	n.s.	n.s.
Y × S × H	n.s.	n.s.	n.s.
I × S × H	*	n.s.	n.s.
Y × I × S × H	n.s.	n.s.	n.s.

Means followed by the same letter are not significantly different for $p \leq 0.05$ according to Tukey's test; n.s. = not significant; * = significant at $p \leq 0.05$.

The interaction between irrigation factors, substrate and plant habitus determined significant differences for height, height-to-diameter ratio, number of secondary branches and plant fresh weight. Observing the various test treatments, the greatest values for height were obtained when irrigation level I_2 interacted with the control substrate and the biotype with erect habitus, whereas the lowest values for height were found with the interaction I_1-by-S_1-by-H_3. The greatest average values for height-to-diameter ratio were determined with the interaction I_2-by-S_4-by-H_1 and I_2-by-S_2-by-H_1. The highest number of secondary branches, however, was found in the interaction irrigation level I_2 with the control substrate and prostrate habitus biotype, whereas the lowest average values were found with the interaction I_1-by-S_4-by-H_1.

Regarding production parameters, it is worth noting that the interaction I_2-by-S_4-by-H_3 presented higher average values for fresh weight (174.72 g), whilst the interaction I_1-by-S_2-by-H_2 presented the lowest averages (64.57 g).

4. Discussion

In this study, over two years of tests, three Sicilian biotypes of *Rosmarinus officinalis* L. were evaluated growing in pots with differing habitus types and two levels of irrigation. Four substrates were compared containing different ratios of peat and compost to perlite.

All the biotypes showed good adaptation capacities to the irrigation conditions and to the different substrate types in both years, showing significant differences for all of the morphological and production characteristics in the study. The general appearance of the plants was monitored regularly and, for both levels of irrigation over the two years, no signs of lack of water or water stress were noted. Various observations have been made in scientific literature on the effects of irrigation on the morphological and production characteristics of rosemary in open-field conditions. Some authors [59] stated that the use of surface or underground micro-drip irrigation systems does not exert a significant effect on the characteristics of rosemary plants. It was demonstrated [23] that the quality of the irrigation water did not have a significant effect on rosemary. In contrast, other authors [60] observed that irrigation significantly influenced the main characteristics of rosemary plants grown in pots. In our study, it was noted that the interval of time between irrigation events significantly influenced the characteristics of rosemary. In general, plant growth was negatively affected by water stress, probably due to a decrease in the stomatal aperture, which limits the circulation of CO_2 in the leaves and reduces photosynthetic activity, as reported in the literature [61].

Although little is known about the effects of cultivation techniques on potted rosemary, it should be emphasized that almost all the studies focus on investigating the influence of substrate on plant growth and production. The composition of the substrate and, in particular, a decrease in the peat content in the mix, had significant effects on the growth and on the production parameters of the plants. Peat represents the most frequently used growing substrate due to its excellent chemical and physical properties for plants grown in pots and its stability over time. Gruda [62] affirms that peat is the standard constituent of substrates used in the production of ornamental plants in pots and that other constituents may vary by 20% to 50%. However, several authors note that high-quality peat is expensive and can cause environmental problems due to the depletion of unrenewable resources [38,55]. Compost represents an interesting substrate alternative to peat and can improve the physical, chemical and biological properties of the substrate [56,63]. Raviv [64] highlights that compost is a bioresource and can be a valid alternative to peat, despite the fact it is potentially a waste material. However, the qualitative properties of compost are strongly linked to its maturity and stability. Rinaldi et al. [38] tested a number of substrates by mixing increasingly greater amounts of eight different composts in place of peat, with a fixed inert material. The authors observed that the most suitable substrates for rosemary growth contained compost at a rate of up to 70%. De Lucia et al. [36] studied four composts, obtained from agro-industrial, urban and green wastes, as growing media components in *Rosmarinus officinalis* L. and obtained improved quality rosemary plants by substituting peat with 30% compost. Our study also assessed the chemical–physical properties of the substrates as these characteristics tend to change as the compost content increases and the peat content decreases. In fact, the particle size of the growing substrate and the geometry of the pot need to be carefully considered in order to balance water availability and root aeration [65]. As reported in some studies [24,66], the decrease in plant height in rosemary grown in pots, using substrates with decreasing quantities of peat, is of interest in order to increase the ornamental value of the rosemary as a reduction in height is a desirable ornamental characteristic. In this study, the greater length and width of the main branches was obtained using a substrate with the greatest compost content. According to Rinaldi et al. [38], compost from pruning residues or mixes with agro-industrial or urban waste are of most interest in rosemary cultivation.

In our study, the factor year did not have significant effects on the rosemary characteristics. This result was also confirmed by some authors [67], who state that the composition and variability of rosemary mainly depend on the genetic background and origin rather than on the environmental conditions and geographic location.

The factor irrigation, however, did have a significant effect on the percent content of the EO. As highlighted in the literature, irrigation influences the morphological and physiological characteristics determining the yield of the plants and also has a bearing on the quantity of some of the principal components of the essential oils (EOs) [22,68–71]. In particular, the percentage of EO increased following limited water availability, in agreement with the results of Pirzad and Mohammadzadeh [27].

The factor substrate affected the percent content of EO. This was confirmed by a number of studies that demonstrated that the quantitative characteristics of EOs of various species in the *Lamiaceae* family, such as *Thymus caespititus* Brot. [72], *Ocimum basilicum* L. [73] and *Lavandula agustifolia* Mill. [15], are greatly affected by the composition of the substrate. It is important to note that the production of EO depends on genetic characteristics, as revealed by this study and in agreement with some authors [67,74–79] who stated that the genetic pool contributes to a greater degree than other factors to determine both the quality and the quantity of EOs.

Rosemary presents a variety of different habitus types, morphological traits, flower colours and aromatic properties [80]. Our study confirms results reported by Flamini et al. [81], who, in a recent study on the evaluation of the agronomic and production characteristics of two rosemary biotypes, state that the best performances are produced by biotypes with an erect habitus. Plant height, in particular, is a morphological characteristic under genetic control; however, its manifestation could depend on environmental factors, such as altitude, air temperature and solar radiation, but also growth techniques, such as irrigation, fertilization and, in the case of pot plants, also the substrate used. The high significance of the interactions substrate-biotype and irrigation-substrate-biotype on the agronomic and production parameters could play an important role in the production of plants with high ornamental value but also in the production of EOs, confirming results of other authors [15,70,81–83].

5. Conclusions

The results of this study show that the use of peat-alternative substrates can represent a valid opportunity for the cultivation of rosemary in pots, and, in general, for the cultivation of numerous medicinal and aromatic plants. The use of compost could allow a partial or total replacement of peat, leading to environmental benefits (the harvesting of peat as a substrate has a strong environmental impact) energy and economic benefits (peat is an expensive and functional material). It could also foster the use of biomass from agro-industrial activities and recycled municipal organic waste. In this study, although the best agronomic results were obtained using the substrate with greater peat content, it is worth highlighting that the substrates with 20% and 30% compost also gave excellent performance results, confirming the idea of a partial replacement of peat. The erect habitus biotype showed the best adaptation capacity to the various treatments. Considering the various interactions between the main factors, the substrate-by-irrigation-by-biotype interaction had significant effects on the morphological and production characteristics of the rosemary plants.

Further research is required, however, to assess both the performance of the various components of the substrate and, more specifically, exactly when the various agronomic factors interact with the substrate. A longer period is also needed to evaluate the morphological and production characteristics. Finally, the various biometric and production characteristics of the genotypes, together with the substrate types, could be of use when selecting rosemary biotypes with good adaptability to cultivation in pots for ornamental purposes, thus favouring the expansion of this species.

Supplementary Materials: The following are available online at https://www.mdpi.com/2077-0472/11/1/13/s1, Table S1: Interactions between the various treatments of the main factors in the study.

Author Contributions: Conceptualization, S.L.B., G.V. and T.T.; methodology, S.L.B., G.V., M.L. and T.T.; software, N.I., L.S. and B.B.C.; validation, L.S., B.B.C. and C.L.; formal analysis, N.I., L.S. and B.B.C.; investigation, S.L.B., G.V. and M.L.; resources, S.L.B. and T.T.; data curation, G.V., N.I., M.L., L.S. and B.B.C.; writing—original draft preparation, S.L.B., N.I. and M.L.; writing—review and editing, S.L.B., G.V. and M.L.; visualization, L.S., C.L. and T.T.; supervision, C.L. and T.T.; project administration, C.L.; funding acquisition, C.L. All authors have read and agreed to the published version of the manuscript.

Funding: This research was funded by Sicilian Regional Ministry of Agriculture and Food Resources (Italy), research project "Environmental and plant resources in the Mediterranean: study, valorisation and defence", grant number 2309/2005.

Institutional Review Board Statement: Not applicable.

Informed Consent Statement: Not applicable.

Data Availability Statement: Data are available by contacting the authors.

Acknowledgments: The authors would like to thank the Sicilian Regional Ministry of Agriculture and Food Resources (Italy), funding the "Environmental and plant resources in the Mediterranean: study, valorisation and defence" research project. Special thanks go to Lucie Branwen Hornsby for her linguistic assistance.

Conflicts of Interest: The authors declare no conflict of interest. The funders had no role in the design of the study; in the collection, analyses, or interpretation of data; in the writing of the manuscript, or in the decision to publish the results.

References

1. De Pasquale, C.; La Bella, S.; Cammalleri, I.; Gennaro, M.C.; Licata, M.; Leto, C.; Tuttolomondo, T. Agronomical and postharvest evaluation of the essential oils of Sicilian rosemary (*Rosmarinus officinalis* L.) biotypes. *Acta Hortic.* **2019**, *1255*, 139–144. [CrossRef]
2. Begum, A.; Sandhya, S.; Ali, S.S.; Vinod, K.R.; Swapna, R.; Banji, D. An in-depth review on the medicinal flora *Rosmarinus officinalis* (Lamiaceae). *Acta Sci. Pol. Technol. Aliment.* **2013**, *12*, 61–74. [PubMed]
3. Heinrich, M.; Kufer, K.; Leonti, M.; Pardo-de-Santayana, M. Ethnobotany and ethnopharmacology—Interdisciplinary links with the historical sciences. *J. Ethnopharmacol.* **2006**, *107*, 157–160. [CrossRef] [PubMed]
4. Moreno, S.; Ojeda Sana, A.M.; Gaya, M.; Barni, M.V.; Castro, A.O.; van Baren, C. Rosemary compounds as nutraceutical health products. In *Food Additives*, 1st ed.; El-Samragy, Y., Ed.; IntechOpen Science: Rijeka, Croatia, 2012; pp. 157–174.
5. Tuttolomondo, T.; La Bella, S.; Leto, C.; Gennaro, M.C.; Calvo, R.; D'Asaro, F. Biotechnical characteristics of root systems in erect and prostrate habit *Rosmarinus officinalis* L. accessions grown in a Mediterranean climate. *Chem. Eng. Trans.* **2017**, *58*, 769–774. [CrossRef]
6. Sarmoum, R.; Haid, S.; Biche, M.; Djazouli, Z.; Zebib, B.; Merah, O. Effect of salinity and water stress on the essential oil components of rosemary (*Rosmarinus officinalis* L.). *Agronomy* **2019**, *9*, 214. [CrossRef]
7. Durán Zuazo, V.H.; Rodriguez Pleguezelo, C.R. Soil-erosion and runoff prevention by plant covers. A review. *Agron. Sustain. Dev.* **2008**, *28*, 65–86. [CrossRef]
8. Napoli, E.M.; Siracusa, L.; Saija, A.; Speciale, A.; Trombetta, D.; Tuttolomondo, T.; La Bella, S.; Licata, M.; Virga, G.; Leone, R.; et al. Wild Sicilian rosemary: Phytochemical and morphological screening and antioxidant activity evaluation of extracts and essential oils. *Chem. Biodivers.* **2015**, *12*, 1075–1094. [CrossRef]
9. Napoli, E.M.; Curcuruto, G.; Ruberto, G. Screening of the essential oil composition of wild Sicilian rosemary. *Biochem. Syst. Ecol.* **2010**, *38*, 659–670. [CrossRef]
10. Tuttolomondo, T.; Dugo, G.; Ruberto, G.; Leto, C.; Napoli, E.M.; Cicero, N.; Gervasi, T.; Virga, G.; Leone, R.; Licata, M.; et al. Study of quantitative and qualitative variations in essential oils of Sicilian *Rosmarinus officinalis* L. *Nat. Prod. Res.* **2015**, *29*, 1928–1934. [CrossRef]
11. Sánchez-Camargo, A.P.; Herrero, M. Rosemary (*Rosmarinus officinalis*) as a functional ingredient: Recent scientific evidence. *Curr. Opin. Food Sci.* **2017**, *14*, 13–19. [CrossRef]
12. Andrade, M.A.; Ribeiro-Santos, R.; Costa Bonito, M.C.; Saraiva, M.; Sanches-Silva, A. Characterization of rosemary and thyme extracts for incorporation into a whey protein based film. *LWT-Food Sci. Technol.* **2018**, *92*, 497–508. [CrossRef]
13. Yosr, Z.; Hnia, C.; Rim, T.; Mohamed, B. Changes in essential oil composition and phenolic fraction in *Rosmarinus officinalis* L. var. *typicus* Batt. organs during growth and incidence on the antioxidant activity. *Ind. Crop. Prod.* **2013**, *43*, 412–419. [CrossRef]
14. Alipour, M.; Saharkhiz, M.J. Phytotoxic activity and variation in essential oil content and composition of rosemary (*Rosmarinus officinalis* L.) during different phenological growth stages. *Biocatal. Agric. Biotechnol.* **2016**, *7*, 271–278. [CrossRef]
15. Najar, B.; Demasi, S.; Caser, M.; Gaino, W.; Cioni, P.L.; Pistelli, L.; Scariot, V. Cultivation substrate composition influences morphology, volatilome and essential oil of *Lavandula angustifolia* Mill. *Agronomy* **2019**, *9*, 411. [CrossRef]
16. Carrubba, A.; Abbate, L.; Sarno, M.; Sunseri, F.; Mauceri, A.; Lupini, A.; Mercati, F. Characterization of Sicilian rosemary (*Rosmarinus officinalis* L.) germplasm through a multidisciplinary approach. *Planta* **2020**, *251*, 37. [CrossRef] [PubMed]
17. Figueiredo, A.C.; Barroso, J.G.; Pedro, L.G.; Scheffer, J.J.C. Factors affecting secondary metabolite production in plants: Volatile components and essential oils. *Flavour Fragr. J.* **2008**, *23*, 213–226. [CrossRef]
18. Franz, C.; Novak, J. Sources of essential oils. In *Handbook of Essential Oils: Science, Technology, and Application*, 3rd ed.; Başer, K.H.C., Buchbauer, G., Eds.; CRC Press: Boca Raton, FL, USA, 2020; Volume 1, pp. 1–43.
19. Farouk, S.; Al-Amri, S.M. Exogenous melatonin-mediated modulation of arsenic tolerance with improved accretion of secondary metabolite production, activating antioxidant capacity and improved chloroplast ultrastructure in rosemary herb. *Ecotox. Environ. Safe* **2019**, *180*, 333–347. [CrossRef]

20. Raffo, A.; Mozzanini, E.; Ferrari Nicoli, S.; Lupotto, E.; Cervelli, C. Effect of light intensity and water availability on plant growth, essential oil production and composition in *Rosmarinus officinalis* L. *Eur. Food Res. Technol.* **2020**, *246*, 167–177. [CrossRef]
21. Böszörményi, A.; Dobi, A.; Skribanek, A.; Pávai, M.; Solymosi, K. The effect of light on plastid differentiation, chlorophyll biosynthesis, and essential oil composition in rosemary (*Rosmarinus officinalis*) leaves and cotyledons. *Front. Plant Sci.* **2020**, *11*, 196. [CrossRef]
22. Sánchez-Blanco, J.M.; Ferrández, T.; Navarro, A.; Bañon, S.; Alarcón, J. Effects of irrigation and air humidity preconditioning on water relations, growth and survival of *Rosmarinus officinalis* plants during and after transplanting. *J. Plant Physiol.* **2004**, *161*, 1133–1142. [CrossRef]
23. Bernstein, N.; Chaimovitch, D.; Dudai, N. Effect of irrigation with secondary treated effluent on essential oil, antioxidant activity, and phenolic compounds in oregano and rosemary. *Agron. J.* **2009**, *101*, 1–10. [CrossRef]
24. Singh, M.; Guleria, N. Influence of harvesting stage and inorganic and organic fertilizers on yield and oil composition of rosemary (*Rosmarinus officinalis* L.) in a semi-arid tropical climate. *Ind. Crop. Prod.* **2013**, *42*, 37–40. [CrossRef]
25. Khalil, S.E.; Khalil, A.M. Effect of water irrigation intervals, compost and dry yeast on growth, yield and oil content of *Rosmarinus officinalis* L. plant. *Am. Eurasian J. Sustain. Agric.* **2015**, *9*, 36–51.
26. Ganjali, A.; Kaykhaii, M. Investigating the essential oil composition of *Rosmarinus officinalis* before and after fertilizzating with vermicompost. *J. Essent. Oil Bear. Plants* **2017**, *20*, 1413–1417. [CrossRef]
27. Pirzad, A.; Mohammadzadeh, S. Water use efficiency of three mycorrhizal Lamiaceae species (*Lavandula officinalis*, *Rosmarinus officinalis* and *Thymus vulgaris*). *Agric. Water Manag.* **2018**, *204*, 1–10. [CrossRef]
28. Tawfeeq, A.; Culham, A.; Davis, F.; Reeves, M. Does fertilizer type and method of application cause significant differences in essential oil yield and composition in rosemary (*Rosmarinus officinalis* L.)? *Ind. Crop. Prod.* **2016**, *88*, 17–22. [CrossRef]
29. Singh, M.; Wasnik, K. Effect of vermicompost and chemical fertilizer on growth, herb, oil yield, nutrient uptake, soil fertility, and oil quality of rosemary. *Commun. Soil Sci. Plant Anal.* **2013**, *44*, 2691–2700. [CrossRef]
30. Bustamante, M.A.; Nogués, I.; Jones, S.; Allison, G.G. The effect of anaerobic digestate derived composts on the metabolite composition and thermal behaviour of rosemary. *Sci. Rep.* **2019**, *9*, 1–15. [CrossRef]
31. Singh, M. Effects of plant spacing, fertilizer, modified urea material and irrigation regime on herbage, oil yield and oil quality of rosemary in semi-arid tropical conditions. *J. Hortic. Sci. Biotechnol.* **2004**, *79*, 411–415. [CrossRef]
32. Kiuru, P.; Muriuki, S.J.N.; Wepukhulu, S.B.; Muriuki, S.J.M. Influence of growth media and regulators on vegetative propagation of rosemary (*Rosmarinus officinalis* L.). *East Afr. Agric. For. J.* **2015**, *81*, 105–111. [CrossRef]
33. Martinetti, L.; Quattrini, E.; Bononi, M.; Tateo, F. Effect of the mineral fertilization on the yield and the oil content of two cultivars of rosemary. *Acta Hortic.* **2006**, *723*, 399–404. [CrossRef]
34. Boyle, T.H.; Craker, L.E.; Simon, J.E. Growing medium and fertilization regime influence growth and essential oil content of rosemary. *Hortscience* **1991**, *26*, 33–34. [CrossRef]
35. Fornes, F.; Liu-Xu, L.; Lidón, A.; Sánchez-García, M.; Cayuela, M.L.; Sánchez-Monedero, M.A.; Belda, R.M. Biochar improves the properties of poultry manure compost as growing media for rosemary production. *Agronomy* **2020**, *10*, 261. [CrossRef]
36. De Lucia, B.; Vecchietti, L.; Rinaldi, S.; Rivera, C.M.; Trinchera, A.; Rea, E. Effect of peat-reduced and peat-free substrates on rosemary growth. *J. Plant Nutr.* **2013**, *36*, 863–876. [CrossRef]
37. Mendoza-Hernández, D.; Fornes, F.; Belda, R.M. Compost and vermicompost of horticultural waste as substrates for cutting rooting and growth of rosemary. *Sci. Hortic.* **2014**, *178*, 192–202. [CrossRef]
38. Rinaldi, S.; De Lucia, B.; Salvati, L.; Rea, E. Understanding complexity in the response of ornamental rosemary to different substrates: A multivariate analysis. *Sci. Hortic.* **2014**, *176*, 218–224. [CrossRef]
39. Kern, J.; Tammeorg, P.; Shanskiy, M.; Sakrabani, R.; Knicker, H.; Kammann, C.; Tuhkanen, E.M.; Smidt, G.; Prasad, M.; Tiilikkala, K.; et al. Synergistic use of peat and charred material in growing media—An option to reduce the pressure on peatlands? *J. Environ. Eng. Landsc. Manag.* **2017**, *25*, 160–174. [CrossRef]
40. Hammond, R.F. The origin, formation and distribution of peatland resources. In *Peat in Horticulture*, 1st ed.; Robinson, D.W., Lamb, J.G.D., Eds.; Academic Press: London, UK, 1975; pp. 1–22.
41. Rydin, H.; Jeglum, J.K. *The Bology of Peatlands*, 2nd ed.; Oxford University Press: Oxford, UK, 2013.
42. Zulfiqar, F.; Allaire, S.E.; Akram, N.A.; Méndez, A.; Younis, A.; Peerzada, A.M.; Shaukat, N.; Wright, S.R. Challenges in organic component selection and biochar as an opportunity in potting substrates: A review. *J. Plant Nutr.* **2019**, *24*, 1–6. [CrossRef]
43. Sannazaro, F.M. Valutazione di Substrati Alternativi alla Torba: Caratterizzazione Chimica, Fisica ed Agronomica di Lolla di Riso. Ph.D. Thesis, Università degli Studi di Padova, Padova, Italy, 31 January 2008.
44. Barrett, G.E.; Alexander, P.D.; Robinson, J.S.; Bragg, N.C. Achieving environmentally sustainable growing media for soilless plant cultivation systems—A review. *Sci. Hortic.* **2016**, *212*, 220–234. [CrossRef]
45. Fenner, N.; Freeman, C. Woody litter protects peat carbon stocks during drought. *Nat. Clím. Chang.* **2020**, *10*, 363–369. [CrossRef]
46. Lazcano, C.; Arnold, J.; Salgado, A.T.; Zaller, J.G.; Martin, J.D. Compost and vermicompost as nursery pot components: Effects on tomato plant growth and morphology. *Span. J. Agric. Res.* **2009**, *7*, 944–951. [CrossRef]
47. Abad, M.; Noguera, P.; Burés, S. National inventory of organic wastes for use as growing media for ornamental potted plant production: Case study in Spain. *Bioresour. Technol.* **2001**, *77*, 197–200. [CrossRef]
48. Abad, M.; Martínez, P.F.; Martínez, M.D.; Martínez, J. Evaluación agronómica de los sustratos de cultivo. *Actas Hortic.* **1992**, *11*, 141–154.

49. Ansorena Miner, J. *Sustratos. Propiedades y Caracterización*; Ediciones Mundi-Prensa: Madrid, Spain, 1994.
50. Lemaire, F.; Rivière, L.; Stievenard, S.; Marfa, O.; Gschwander, S.; Giuffrida, F. Consequences of organic matter biodegradability on the physical, chemical parameters of substrates. *Acta Hortic.* **1998**, *469*, 129–138. [CrossRef]
51. Cabrera, F.; Clemente, L.; Díiaz Barrientos, E.; López, R.; Murillo, J.M. Heavy metal pollutions of soils affected by the Guadiamar toxic flood. *Sci. Total Environ.* **1999**, *242*, 117–129. [CrossRef]
52. Chen, J.; Mc Connell, D.B.; Robinson, C.A.; Caldwell, R.D.; Huang, Y. Production and interior performances of tropical ornamental foliage plants grown in container substrates amended with composts. *Comp. Sci. Util.* **2002**, *10*, 217–225. [CrossRef]
53. Benito, M.; Masaguer, A.; Moliner, A.; Antonio, R.D. Chemical and physical properties of pruning waste compost and their seasonal variability. *Bioresour. Technol.* **2006**, *97*, 2071–2076. [CrossRef]
54. Tittarelli, F.; Rea, E.; Verrastro, V.; Pascual, J.A.; Canali, S.; Ceglie, F.G.; Trinchera, A.; Rivera, C.M. Compost-based nursery substrates: Effect of peat substitution on organic melon seedlings. *Comp. Sci. Util.* **2009**, *17*, 220–228. [CrossRef]
55. De Lucia, B.; Vecchietti, L.; Leone, A. Italian buckthorn response to compost based substrates. *Acta Hortic.* **2011**, *891*, 231–236. [CrossRef]
56. Stellacci, A.M.; Cristiano, G.; Rubino, P.; De Lucia, B.; Cazzato, E. Nitrogen uptake, nitrogen partitioning and N-use efficiency of container-grown Holm oak (Quercus ilex L.) under different nitrogen levels and fertilizer sources. *Int. J. Food Agric. Environ.* **2013**, *11*, 132–137.
57. European Pharmacopoeia. *Determination of Essential Oils in Herbal Drugs*, 6th ed.; Council of Europe European, European Directorate for the Quality of Medicines: Strasbourg, France, 2008; pp. 251–252.
58. Servizio Informativo Agrometeorologico Siciliano. Available online: www.sias.regione.sicilia.it (accessed on 20 July 2020).
59. Omer, E.; Hendawy, S.; El Gendy, A.N.; Mannu, A.; Petretto, G.L.; Pintore, G. Effect of irrigation systems and soil conditioners on the growth and essential oil composition of *Rosmarinus officinalis* L. cultivated in Egypt. *Sustainability* **2020**, *12*, 6611. [CrossRef]
60. Tuttolomondo, T.; Virga, G.; Licata, M.; Leto, C.; La Bella, S. Constructed wetlands as sustainable technology for the treatment and reuse of the first-flush stormwater in agriculture—A case study in Sicily (Italy). *Water* **2020**, *12*, 2542. [CrossRef]
61. Osakabe, Y.; Osakabe, K.; Shinozaki, K.; Tran, L.-S.P. Response of plants to water stress. *Plant Sci.* **2014**, *5*, 1–8. [CrossRef] [PubMed]
62. Gruda, N.S. Increasing sustainability of growing media constituents and stand-alone substrates in soilless culture systems. *Agronomy* **2019**, *9*, 298. [CrossRef]
63. Garcia-Gomez, A.; Bernal, M.P.; Roig, A. Growth of ornamental plants in two composts prepared from agro-industrial wastes. *Bioresour. Technol.* **2002**, *83*, 81–87. [CrossRef]
64. Raviv, M. Can compost improve sustainability of plant production in growing media? *Acta Hortic.* **2017**, *1168*, 119–133. [CrossRef]
65. Savvas, D.; Gruda, N. Application of soilless culture technologies in the modern greenhouse industry—A review. *Eur. J. Hortic. Sci.* **2018**, *83*, 280–293. [CrossRef]
66. Han, S.; Kim, K. Effects of growth retardans on growth, flowering, and germination of harvested seed in *Clinopodium chinense* var. *parviflorum*. *J. Korean Soc. Hortic. Sci.* **1999**, *40*, 765–768.
67. Lì, Z.; Wu, N.; Liu, T.; Chen, H.; Tang, M. Sex-related responses of *Populus cathayana* shoots and roots to AM fungi and drought stress. *PLoS ONE* **2015**, *10*, e0142356. [CrossRef]
68. Llorens-Molina, J.A.; Vacas, S. Effect of drought stress on essential oil composition of *Thymus vulgaris* L. (Chemotype 1, 8-cineole) from wild populations of Eastern Iberian Peninsula. *J. Essent. Oil Res.* **2017**, *29*, 144–155. [CrossRef]
69. An, Y.Y.; Liang, Z.S. Drought tolerance of *Periploca sepium* during seed germination: Antioxidant defense and compatible solutes accumulation. *Acta Physiol. Plant.* **2013**, *35*, 959–967. [CrossRef]
70. Mathobo, R.; Marais, D.; Steyn, J.M. The effect of drought stress on yield, leaf gaseous exchange and chlorophyll fluorescence of dry beans (*Phaseolus vulgaris* L.). *Agric. Water Manag.* **2017**, *180*, 118–125. [CrossRef]
71. Rioba, N.B.; Itulya, F.M.; Saidi, M.; Dudai, N.; Bernstein, N. Effects of nitrogen, phosphorus and irrigation frequency on essential oil content and composition of sage (*Salvia officinalis* L.). *J. Appl. Res. Med. Aromat. Plants* **2015**, *2*, 21–29. [CrossRef]
72. Pereira, S.I.; Santos, P.A.G.; Barroso, J.G.; Figueiredo, A.C.; Pedro, L.G.; Salgueiro, L.R.; Deans, S.G.; Scheffer, J.J.C. Chemical polymorphism of the essential oils from populations of *Thymus caespititius* grown on the island S. Jorge (Azores). *Phytochemistry* **2000**, *55*, 241–246. [CrossRef]
73. Burdina, I.; Priss, O. Effect of the substrate composition on yield and quality of basil (*Ocimum basilicum* L.). *J. Hortic. Res.* **2016**, *24*, 109–118. [CrossRef]
74. Sadeh, D.; Nitzan, N.; Chaimovitsh, D.; Shachter, A.; Ghanim, M.; Dudai, N. Interactive effects of genotype, seasonality and extraction method on chemical compositions and yield of essential oil from rosemary (*Rosmarinus officinalis* L.). *Ind. Crop Prod.* **2019**, *138*, 1–7. [CrossRef]
75. La Bella, S.; Tuttolomondo, T.; Dugo, G.; Ruberto, G.; Leto, C.; Napoli, E.M.; Potortì, A.G.; Fede, M.R.; Virga, G.; Leone, R.; et al. Composition and variability of the essential oil of the flowers of *Lavandula stoechas* from various geographical sources. *Nat. Prod. Commun.* **2015**, *10*, 2001–2004. [CrossRef]
76. Tuttolomondo, T.; Dugo, G.; Ruberto, G.; Leto, C.; Napoli, E.M.; Potortì, A.G.; Fede, M.R.; Virga, G.; Leone, R.; D'Anna, E.; et al. Agronomical evaluation of Sicilian biotypes of *Lavandula stoechas* L. spp. *stoechas* and analysis of the essential oils. *J. Essent. Oil Res.* **2015**, *27*, 115–124. [CrossRef]

77. Tuttolomondo, T.; Dugo, G.; Leto, C.; Cicero, N.; Tropea, A.; Virga, G.; Leone, R.; Licata, M.; La Bella, S. Agronomical and chemical characterisation of *Thymbra capitata* (L.) Cav. biotypes from Sicily, Italy. *Nat. Prod. Res.* **2015**, *29*, 1289–1299. [CrossRef]
78. Saija, A.; Speciale, A.; Trombetta, D.; Leto, C.; Tuttolomondo, T.; La Bella, S.; Licata, M.; Virga, G.; Bonsangue, G.; Gennaro, M.C.; et al. Phytochemical, ecological and antioxidant evaluation of wild Sicilian thyme: *Thymbra capitata* (L.) Cav. *Chem. Biodivers.* **2016**, *13*, 1641–1655. [CrossRef]
79. Tuttolomondo, T.; Iapichino, G.; Licata, M.; Virga, G.; Leto, C.; La Bella, S. Agronomic evaluation and chemical characterization of Sicilian *Salvia sclarea* L. accessions. *Agronomy* **2020**, *10*, 1114. [CrossRef]
80. Nunziata, A.; De Benedetti, L.; Marchioni, I.; Cervelli, C. High resolution melting profiles of 364 genotypes of *Salvia rosmarinus* in 16 microsatellite loci. *Ecol. Evol.* **2019**, *9*, 3728–3739. [CrossRef] [PubMed]
81. Flamini, G.; Najar, B.; Leonardi, M.; Ambryszewska, K.E.; Cioni, P.G.; Parri, F.; Melai, B.; Pistelli, L. Essential oil composition of *Salvia rosmarinus* spenn. wild samples collected from six sites and different seasonal periods in Elba Island (Tuscan Archipelago, Italy). *Nat. Prod. Res.* **2020**, 1–8. [CrossRef] [PubMed]
82. Bolechowski, A.; Moral, M.A.; Bustamante, M.A.; Bartual, J.; Paredes, C.; Pérez-Murcia, M.A.; Carbonell-Barrachina, A.A. Winery–distillery composts as partial substitutes of traditional growing media: Effect on the volatile composition of thyme essential oils. *Sci. Hortic.* **2015**, *193*, 69–76. [CrossRef]
83. Bolechowski, A.; Moral, R.; Bustamante, M.A.; Paredes, C.; Agulló, E.; Bartual, J.; Carbonell-Barrachina, A.A. Composition of oregano essential oil (*Origanum vulgare*) as affected by the use of winery-distillery composts. *J. Essent. Oil Res.* **2011**, *23*, 32–38. [CrossRef]

Article

Antiarthritic Potential of *Calotropis procera* Leaf Fractions in FCA-Induced Arthritic Rats: Involvement of Cellular Inflammatory Mediators and Other Biomarkers

Vandana S. Singh [1], Shashikant C. Dhawale [1], Faiyaz Shakeel [2], Md. Faiyazuddin [3,4] and Sultan Alshehri [2,*]

[1] School of Pharmacy, S.R.T.M. University, Nanded 431606, Maharashtra, India; vandana27011990@gmail.com (V.S.S.); sashiprathmesh@gmail.com (S.C.D.)
[2] Department of Pharmaceutics, College of Pharmacy, King Saud University, P.O. Box 2457, Riyadh 11451, Saudi Arabia; faiyazs@fastmail.fm
[3] School of Pharmacy, Alkarim University, Katihar 854106, Bihar, India; md.faiyazuddin@gmail.com
[4] Nano Drug Delivery®, Raleigh-Durham, NC 27705, USA
* Correspondence: salshehri1@ksu.edu.sa

Abstract: *Calotropis procera* (commonly known as Swallow wort) is described in the Ayurvedic literature for the treatment of inflammation and arthritic disorders. Therefore, in the present work, the antiarthritic activity of potential fractions of Swallow wort leaf was evaluated and compared with standards (indomethacin and ibuprofen). This study was designed in Wistar rats for the investigation of antiarthritic activity and acute toxicity of Swallow wort. Arthritis was induced in Wistar rats by injecting 0.1 mL of Freund's complete adjuvant (FCA) on the 1st and 7th days subcutaneously into the subplantar region of the left hind paw. Evaluation of our experimental findings suggested that antiarthritic activity of methanol fraction of Swallow wort (MFCP) was greater than ethyl acetate fraction of Swallow wort (EAFCP), equal to standard ibuprofen, and slightly lower than standard indomethacin. MFCP significantly reduced paw edema on the 17th, 21st, 24th, and 28th days. It also showed significant effect ($p < 0.01$) on arthritic score, paw withdrawal latency, and body weight. The inhibition of serum lysosomal enzymes and proinflammatory cytokines along with improvement of radiographic features of hind legs was also recorded with MFCP. Finally, it was concluded that MFCP can be a feasible therapeutic candidate for the treatment of inflammatory arthritis.

Keywords: *Calotropis procera* leaves; chronic inflammatory model; cytokines; Freund's complete adjuvant; indomethacin

1. Introduction

Rheumatoid arthritis (RA) is a chronic autoimmune disease which is characterized by nonspecific inflammation of peripheral joints, destruction of articular tissues, and deformities in the joints [1,2]. An epidemiology of RA in terms of males to females has been reported as 1:3, and the prevalence was found to be 1% around the globe [3]. The inflammatory cytokines such as "tumor necrosis factor-α (TNF-α), interleukin-1β (IL-1β), and interleukin-6 (IL-6)" are the major biomarkers responsible for the inflammation and joint damage during the development of RA [4–6].

Presently, the treatment strategies focus on the reduction in inflammation in the joints, as no proper treatments are available. Conventional medicines including analgesics, steroids, nonsteroidal anti-inflammatory drugs (NSAIDs), glucocorticoids, disease-modifying antirheumatic drugs (DMARDs), and anticytokines are reported to show very limited success rates in the treatment of RA, although these medicines are helpful in controlling the symptoms of acute RA [4,7]. Mild to moderate arthritis is relieved by the use of different analgesics. However, these analgesics did not offer anti-inflammatory efficacy and hence these are usually administered in combination with other drugs [8]. Most of the NSAIDs are reported to possess both analgesic and anti-inflammatory efficacy but are not

capable of preventing the destruction of joints [9]. Ultimately, the treatments with NSAIDs fail to produce long-term benefits and produce serious adverse effects such as gastric toxicity and cardiotoxicity, which limit their application in the treatment of RA [10,11]. Therefore, the search for drugs with low or no adverse effects are of prime importance to treat RA. Herbal drugs constitute a major part of all the traditional systems of medicine. According to the WHO, 80% of the world's population use herbal medicines for primary health care. Herbal medicine is a triumph of popular therapeutic diversity and is also used to derive a number of synthetic medicines such as aspirin, paclitaxel, digoxin, and morphine, etc. [12,13]. Several plant-based materials showed potent anti-inflammatory activity in animal models which could also be useful in the treatment of RA [14,15]. These plant-based medicines are supposed to have low incidence of serious adverse effects, low cost, and be easily accessible to the consumers [16,17].

Calotropis procera Linn. (commonly known as Swallow wort), which belongs to the Asclepiadaceae family, is an Ayurvedic plant with important medicinal properties [18,19]. It is represented in India by two different species—viz., *C. procera* and *C. gigantean* [19,20]. Traditionally, it is used in the treatment of ulcers, leprosy, cancers, piles, and liver diseases [21,22]. It has also been reported to have purgative, anthelmintic, analgesic, anti-inflammatory, protective, anticoagulant, antipyretic, and antimicrobial activities [18,19,22,23]. *C. procera* (Swallow wort) latex has been used as a hepatoprotective [21], anti-inflammatory [24], anthelmintic [25], anticonvulsant [26], and antimicrobial agent [27].

A traditional claim made about Swallow wort includes its beneficial effects in inflammation and arthritis. Anti-inflammatory activity of the latex of this plant has been demonstrated in various animal models [28]. The protective effects of different extracts of Swallow wort latex have also been evaluated against inflammation and oxidative stress in monoarthritis induced by different inflammatory mediators [29–34]. Although the leaves of this plant are reported to have traditional uses regarding its beneficial effects in inflammatory conditions, these claims are not experimentally verified. Previous studies have shown the significant anti-inflammatory activity of this plant both in in vitro and in vivo models [24]. Therefore, the current investigations were planned to explore the antiarthritic potential of different fractions of Swallow wort for the evaluation of antiarthritic potential using Freund's complete adjuvant (FCA) model on Wistar rats, because no such study has been carried out in the past.

2. Materials and Methods

2.1. Procurement and Authentication of the Plant

The leaves of Swallow wort were procured from the herbal garden of Swami Ramanand Teerth Marathwada University, Nanded, in the month of January 2017. The Swallow wort plant specimens were identified and authenticated at Botanical Survey of India, Pune, India, where a voucher specimen is preserved (BSI/WC/100-1/Tech./2017/1). The fresh leaves were washed properly with tap water and shade dried for one week. The dried leaves were ground coarsely using an electrical mixer and stored in airtight plastic bags in a cool and dark place until further use.

2.2. Materials

FCA was procured from "Sigma Aldrich (Bangalore, India)". Indomethacin and ibuprofen were received as gift samples kindly provided by "Wockhardt Research Center (Aurangabad, India)". Petroleum ether, chloroform, ethyl acetate, acetone, ethanol, and methanol were procured from "S.D. Fine Chemicals (Mumbai, India)". All the reagents used were of analytical grade.

2.3. Plant Extraction and Characterization

2.3.1. Extraction Method

The successive extraction of dried powder of Swallow wort leaves was carried out using Soxhlet extractor. The solvents used to obtain different fractions of Swallow wort leaves were petroleum ether, chloroform, ethyl acetate, acetone, ethanol, and methanol. The fractions were concentrated by evaporating their respective solvents under reduced pressure using rotary vacuum evaporator at 42–45 °C. The obtained fractions were further dried at room temperature in a petri dish and stored in refrigerator at 4–6 °C till further evaluation.

2.3.2. Qualitative Phytochemical Screening

The bioactive agents present in Swallow wort were detected by a standard phytochemical screening procedure and inference was based on visual observations of color change or precipitate formation [35]. The stock solution of the plant extracts was prepared with their respective solvents. Mayer's reagent (potassium mercuric iodide solution) was added to 1 mL of plant fraction, and the appearance of a cream precipitate was indicated as the end point for the presence of alkaloids. In total, 1 mL of glacial acetic acid was added to 1 mL of plant fraction, which was dissolved and then cooled, followed by addition of 2–3 drops of ferric chloride. Then, 2 mL of conc. H_2SO_4 was carefully added along the walls of test tube; the appearance of reddish brown ring at the junction of two layers indicated the presence of glycosides. Steroids were investigated by adding 5 mL of chloroform and a few drops of conc. H_2SO_4; this mixture was allowed to stand for some time, the reddish precipitate in the lower layer indicates the presence of steroids. The formation of foam upon vigorous shaking of plant fraction solution indicates the presence of saponins. Flavonoids were investigated by adding basic lead acetate separately to 1 mL of plant fraction. A bulky reddish brown precipitate indicates the presence of flavonoids.

2.4. Biological Evaluation

2.4.1. Prediction of Biological Activity

The prediction of the activity spectra for substance (PASS) is an online software database program used to predict the various biological properties of compounds [36]. PASS software helps estimate the probable biological activity of drugs such as organic compounds (pursuing molecular weights of 50 to 1250 Da) or phytoconstituents based on structural activity relationship analysis of training set consisting of information on the structures of more than 205,000 organic compounds which exhibit more than 3750 kinds of biological activity [37,38]. The MDL mole file [V 3000] (*mol) structure of desired phytoconstituent drawn with the help of ACD/Labs chemsketch software 2015 release (file version C10E41) was fed in PASS online Way2drug online software. The software gave the Pa and Pi values (active and inactive) [38]. The probability of experimental biological and pharmacological activities is high if the value of Pa is greater than 0.7 and less if Pa value is $0.5 < Pa > 0.7$ [39].

2.4.2. Experimental Animals

Male Albino Wistar rats (weighing 180–220 g) were purchased from "Wockhardt Research Center (Aurangabad, India)". The animals were housed under standard animal housing facility with temperature (24 ± 1 °C), relative humidity (45–50%), and 12 h light/dark cycle. The animals were fed with standard pellet chow diet and water ad libitum. All rats were allowed to adopt laboratory conditions before starting experimental studies. The study protocol was approved by the Institutional Animal Ethics Committee (Reg. No. 731/PO/Re/2002/CPCSEA), Approval No. CPCSEA/CBPL/AH-11 under the Committee for "Purpose of Control and Supervision of Experiments on Animals (CPCSEA)". The ethical protocol and guidelines were strictly followed throughout the experimental studies.

2.4.3. Acute Toxicity Studies

The Swallow wort plant fractions were evaluated for acute toxicity as per the organization for economic co-operation and development (OECD) guidelines 425. The single oral dose of 2000 mg/kg of plant fraction was administered to overnight fasted rats and observations were continuously recorded for behavioral profiles for 2 h and for mortality up to 72 h.

2.4.4. FCA-Induced Arthritis

For the evaluation of FCA-induced arthritis, the rats were divided into six different groups (each containing six rats) as follows:

Group I—Vehicle control, 1% w/v suspension of sodium carboxymethyl cellulose (SCMC) was administered orally;

Group II—Arthritic control, 1% w/v suspension of SCMC was administered orally;

Group III—Arthritic animals treated with oral administration of standard indomethacin at 10 mg/kg dose;

Group IV—Arthritic animals treated with oral administration of standard ibuprofen at 15 mg/kg dose;

Group V—Arthritic animals treated with oral administration of methanolic fraction of Swallow wort (MFCP) at a 300 mg/kg dose;

Group VI—Arthritic animals treated with oral administration of ethyl acetate fraction of Swallow wort (EAFCP) at a 300 mg/kg dose.

Around 0.1 mL of FCA was administered subcutaneously into the subplantar region of the left hind paw on 1st and 7th days to all the animals of all groups except vehicle control [40]. Then, 300 mg/kg of each of MFCP and EAFCP were orally administered once daily from day 12 to day 28 [41]. The standards (indomethacin and ibuprofen) at the dose of 10 and 15 mg/kg, respectively, were also orally administered once daily from day 12 to day 28 for the evaluation of antiarthritic potential of these standards. The antiarthritic activities of each plant fraction of Swallow wort and standards were determined using a "Plethysmometer (Ugo, Basile, Italy)" by the hind paw method on days 1, 4, 7, 10, 12, 14, 17, 21, 24, and 28 [42]. The percentage inhibition value of each fraction of plant was calculated using its standard formula, described previously [14,16].

Arthritic index was obtained using different pharmacological parameters such as hind paw edema, mechanical withdrawal threshold, body weight, and arthritic score. On the 28th day, the rats were anaesthetized using anesthetic ketamine and blood samples were withdrawn from the retro-orbital puncture for the determination of different biochemical parameters. At the end of the experiment, X-rays of the joints of the hind paws of animals were recorded under mild diethyl ether anesthesia for the evaluation of possible bone, cartilage, and other structural degeneration.

2.4.5. Paw Volume Evaluation

The left paw volume was measured up to the lateral malleolus by the mercury displacement method just before FCA injection on 1st day and subsequently at various time intervals until the 28th day using a "Plethysmometer (Ugo, Basile, Italy)" [43]. The alterations in the paw volume were considered as the difference between the final and initial paw volumes.

2.4.6. Visual Arthritis Scoring System

The visual arthritis scoring systems were used to assess the severity of the arthritis, as described previously [44]. The arthritis score was ranged from 0 to 4 which is graded as follows: 0 = normal paw; 1 = mild swelling and erythema; 2 = swelling and erythema; 3 = severe swelling and erythema; 4 = gross deformity and inability.

2.4.7. Evaluation of Thermal Hyperalgesia

The thermal hyperalgesia/paw withdrawal latency of injected paw was evaluated using hot plate method just before FCA injections on the 1st day and subsequently at various time intervals until the 28th day. The paw was placed on the flat surface of the hot plate which was maintained at $55 \pm 5\,°C$. The reaction time to heat stimulus in terms of paw licking or jumping was recorded as the end point of the pain threshold [45]. A cut off time was selected as 20 s to avoid tissue damage.

2.4.8. Body Weight Recording

The body weight was recorded during the experimental period using a Digital Weighing Balance (Sartorius 1413, MP 8/8-1, Bohemia, NY, USA) just before FCA injections on the 1st day and subsequently at various time intervals until the 28th day [46].

2.4.9. Determination of the Weight of Spleen and Thymus

At the end of experimental studies, the rats were sacrificed using ketamine anesthesia. Then, the thymuses and spleens of all rats were taken and weights were recorded.

2.4.10. Biochemical Estimation

On the 28th day, the blood samples of the rats were withdrawn from the retro-orbital puncture of all the animal groups and various biochemical parameters such as aspartate aminotransferase (AST), alanine transaminase (ALT), and alkaline phosphatase (ALP) were estimated using a standard kit (Sigma-Aldrich assay kit) [47].

2.4.11. Estimation of Serum Parameters

Various hematological/serum parameters were estimated using routine laboratory techniques. The levels of serum C-reactive protein (CRP) and rheumatoid factor (RF) were estimated using commercial kits (Aspen laboratories) following the manufacturer's instructions.

2.4.12. Proinflammatory Biomarkers (TNF-α and IL-6)

For the estimation of proinflammatory biomarkers such as TNF-α and IL-6, the blood samples were left to stand for about 30 min. The serum was separated from the blood by centrifugation at 3000 rpm for about 10 min. The obtained serum samples were stored at $-20\,^\circ$C until further evaluation. Proinflammatory biomarkers (TNF-α and IL-6) were estimated using readymade ELISA reagent kits [40].

2.4.13. Radiological Analysis of Ankle Joints

On the 28th day, the animals were anaesthetized using ketamine and radiographs of FCA-injected hind paws were recorded using X-ray (GE DX-300). Radiographic analysis of hind paws was carried out at 40 kV peak and 12 Ms. The X-ray image was interpreted for the radiographic changes.

2.5. Data and Statistical Analysis

The values are expressed as mean $\pm$ SEM for 6 animals. The experimental results were analyzed statistically using one-way ANOVA followed by Dunnett test using Graphpad instant software. $p < 0.05$ was considered a statistically significant value.

3. Results
3.1. Plant Extraction and Characterization
3.1.1. Extractive Value of Fraction

After extracting 702 g of powder Swallow wort leaves with petroleum ether, chloroform, ethyl acetate, acetone, and methanol, the fraction yields were found to be 12.25, 12.51, 5.87, 5.43, and 42.66 g of the fraction, respectively. The yield was recorded highest in methanol fraction of Swallow wort leaves, while the lowest one was found in acetone fraction. The yield was negligible for ethanol fraction. The yield of above mentioned fractions of Swallow wort leaves was not found in the literature. However, the yield of methanol fraction of Swallow wort latex was recorded as 25% (dry weight) in the literature [30]. In the present study, the yield of methanol fraction of Swallow wort leaves was 42.66 g out of

702 g of powder, which was approximately 6.10%. These observations suggested that latex of Swallow wort is better than its leaves in terms of yield.

3.1.2. Phytochemical Screening

The preliminary phytochemical screening of MFCP and EAFCP demonstrated the presence of alkaloids, glycosides, flavonoids, and steroids. Other studied fractions did not show the presence of all these compounds.

3.2. Biological Evaluation

3.2.1. PASS Prediction

We performed the PASS prediction of various phytoconstituents of Swallow wort leaves for the prediction of various biological activities such as anti-inflammatory and antiarthritic activity. From this study, we found most of the constituents such as α-amyrin, β-amyrin, β- sitosterol, and stigmasterol have shown good anti-inflammatory activity (Table 1).

Table 1. Prediction of activity spectra for substance (PASS) of Swallow wort for antiarthritic activity.

Phytoconstituent	Pa	Pi	Activity
α-amyrin	0.889	0.004	Anti-inflammatory
	0.835	0.002	Nitric oxide antagonist
	0.522	0.043	Antiarthritic
β-amyrin	0.411	0.011	Antioxidant
	0.843	0.005	Anti-inflammatory
	0.793	0.003	Nitric oxide antagonist
Calitropigenin	0.405	0.012	Antioxidant
	0.357	0.119	Anti-inflammatory
	0.490	0.060	Anti-inflammatory
Asclepin	0.357	0.119	Anti-inflammatory
	0.490	0.060	Anti-inflammatory
	0.608	0.005	Calcium regulator
β-sitosterol	0.572	0.038	Anti-inflammatory
	0.482	0.004	Anti-inflammatory, Ophthalmic
	0.740	0.011	Anti-inflammatory
Stigmasterol	0.669	0.004	Calcium regulator
	0.662	0.006	Bone diseases treatment
	0.373	0.017	Anti-inflammatory, ophthalmic

3.2.2. FCA-Induced Arthritis

The subplantar injection of FCA in the left hind paw of rats resulted in the progressive increase in the volume of the ipsilateral (injected) paw as well as contralateral (noninjected) paw.

3.2.3. Effect of Swallow Wort Fractions on Paw Volume

FCA was administered on 1st and 7th days, which resulted in the progressive increase in paw volume. The treatment with standards (indomethacin and ibuprofen) and plant fractions (MFCP and EAFCP) started from the day 12 to day 28. As presented in Table 2, it can be seen that both of the standards as well as fractions caused significant abatement of paw volume which was noticed from day 17 to day 28. The MFCP demonstrated the same level of antiarthritic effects (46.42%) with that of ibuprofen, which was slightly lower than that of indomethacin (51.78%). However, the antiarthritic effects of EAFCP (26.78%) were found to be significantly lower than both of the standards (indomethacin and ibuprofen) and MFCP ($p < 0.01$). Hence, MFCP could be used as an alternative to indomethacin and ibuprofen in the treatment of RA.

Table 2. Effect of Swallow wort fractions on Freund's complete adjuvant (FCA)-induced paw volume of rats.

Groups	Paw Volume on Different Days (mL)									Inhibition (%)
	4th	7th	10th	12th	14th	17th	21st	24th	28th	
Normal	0.18 ± 0.009	0.19 ± 0.012	0.2 ± 0.008	0.20 ± 0.008	0.2 ± 0.08	0.19 ± 0.01	0.2 ± 0.02	0.21 ± 0.01	0.21 ± 0.008	–
Arthritic control	0.38 ± 0.011 [#]	0.50 ± 0.012 [#]	0.59 ± 0.008 [#]	0.61 ± 0.01 [#]	0.59 ± 0.008 [#]	0.59 ± 0.008 [#]	0.58 ± 0.005 [#]	0.57 ± 0.005 [#]	0.56 ± 0.004 [#]	–
Indomethacin	0.38 ± 0.009	0.51 ± 0.012	0.59 ± 0.014	0.61 ± 0.01	0.58 ± 0.004	0.42 ± 0.005 [**]	0.38 ± 0.005 [**]	0.34 ± 0.006 [**]	0.27 ± 0.01 [**]	51.78
Ibuprofen	0.39 ± 0.011	0.51 ± 0.012	0.59 ± 0.008	0.60 ± 0.012	0.58 ± 0.012	0.44 ± 0.01 [**]	0.42 ± 0.008 [**]	0.37 ± 0.008 [**]	0.3 ± 0.008 [**]	46.42
MFCP	0.38 ± 0.019	0.51 ± 0.012	0.58 ± 0.006	0.59 ± 0.012	0.58 ± 0.009	0.44 ± 0.005 [**]	0.42 ± 0.007 [**]	0.38 ± 0.01 [**]	0.30 ± 0.008 [**]	46.42
EAFCP	0.39 ± 0.009	0.51 ± 0.012	0.59 ± 0.012	0.60 ± 0.011	0.59 ± 0.008	0.58 ± 0.007	0.56 ± 0.008 [*]	0.49 ± 0.007 [**]	0.41 ± 0.008 [**]	26.78

Values are mean ± SEM for 6 animals. * $p < 0.05$, ** $p < 0.01$ vs. control group # $p < 0.01$ when compared to normal control. One-way ANOVA followed by Dunnett test were used. Methanol fraction of Swallow wort (MFCP) showed similar percentages inhibition with that of ibuprofen.

3.2.4. Arthritic Scoring System

The effect of the administration of different fractions (MFCP and EAFCP) and standards (indomethacin and ibuprofen) on arthritic score was assessed through visual observation. The different grading systems as described in the experimental section were used to assess the arthritic scores. On the 28th day, the MFCP was found to reduce the arthritic score significantly compared to the control and arthritic control ($p < 0.01$). The effect of MFCP was similar to that of standard ibuprofen (Figure 1). Although EAFCP demonstrated lesser effects than MFCP, a significant reduction in arthritic score was recorded compared to the control and arthritic control ($p < 0.01$). Overall, MFCP was found to be better than EAFCP in reducing the arthritic scores in the rats.

3.2.5. Measurement of Paw Withdrawal Latency

The significant increment in the paw withdrawal latency was witnessed on day 21 and day 28 in animals treated with MFCP, EAFCP, and standards (Figure 2). The paw withdrawal latency of MFCP (2.11 ± 0.17) was found to be similar to that of standard ibuprofen (2.33 ± 0.18) on the 28th day. However, the paw withdrawal latency of EAFCP (1.74 ± 0.09) was found to be significantly lower than MFCP and both of the standards (indomethacin and ibuprofen) on the 28th day ($p < 0.05$). These observations indicate the superiority of MFCP over EAFCP in the increment of paw withdrawal latency.

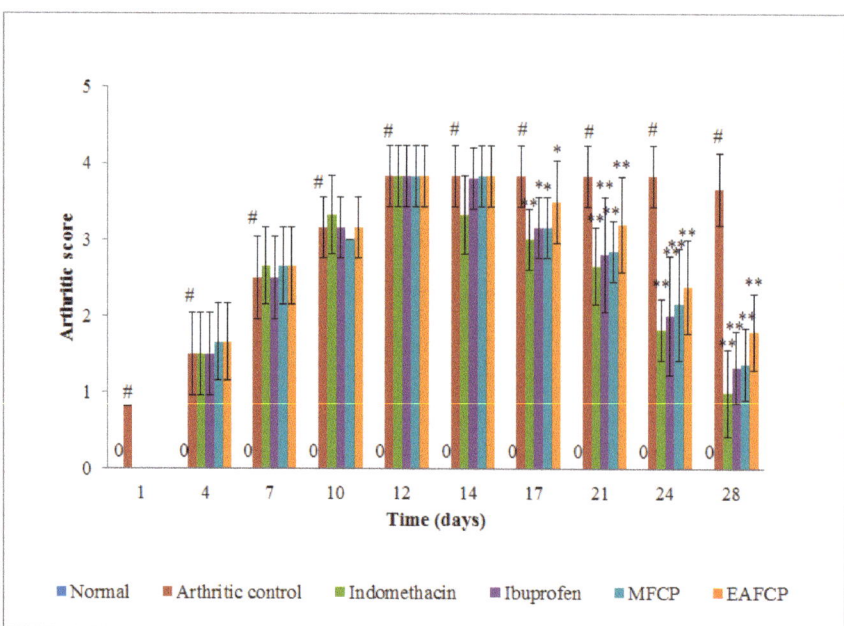

Figure 1. Effect of Swallow wort fractions on arthritic score. Values are mean ± SEM for 6 animals; statistical analysis by one-way ANOVA followed by Dunnett test using Graphpad Instat software; * $p < 0.05$, ** $p < 0.01$ when compared to arthritic control, and # $p < 0.01$ when compared to normal control.

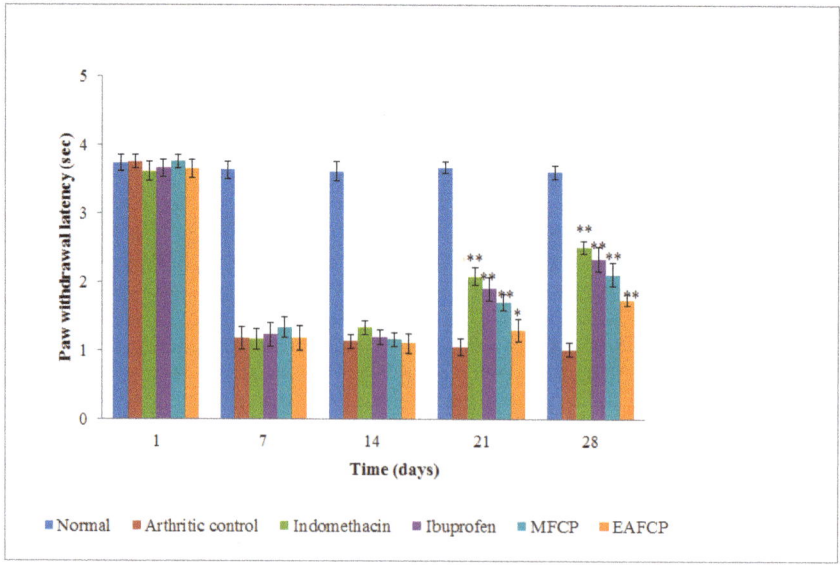

Figure 2. Measurement of paw withdrawal latency. Values are mean ± SEM for 6 animals; statistical analysis by one-way ANOVA followed by Dunnett test using Graphpad Instat software; * $p < 0.05$ and ** $p < 0.01$ when compared to arthritic control normal control.

3.2.6. Body Weight

All animals injected with FCA showed the reduction in the body weight which might be due to the decreased absorption of nutrients of Swallow wort through the intestine [45]. However, the treatment with both of the standards (indomethacin and ibuprofen) and plant fractions showed increase in the body weight from the 12th day onwards. The MFCP and EAFCP were found to almost restore the body weight in the progressive manner such as in case of indomethacin and ibuprofen (Figure 3). Overall, both MFCP and EAFCP were found to have a good impact on the body weight of the rats.

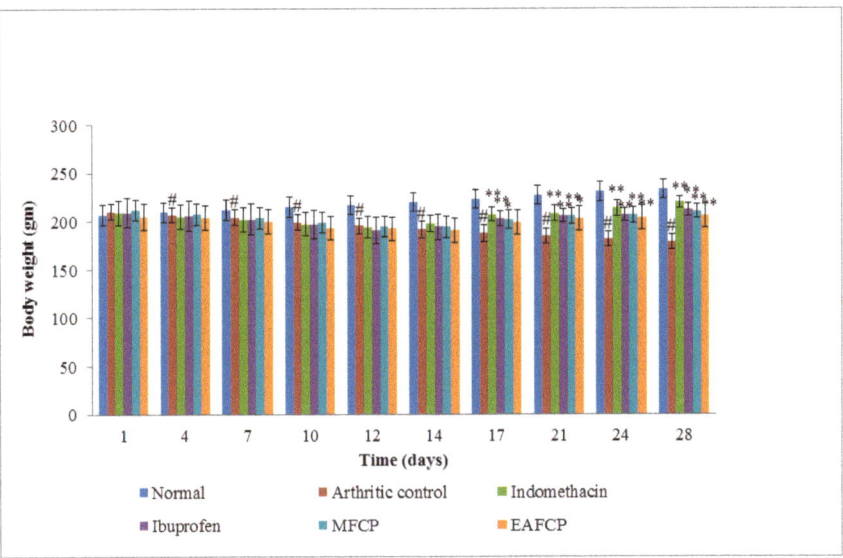

Figure 3. Effect of Swallow wort fractions on body weight. Values are mean ± SEM for 6 animals; statistical analysis by one-way ANOVA followed by Dunnett test using GraphpadInstat software; * $p < 0.05$, ** $p < 0.01$ when compared to arthritic control, and # $p < 0.01$ when compared to normal control.

3.2.7. Measurement of Spleen and Thymus Weights

Immunological functions are related to the thymus and spleen indexes. On the 28th day, the rats were sacrificed and thymus index and spleen index were determined. As presented in Table 3, the weights of spleen and thymus for MFCP group animals were found to be significantly lower than arthritic control group ($p < 0.01$). The MFCP also showed greater effect in thymus weight (0.13 ± 0.017 g) than ibuprofen (0.14 ± 0.017 g) but slightly lower effect in spleen weight (0.33 ± 0.014 g) than ibuprofen (0.37 ± 0.01 g). The weight of spleen and thymus for EAFCP group animals was also found to be significantly lower than the arthritic control ($p < 0.01$). Based on these observations, MFCP was found to be much better than EAFCP in reducing the thymus and spleen weights.

Table 3. Measurement of spleen and thymus weights in FCA-induced arthritis.

Groups	Organs Weight (g)	
	Spleen	Thymus
Normal	0.21 ± 0.01	0.11 ± 0.01
Arthritic control	0.44 ± 0.012 #	0.2 ± 0.011 #
Indomethacin 10 mg/kg	0.29 ± 0.012 **	0.13 ± 0.016 **
Ibuprofen 15 mg/kg	0.37 ± 0.01 **	0.14 ± 0.017 **
MFCP 300 mg/kg	0.33 ± 0.014 **	0.13 ± 0.017 **
EAFCP 300 mg/kg	0.41 ± 0.024 **	0.18 ± 0.008 **

Values are mean ± SEM for 6 animals. ** $p < 0.01$ vs. control group and # $p < 0.01$ when compared to normal control. One-way ANOVA followed by Dunnett test were used.

3.2.8. Serum Lysosomal Enzymes in FCA-Induced Arthritis

Enzymes such as AST, ALT, and ALP have significant roles in the formation of biologically active chemical mediators such as bradykinins in the inflammatory process [46]. In addition, serum AST, ALT, and ALP are the specific biomarkers which are useful in the evaluation of liver damage [13]. Therefore, the levels of AST, ALT, and ALP were measured in this study. All group animals treated with FCA caused an increase in the level of these enzymes. However, the treatment of animals with MFCP (300 mg/kg), EAFCP (300 mg/kg), indomethacin (10 mg/kg), and ibuprofen (15 mg/kg) significantly reduced the elevated levels of serum enzymes (Table 4), while the animals treated with MFCP showed greater effect than EAFCP.

Table 4. Serum lysosomal enzyme in FCA-induced arthritis.

Groups	Serum Enzymes on Different Days		
	AST (U/mL)	ALT (U/mL)	ALP (U/mL)
Normal	32.33 ± 0.81	26.16 ± 0.98	42.83 ± 1.72
Arthritic control	76.66 ± 1.63 #	74 ± 1.6 #	122.66 ± 1.63 #
Indomethacin 10 mg/kg	38.16 ± 1.72 **	35.66 ± 1.3 **	42.33 ± 1.36 **
Ibuprofen 15 mg/kg	42.33 ± 1.36 **	38.66 ± 1.03 **	55.83 ± 1.16 **
MFCP 300 mg/kg	45.16 ± 0.75 **	41.83 ± 0.75 **	56 ± 1.78 **
EAFCP 300 mg/kg	49.66 ± 1.5 **	47.5 ± 1.04 **	62.16 ± 1.47 **

Values are mean ± SEM for 6 animals. ** $p < 0.01$ vs. control group and # $p < 0.01$ when compared to normal control. One-way ANOVA followed by Dunnett test were used.

3.2.9. Alterations in CRP and RF in FCA-Induced Arthritis in Rats

The serum CRP and RF are the markers of the inflammation and antibody production against the injected FCA. The high levels of CRP (7.1 mg/L) and RF (59.71 IU/L) were recorded in the FCA control group animals. However, the MFCP, EAFCP, indomethacin, and ibuprofen treatments significantly reduced the levels of CRP and RF ($p < 0.01$) as shown in Table 5. The effects of MFCP were better than EAFCP. Hence, MFCP can be effectively utilized in reducing the serum levels of CRP and RF compared to EAFCP.

3.2.10. TNF-α and IL-6

Proinflammatory cytokines such as TNF-α and IL-6 play essential roles in the pathogenesis of RA. Therefore, the levels of TNF-α and IL-6 cytokines in the serum of arthritic rats were analyzed and results are shown in Figure 4. The levels of TNF-α and IL-6 were significantly elevated in FCA-induced arthritic rats ($p < 0.01$). However, the treatment of arthritic rats with MFCP reduced the elevated levels of serum TNF-α and IL-6. This effect was almost equivalent to that of standard ibuprofen. The treatment of arthritic rats with EAFCP also reduced the elevated levels of serum TNF-α and IL-6 compared to arthritic control but its level was significantly higher than MFCP and both of the standards ($p < 0.05$).

Table 5. Alterations in C-reactive protein (CRP) and rheumatoid factor (RF) in FCA-induced arthritis in rats.

Groups	RF (IU/L)	CRP (mg/L)
Normal control	-	1.19 ± 0.3
Arthritic control	59.71 ± 1.18 #	7.1 ± 0.48 #
Indomethacin (10 mg/kg)	34.55 ± 0.96 **	2.41 ± 0.51 **
Ibuprofen (15 mg/kg)	37.08 ± 0.91 **	2.93 ± 0.51 **
MFCP 300 mg/kg	46.03 ± 1.25 **	4.1 ± 0.49 **
EAFCP 300 mg/kg	49.34 ± 1.21 **	6.38 ± 0.34 **

Values are mean ± SEM for 6 animals. ** $p < 0.01$ vs. control group and # $p < 0.01$ when compared to normal control. One-way ANOVA followed by Dunnett test were used.

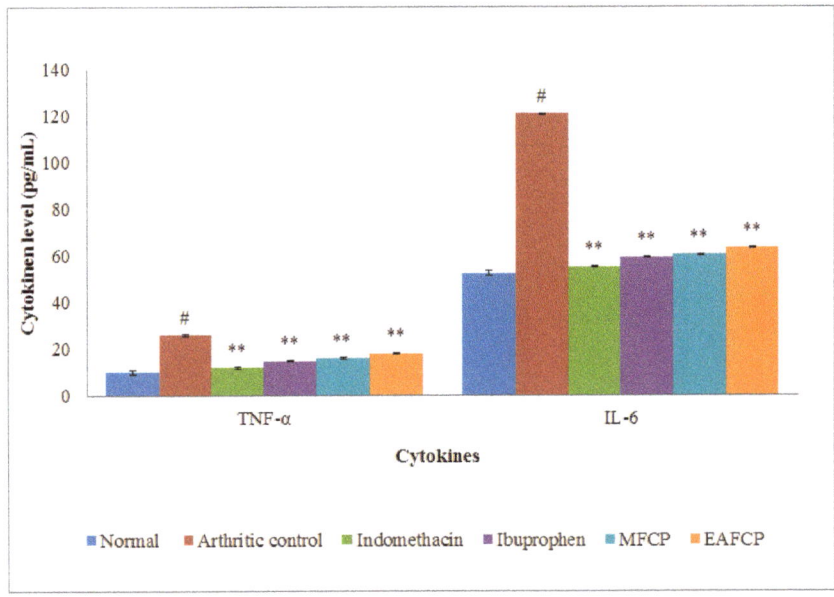

Figure 4. Effects of Swallow wort fractions on cytokine production in serum. Values are mean ± SEM for 6 animals; statistical analysis by one-way ANOVA followed by Dunnett test using Graphpad Instat software; ** $p < 0.01$ when compared to arthritic control and # $p < 0.01$ when compared to normal control.

3.2.11. Radiological Analysis of Ankle Joints

Figure 5A presents the photographs of the tarsotibial joint swelling of the left hind paws from different groups on the 14th day after CFA injection. Figure 5B presents the X-ray radiographs of the same paws recorded on the 28th day. The radiographic analysis of the hind legs for FCA-induced arthritic rats presented soft tissue swelling (phalangeal region) along with joint space narrowing (intertarsal joints), cystic enlargement of the bone, and extensive erosions which clearly indicated the degeneration of the cartilage (Figure 5). However, the treatment with MFCP, EAFCP, indomethacin, and ibuprofen reduced the narrowing of joint space and improved the radiographic pattern of the joints. The radiographic patterns of the joints were found to be better in MFCP treated animals than EAFCP threated animals.

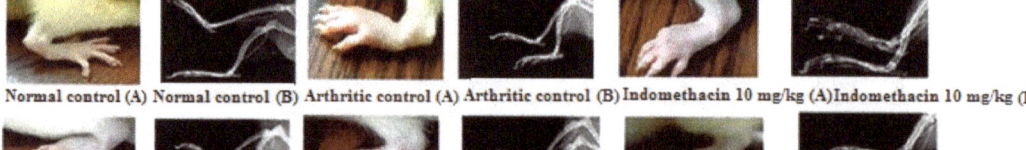

Normal control (A) Normal control (B) Arthritic control (A) Arthritic control (B) Indomethacin 10 mg/kg (A) Indomethacin 10 mg/kg (B)
Ibuprofen 15 mg/kg (A) Ibuprofen 15 mg/kg (B) MFCP 300 mg/kg (A) MFCP 300 mg/kg (B) EAFCP 300 mg/kg (A) EAFCP 300 mg/kg (B)

Figure 5. Photographic and radiographic analysis of FCA-induced arthritis in rats. (**A**) Photographic of the left hind paw taken 14 days after FCA injection and (**B**) radiographic analysis of the left and right hind paws at day 21 after FCA injection.

4. Discussion

Swallow wort is growing wildly as an Ayurvedic plant with important medicinal properties. The leaves, latex, flowers, stems, and roots of Swallow wort are used traditionally to treat complications such as ulcers, leprosy, tumors, inflammatory disorders, piles, and liver diseases [22]. Anti-inflammatory efficacy of the latex of Swallow wort has been demonstrated in various animal models [28]. The warm leaf paste of Swallow wort is usually applied on the swollen part to minimize the inflammation and pain [48]. Although the leaves of this plant are reported to have traditional uses regarding its beneficial effects in inflammatory conditions, these claims are not experimentally verified. In our earlier studies, we found significant activity of Swallow wort (extract and fraction) in in vitro, carrageenan-induced paw edema, and formalin-induced paw edema rat models [49]. The PASS studies were performed to predict the biological activities of various phytoconstituents present in Swallow wort leaves. Based on current phytochemical analysis and reported gas-chromatography mass-spectrometry data of Swallow wort leaves, several reported phytoconstituents of Swallow wort leaves were utilized for the prediction of biological activities [50]. The PASS studies predicted the anti-inflammatory potentials of most of its constituents such as α-amyrin, β-amyrin, stigmasterol, and β-sitosterol. These compounds also showed good anti-inflammatory activities in experimental animal models [51,52]. So, in the present study, we have investigated the antiarthritic potential of different fractions of Swallow wort leaves using an FCA-induced arthritis model. Due to unavailability of female rats, these studies were carried out on male rats. In addition, no differences were recorded in the susceptibility to RA between the genders previously in the literature [53]. Hence, either sex of the rats can be selected for these studies.

FCA-induced arthritis model is the most common chronic model which is comparable to the human RA model [54]. This model is predicted as a progressive increase in the volume of the injected paw of rats and used to evaluate the pathogenesis of RA for the screening of different therapeutics [55]. In this model, the primary reaction of edema and soft-tissue thickening at the depot site is caused by the FCA, which has an irritant effect [56]. However, the secondary lesion is associated with the formation of antibodies [57].

An acute toxicity evaluation demonstrated the nontoxic nature of both of the extracts at the dose of 2000 mg/kg. The selected dose of 300 mg/kg for both of the extracts was much lower than studied dose of acute toxicity evaluation. Accordingly, a dose of 300 mg/kg was selected in this study. Although FCA was administered on the 1st and the 6th days of the treatment, the signs of inflammation appeared from the 2nd day, which might be due to fluid exudation, neutrophil infiltration, and mast cell activation. It was followed by a slow regression and the joint swellings [29]. Antiarthritic activity of MFCP was greater than EAFCP, equal to standard ibuprofen, and slightly lower than standard indomethacin, which is indicated by the reduction in paw volume of inhibiting the FCA-challenged rats. Indomethacin, a nonselective COX inhibitor, is a more potent anti-inflammatory, analgesic, and antipyretic than aspirin [58].

The swelling of joints and the increased arthritic score are the real index for chronic inflammation of FCA-induced arthritic model [40]. MFCP was found to reduce the arthritic score and paw swelling significantly, which was similar to that of standard ibuprofen (Table 2 and Figure 1). These observations suggest the possible immunosuppressant effects of MFCP [54]. The animal groups that underwent MFCA and EAFCP treatments showed a marked enhancement in rat body weight compared to the arthritic control group. A significant reduction in rat body weight was witnessed due to the decreased absorption of nutrients from the intestine in FCA-challenged rats. Similar to that of standards indomethacin and ibuprofen, MFCP significantly restored the body weight of the rats which may be due to the improvement in the arthritic condition, which in turn, may normalize the absorption of nutrients from the intestine. These result also suggested that there is a close relationship between the extent of inflammation and the loss of body weight [59]. These results were in good agreement with those reported for the treatment of monoarthritis in rats using a latex extract of Swallow wort [29].

In adjuvant arthritis, the spleen is an essential organ for the formation of cells and antibodies, which are responsible for the immunological effects. The increase in the cellularity occurs in the spleen of the arthritic rats [10]. The organ weight of the spleen and thymus in FCA-induced rats was evidently increased in the present study [60]. The MFCP and EAFCP resulted in the increased weight of the spleen and thymus, probably by suppression of splenic lymphocytes and inhibition of the infiltration lymphocytes into the synovium [11].

Additional verifications of the antiarthritic potentials of MFCP, EAFCP, indomethacin, and ibuprofen were assessed by the estimation of the various biochemical parameters of rat serum including AST, ALT, ALP, CRP, and RF, which are important tools to study the antiarthritic potential of the drugs. In the inflammatory process, AST and ALT are responsible for the formation of inflammatory mediators such as bradykinins [61]. In the FCA model, the liver function indicative enzymes may be altered due to the enhancement in liver and bone fractions. This is associated with localized bone erosion and periarticular osteopenia. The CRP levels increased during inflammation which might be due to the rise in the IL-6 levels. This is produced by the macrophages and the adipocytes [62]. FCA-induced arthritis is linked with an increase in the RF and CRP levels [54]. In this study, the treatment of arthritic rats with MFCP, EAFCP, indomethacin, and ibuprofen significantly reduced the elevated levels of AST, ALT, ALP, CRP, and RF in rat serum. The animals treated with MFCP showed a greater effect than EAFCP. These studies suggested the potential of studied fractions of Swallow wort in the treatment of RA. The significant reduction in the elevated serum levels of AST, ALT, and ALP also suggested that the studied fractions did not cause hepatic injury to the rats [13]. Overall, the results of lysosomal enzyme measurements were found to be in good agreement with those reported previously in the literature for the *Solanum xanthocarpum* fruit extract [13].

Cytokines are the mediators of inflammatory conditions. The synovial macrophages and fibroblasts produce an excess amount of ILs, TNF-α, and other cytokines. Moreover, the activated neutrophils are also the sources of leukotrienes and prostaglandins. Mediators such as IL, IL-1β, IL-6, and TNF-α are responsible for the pathogenesis and progression of RA [40]. In the present study, the serum TNF-α and IL-6 levels were significantly increased in FCA-induced arthritic rats. However, the treatment with MFCP showed the maximum inhibition of TNF-α and IL-6 levels compared to EAFCP. These results were in accordance with those reported for latex extraction of Swallow wort [29].

For the diagnosis of tissue swelling, erosions, and joint deformity in arthritis patients, radiographic images are used. This gives an idea about soft tissue lesions which are observed as an early sign of the arthritis. The bone erosion and deterioration in trabecular bone are also typical pathologic changes of human arthritis [63]. In FCA-induced arthritic rats, soft tissue swelling along with a narrowing of the joint spaces was recorded, which implies bone destruction in arthritic conditions. The radiographic observations of the treatments group with MFCP and EAFCP inhibited the arthritis-associated joint changes (Figure 4).

The antiarthritic effects of MFCP and EAFCP could be due to the presence of different classes of phytochemicals including flavonoids, alkaloids, terpenoids, glycosides, saponins, tannins, and steroids, which were identified using preliminary phytochemical screening. Accordingly, the current investigation showed that the leaf of Swallow wort markedly reduced the paw inflammation, serum enzyme levels, and release of proinflammatory mediators associated with RA and hence it has great potential in the treatment of RA.

5. Conclusions

The results of the current investigation contribute toward the exploration of Swallow wort leaf extracts in the treatment of RA. The obtained data suggested that the antiarthritic potentials of MFCP and EAFCP might be due to the protection of vascular permeability, synovial membrane, and prevention of cartilage destruction, which could finally result in improved health status. It also demonstrates its beneficial effects during recovery from RA by including body weight, arthritic score, and organ weights along with clinical signs including paw edema, paw withdrawal latency, and radiological pattern. This research established the antiarthritic potential of MFCP in Wistar rats. However, further investigations are required to identify and isolate the potential phytoconstituents responsible for the antiarthritic efficacy, which could facilitate the utilization of Swallow wort in arthritic disorders.

Author Contributions: Conceptualization, S.C.D. and M.F.; methodology, V.S.S., F.S., S.A. and M.F.; software, V.S.S.; validation, S.C.D., S.A. and F.S.; formal analysis, V.S.S. and F.S.; investigation, M.F.; resources, S.C.D.; data curation, F.S.; writing—original draft preparation, V.S.S.; writing—review and editing, S.C.D., F.S., M.F. and S.A.; visualization, V.S.S.; supervision, S.C.D.; project administration, S.C.D.; funding acquisition, S.A. All authors have read and agreed to the published version of the manuscript.

Funding: This research was funded by the Researchers Supporting Project (number RSP-2020/146) at King Saud University, Riyadh, Saudi Arabia and the APC was also funded by the Researchers Supporting Project.

Institutional Review Board Statement: The study was conducted according to the guidelines of the Declarations of Helsinki, and approved by the Institutional Animal Ethics Committee (Reg. No. 731/PO/Re/2002/CPCSEA), Approval No. CPCSEA/CBPL/AH-11 under the Committee for "Purpose of Control and Supervision of Experiments on Animals (CPCSEA)".

Informed Consent Statement: Not applicable.

Data Availability Statement: Not applicable.

Acknowledgments: The authors are thankful to the Researchers Supporting Project (number RSP-2020/146) at King Saud University, Riyadh, Saudi Arabia for funding this project. The authors are also thankful to S.R.T. University for providing facilities to carry out these studies.

Conflicts of Interest: The authors declare no conflict of interest associated with this manuscript.

References

1. Putterman, C. Introduction: New treatment paradigms in rheumatoid arthritis. *Am. J. Orthopaed.* **2006**, *36*, S2–S3.
2. Tanaka, T. Introduction for inflammation and cancer. *Sem. Immunopathol.* **2013**, *35*, 121–122. [CrossRef] [PubMed]
3. Bihania, G.V.; Rojatkar, S.R.; Bodhankar, S.L. Anti-arthritic activity of methanol extract of *Cyathocline purpurea* (whole plant) in Freund's complete adjuvant-induced arthritis in rats. *Biomed. Aging Pathol.* **2014**, *4*, 197–206. [CrossRef]
4. Reddy, D.; Trost, L.W.; Lee, T.; Baluch, A.R.; Kaye, A.D. Rheumatoid arthritis: Current pharmacologic treatment and anesthetic considerations. *Middle East J. Anesthesiol.* **2007**, *19*, 311–335.
5. McInnes, I.B.; Schett, G. The pathogenesis of rheumatoid arthritis. *N. Eng. J. Med.* **2011**, *365*, 2205–2219. [CrossRef]
6. Bhardwaj, L.K.; Chandrul, K.K.; Sharma, U.S. Evaluation of anti-arthritic activity of *Ficus benghalensis* Linn. root extracts on Freund's adjuvant induced arthritis. *J. Phytopharmacol.* **2006**, *5*, 10–14.
7. Fan, A.; Lao, L.; Zhang, R.; Zhou, A.; Wang, L.; Moudgil, K. Effects of an acetone extract of *Boswellia carterii Birdw* (Burseraceae) gum resin on adjuvant-induced arthritis in lewis rats. *J. Ethnopharmacol.* **2005**, *101*, 104–109. [CrossRef]

8. Ofman, J.J.; Badamgarav, E.; Henning, J.M.; Knight, K.; Laine, L. Utilization of nonsteroidal anti-inflammatory drugs and anti secretory agents: A managed care claims analysis. *Am. J. Med.* **2004**, *116*, 835–842. [CrossRef]
9. American college of rheumatology ad hoc Committee on clinical guidelines. Guidelines for the management of rheumatoid arthritis. *Arth. Rheum.* **1996**, *39*, 713–722. [CrossRef]
10. Choudhary, M.; Kumar, V.; Gupta, P.; Singh, S. Investigation of antiarthritic potential of *Plumeria alba* L. leaves in acute and chronic models of arthritis. *BioMed. Res. Int.* **2014**, *2014*, 474616. [CrossRef]
11. Choudhary, M.; Kumar, V.; Gupta, P.K.; Singh, S. Anti-arthritic activity of Barleria prionitis Linn. leaves in acute and chronic models in Sprague Dawley rats. *Bull. Facul. Pharm. Cairo Univ.* **2014**, *52*, 199–209. [CrossRef]
12. Vedpal; Gupta, S.K.; Gupta, A.K.; Pakash, D.; Gupta, A. Anti-arthritic activity of Desmodium gangeticum root. *Int. Res. J. Pharm.* **2013**, *4*, 100–102.
13. Gupta, R.K.; Hussain, T.; Panigrahi, G.; Das, A.; Singh, G.N.; Sweety, K.; Faiyazuddin, M.; Rao, C.V. Hepatoprotective effect of Solanum xanthocarpum fruit extract against CCl4 induced acute liver toxicity in experimental animals. *Asian Pac. J. Trop. Med.* **2011**, *4*, 964–968. [CrossRef]
14. Perera, H.D.S.M.; Samarasekera, J.K.R.R.; Handunnetti, S.M.; Weerasena, O.V.D.S.J. In vitro anti-inflammatory and anti-oxidant activities of Sri Lankan medicinal plants. *Ind. Crops Prod.* **2016**, *94*, 610–620. [CrossRef]
15. Mahboubi, M. *Elaeagnus angustifolia* and its therapeutic applications in osteoarthritis. *Ind. Crops Prod.* **2018**, *121*, 36–45. [CrossRef]
16. Ali, B.; Mujeeb, M.; Aeri, V.; Mir, S.R.; Faiyazuddin, M.; Shakeel, F. Anti-inflammatory and antioxidant activity of Ficus carica Linn. leaves. *Nat. Prod. Res.* **2012**, *26*, 460–465. [CrossRef]
17. Azhar, M.F.; Siddiqui, M.T.; Ishaque, M.; Tanveer, A. Study of ethnobotany and indegenous use of Calotropis procera (Ait.) in Cholistan desert, Punjab, Pakistan. *J. Agric. Res.* **2014**, *52*, 117–126.
18. Meena, A.K.; Yadav, A.; Rao, M.M. Ayurvedic uses and pharmacological activities of *Calotropis procera* Linn. *Asian J. Trad. Med.* **2011**, *6*, 45–53.
19. Panda, P.; Das, B.; Sahu, D.S.; Meher, S.K.; Das, B.K.; Rao, M.M.; Lakshmi, G.C.H.D.N. Important uses of arka (*Calotropis procera* Linn) in Indian system of medicine with pharmacological evidence. *Res. J. Pharmacol. Pharmacodyn.* **2015**, *7*, 46–49. [CrossRef]
20. Biswasroy, P.; Panda, S.; Das, D.; Kar, D.M.; Ghosh, D. Pharmacological investigation of Calotropis gigantea: A benevolent herb of nature. *Res. J. Pharm. Technol.* **2020**, *13*, 461–467. [CrossRef]
21. Ramachandran Setty, S.; Quereshi, A.A.; Viswanath Swamy, A.H.; Patil, T.; Prakash, T.; Prabhu, K.; Veeran Gouda, A. Hepatoprotective activity of *Calatropis procera* flowers against paracetamol-induced hepatic injury in rats. *Fitoterapia* **2007**, *78*, 451–454. [CrossRef] [PubMed]
22. Moustafa, A.M.; Ahmed, S.H.; Nabil, Z.I.; Hussein, A.A.; Omran, M.A. Extraction and phytochemical investigation of *Calotropis procera*: Effect of plant extracts on the activity of diverse muscles. *Pharm. Biol.* **2010**, *48*, 1080–1190. [CrossRef] [PubMed]
23. Nenaah, G.E. Potential of using flavonoids, latex and extracts from *Calotropis procera* (Ait.) as grain protectants against two coleopteran pests of stored rice. *Ind. Crops Prod.* **2013**, *45*, 327–334. [CrossRef]
24. Kumar, V.L.; Basu, N. Anti-inflammatory activity of the latex *Calotropis procera*. *J. Ethnopharmacol.* **1994**, *44*, 123–125. [CrossRef]
25. Iqbal, Z.; Lateef, M.; Jabbar, A.; Muhammad, G.; Khan, M.N. Anthelmintic activity of *Calotropis procera* (Ait.) Ait. F. flowers in sheep. *J. Ethanopharmacol.* **2005**, *102*, 256–261. [CrossRef] [PubMed]
26. Kamath, J.V.; Rana, A.C. Pharmacological activities of ethanolic extract of Calotropis procera roots. *Indian Drugs* **2003**, *40*, 292–295.
27. Kareem, S.O.; Akpan, I.; Ojo, O.P. Antimicrobial activities of *Calotropis procera* on selected pathogenic microorganisms. *Afr. J. Biomed. Res.* **2008**, *11*, 105–110. [CrossRef]
28. Butler, S.H.; Godefroy, F.; Besson, J.M.; Weil-Fugazza, J. A limited arthritic model for chronic pain studies in the rat. *Pain* **1992**, *48*, 73–78. [CrossRef]
29. Kumar, V.L.; Roy, S. *Calotropis procera* Latex extract affords protection against inflammation and oxidative stress in Freund's complete adjuvant-induced monoarthritis in rats. *Med. Inflamm.* **2007**, *2007*, 47523. [CrossRef]
30. Kumar, V.L.; Roy, S. Protective effect of latex of *Calotropis procera* in Freund's complete adjuvant induced monoarthritis. *Phytother. Res.* **2009**, *23*, 1–5. [CrossRef]
31. Sangraula, H.; Dewan, S.; Kumar, V.L. Evaluation of anti-inflammatory activity of latex of *Calotropis procera* in different models of inflammation. *Inflammopharmacology* **2002**, *9*, 257–264. [CrossRef]
32. Kumar, V.L.; Chaudhary, P.; Ramos, M.V.; Moahn, M.; Matos, M.P.V. Protective effects of proteins derived latex of *Calotropis procera* against inflammatory hyperalgesia in monoarthritic rats. *Phytother. Res.* **2011**, *25*, 1336–1341. [CrossRef] [PubMed]
33. Chaudhary, P.; Ramos, M.V.; Vasconcelos, M.D.S.; Kumar, V.L. Protective effects of high molecular weight protein sub-fraction of *Calotropis procera* latex in monoarthritic rats. *Pharmacog. Mag.* **2016**, *112*, S147–S151.
34. Chandrasekar, R.; Chandrasekar, S. Natural herbal treatment for rheumatoid arthritis-a review. *Int. J. Pharm. Sci. Res.* **2017**, *8*, 368–384.
35. Khandelwal, K.R. *Practical Pharmacognosy*; Nirali Prakashan: Pune, India, 2004; pp. 149–155.
36. Parasuraman, S. Computer-aided prediction of biological activity spectra, pharmacological and toxicological properties of cleistanthin A and B. *Int. J. Res. Pharm. Sci.* **2010**, *1*, 333–337.
37. Filimonov, D.A.; Lagunin, A.A.; Gloriozova, T.A.; Rudik, A.V.; Druzhilovskii, D.S.; Pogodin, P.V.; Poroikov, V.V. Prediction of the biological activity spectra of organic compounds using the pass online web resource. *Chem. Het. Comp.* **2014**, *50*, 444–457. [CrossRef]

38. Jamkhande, P.G.; Barde, S.R. Evaluation of anthelmintic activity and in silico PASS assisted prediction of *Cordia dichotoma* (Forst.) root extract. *Anc. Sci. Life* **2014**, *34*, 39–43. [CrossRef]
39. Sambavekar, P.P.; Aitawade, M.M.; Patil, D.R.; Kolekar, G.B.; Deshmukh, M.B.; Anbhule, P.V. In-silico, in-vitro antibacterial activity and toxicity profile of new quinoline derivatives. *Indian J. Chem.* **2013**, *52*, 1521–1526.
40. Thite, A.T.; Patil, R.R.; Naik, S.R. Anti-arthritic activity profile of methanolic extract of *Ficus bengalensis*: Comparison with some clinically effective drugs. *Biomed. Aging Pathol.* **2014**, *4*, 207–217. [CrossRef]
41. Kshirsagar, A.D.; Panchal, P.V.; Harle, U.N.; Nanda, R.K.; Shaikh, H.M. Anti-Inflammatory and antiarthritic activity of anthraquinone derivatives in rodents. *Int. J. Inflamm.* **2014**, *2014*, 690596. [CrossRef]
42. Mali, S.M.; Sinnathambi, A.; Kapase, C.U.; Bodhankar, S.L.; Mahadik, K.R. Anti-arthritic activity of standardised extract of *Phyllanthus amarus* in Freund's complete adjuvant induced arthritis. *Biomed. Aging Pathol.* **2011**, *1*, 185–190. [CrossRef]
43. Arya, S.; Kumar, V.L. Anti-inflammatory efficacy of extracts of latex of *Calotropis procera* against different mediators of inflammation. *Med. Inflamm.* **2005**, *4*, 228–232. [CrossRef] [PubMed]
44. Kumar, V.L.; Roy, S.; Sehgal, R.; Padhy, B.M. A comparative study on the efficacy of rofecoxib in monoarticular arthritis induced by latex of *calotropis procera* and freund's complete adjuvant. *Inflammopharmacology* **2006**, *14*, 17–21. [CrossRef] [PubMed]
45. Zhang, G.Q.; Huang, X.D.; Wang, H.; Leung, A.K.; Chan, C.L.; Fong, D.W.; Yu, Z.L. Anti-inflammatory and analgesic effects of the ethanol extract of *Rosa multiflora Thunb.* hips. *J. Ethnopharmacol.* **2008**, *118*, 290–294. [CrossRef]
46. Asquith, D.L.; Miller, A.M.; McInnes, I.B.; Liew, F.Y. Animal models of rheumatoid arthritis. *Eur. J. Immunol.* **2009**, *39*, 2040–2044. [CrossRef]
47. Mythilypriya, R.; Shanthi, P.; Sachdanandam, P. Salubrious effect of Kalpaamruthaa, a modified indigenous preparation in adjuvant-induced arthritis in rats–a biochemical approach. *Chem. Biol. Inter.* **2008**, *173*, 148–158. [CrossRef]
48. Pawar, S.; Patil, D.A. Folk remedies against rheumatic disorders in Jalgaon district, Maharashtra. *Indian J. Trad. Know.* **2006**, *5*, 314–316.
49. Singh, V.S.; Dhawale, S.C.; Shakeel, F.; Faiyazuddin, M.; Alshehri, S. In vitro and in vivo anti-inflammatory effects of Calotropis procera leave fractions. unpublished work, manuscript in preparation.
50. Sameeh, M.Y.; Mohamed, A.A. Characterization of polyphenols, their antioxidant and GC-MS analysis of wild Calotropis procera leaves and fruit extracts. *Int. J. Chem. Tech. Res.* **2018**, *11*, 319–327.
51. Wu, C.R.; Hseu, Y.C.; Lien, J.C.; Lin, L.W.; Lin, Y.T.; Ching, H. Triterpenoid contents and anti-inflammatory properties of the methanol extracts of *Ligustrum* species leaves. *Molecules* **2011**, *16*, 1–15. [CrossRef]
52. Mallick, S.S.; Dighe, V.V. Detection and estimation of alpha-amyrin, beta-sitosterol, lupeol, and n-triacontane in two medicinal plants by high performance thin layer chromatography. *Adv. Chem.* **2014**, *2014*, 143948. [CrossRef]
53. Tuncel, J.; Haag, S.; Hoffmann, M.H.; Yau, A.C.Y.; Hulqvist, M.; Olofsson, P.; Backlund, J.; Nandakumar, K.S.; Weidner, D.; Fischer, A.; et al. Animal models of rheumatoid arthritis (I): Pristine-induced arthritis in the rat. *PLoS ONE* **2016**, *11*, e0155936. [CrossRef] [PubMed]
54. Patil, K.R.; Patil, C.R.; Jadhav, R.B.; Mahajan, V.K.; Raosaheb, P.; Gaikwad, P.S. Anti-arthritic activity of bartogenic acid isolated from fruits of *Barringtonia racemosa* Roxb. (Lecythidaceae). *Evid.-Based Complement. Altern. Med.* **2011**, *2011*, 785245. [CrossRef] [PubMed]
55. Petchi, R.R.; Parasuraman, S.; Vijaya, C.; Gopala Krishna, S.V.; Kumar, M.K. Antiarthritic activity of a polyherbal formulation against Freund's complete adjuvant induced arthritis in female Wistar rats. *J. Basic Clin. Pharmacol.* **2015**, *6*, 77–83. [CrossRef] [PubMed]
56. Jalalpure, S.S.; Mandavkar, Y.D.; Khalure, P.R.; Shinde, G.S.; Shelar, P.A.; Shah, A.S. Antiarthritic activity of various extracts of *Mesua ferrea* Linn. Seed. *J. Ethnopharmacol.* **2011**, *138*, 700–704. [CrossRef] [PubMed]
57. Kubo, M.; Matsuda, H.; Tanaka, M.; Kimura, Y.; Okuda, H.; Higashino, M.; Tani, T.; Namba, K.; Aaichi, S. Studies on scutellariadix.VII. anti-arthritic and anti-inflammatory action of methanolic extract and flavonoid components from *scutellariae radix*. *Chem. Pharm. Bull.* **1984**, *32*, 2724–2729. [CrossRef] [PubMed]
58. Prasad, P.J. *Conceptual Pharmacology*; Universities Press: Hydrabad, India, 2010; pp. 254–268.
59. Ekambaram, S.; Perumal, S.S.; Subramanian, V. Evaluation of anti-arthritic activity of *Strychnos potatorum* Linn seeds in Freund's adjuvant induced arthritic rat model. *BMC Complement. Altern. Med.* **2010**, *10*, 56. [CrossRef]
60. Zhang, X.; Dong, Y.; Li, F. Inevestigation of the effect of phlomisoside F on complete Freund's adjuvant-induceed arthritis. *Exp. Ther. Med.* **2017**, *12*, 710–718. [CrossRef]
61. Mbiantcha, M.; Almas, J.; Shabana, S.U.; Nida, D.; Aisha, F. Anti-arthritic property of crude extracts of *Piptadeniastrum africanum* (Mimosaceae) in complete Freund's adjuvant-induced arthritis in rats. *BMC Complment. Altern. Med.* **2017**, *17*, E111. [CrossRef]
62. Perumal, S.S.; Ekambaram, P.S.; Dhanam, T. In vivo antiarthritic activity of the ethanol extracts of stem bark and seeds of *Calophyllum inophyllum* in Freund's complete adjuvant induced arthritis. *Pharm. Biol.* **2017**, *55*, 1330–1336. [CrossRef]
63. Almarestani, L.; Fitzcharles, M.A.; Bennett, G.J.; Ribeiro-da-Silva, A. Imaging studies in Freund's complete adjuvant model of regional polyarthritis, a model suitable for the study of pain mechanisms, in the rat. *Arth. Rheum.* **2011**, *63*, 1573–1581. [CrossRef] [PubMed]

Article

Stevia rebaudiana (Bertoni) as a Multifunctional and Sustainable Crop for the Mediterranean Climate

Clarissa Clemente [1], Luciana G. Angelini [1,2], Roberta Ascrizzi [3] and Silvia Tavarini [1,2,*]

[1] Department of Agriculture, Food and Environment, University of Pisa, Via del Borghetto 80, 56124 Pisa, Italy; clarissa.clemente@phd.unipi.it (C.C.); luciana.angelini@unipi.it (L.G.A.)
[2] Interdepartmental Research Center "Nutraceuticals and Food for Health", University of Pisa, Via del Borghetto 80, 56124 Pisa, Italy
[3] Department of Pharmacy, University of Pisa, Via Bonanno 6, 56124 Pisa, Italy; roberta.ascrizzi@for.unipi.it
* Correspondence: silvia.tavarini@unipi.it

Citation: Clemente, C.; Angelini, L.G.; Ascrizzi, R.; Tavarini, S. *Stevia rebaudiana* (Bertoni) as a Multifunctional and Sustainable Crop for the Mediterranean Climate. *Agriculture* **2021**, *11*, 123. https://doi.org/10.3390/agriculture11020123

Academic Editors: Mario Licata, Antonella Maria Maggio, Salvatore La Bella and Teresa Tuttolomondo

Received: 29 December 2020
Accepted: 1 February 2021
Published: 4 February 2021

Publisher's Note: MDPI stays neutral with regard to jurisdictional claims in published maps and institutional affiliations.

Copyright: © 2021 by the authors. Licensee MDPI, Basel, Switzerland. This article is an open access article distributed under the terms and conditions of the Creative Commons Attribution (CC BY) license (https://creativecommons.org/licenses/by/4.0/).

Abstract: *Stevia rebaudiana* (Bertoni) is a promising medicinal and aromatic plant for Mediterranean agroecosystems given its positive agronomic attributes and interesting quality features. It has both food and pharmaceutical applications, since its leaves contain sweet-tasting steviol glycosides (SVglys) and bioactive compounds, such as phenolics, flavonoids, and vitamins. We evaluated the agronomic and qualitative performances of nine stevia genotypes cultivated, in open field conditions, for two consecutive years under the Mediterranean climate of central Italy. Growth, biomass production, and accumulation of bioactive compounds (SVglys, polyphenols, and their related antioxidant activities) were evaluated, considering the effect of harvest time and crop age (first and second year of cultivation). The results showed high variability among genotypes in terms of both morpho-productive and phytochemical characteristics. In general, greater leaf dry yields, polyphenol accumulation, and antioxidant activities were found in the second year of cultivation, harvesting the plants in full vegetative growth. On the other hand, total SVglys leaf content reached the highest values in the first year when plants were at the beginning of the reproductive phase. On the other hand, although the SVglys profile (Rubusoside, Dulcoside A, Stevioside, Rebaudioside A, C, D, E, and M) remained stable over harvest times, it differed significantly depending on the crop age and genotype. Our findings provide useful information on the influence of crop age and harvest time in defining quanti-qualitative traits in stevia, with PL, SL, BR5, and SW30 being the best performing genotypes and thus suitable for breeding programs. Our study highlighted that stevia, in the tested environment, represents a promising semi-perennial crop which offers new solutions in terms of cropping system diversification and marketing opportunities.

Keywords: medicinal and aromatic plants; crop diversification; sustainability; leaf yield; biofunctional products; genotypic variability

1. Introduction

In many Mediterranean areas, high evapotranspiration rates, increased precipitation variability, and intense summer drought are the main environmental constraints to agricultural management and crop production. Under such conditions, sustainable agricultural practices and the introduction of new crops that are able to diversify cropping systems and to mitigate climate changes are very promising strategies for addressing cropping system sustainability.

Compared to continuous cropping systems, crop diversification involves better utilization of land resources, lower risks from pests and diseases, and greater yield stability. In this context, medicinal and aromatic plants (MAPs) can be included in low-input productive systems: they can enhance the multifunctionality traits of the agricultural sector, produce safe final products, respect the environment, and promote rural areas. Various MAPs find optimal growth conditions in Mediterranean environments and, among these, perennial and semi-perennial species can contribute to agro-ecosystem services.

The cultivation of perennial crops reduces soil erosion, minimizes nutrient leaching, sequesters more C in soils, protects water resources, creates a better pest tolerance, and provides a continuous habitat for wildlife [1–3]. In addition, perennial species require less use of farm equipment than annual crops as well as fewer fertilizers and herbicides [4]. Among perennial MAPs, *Stevia rebaudiana* (Bertoni) is grown successfully in a wide range of agro-ecological environments, from semi-humid, subtropical to temperate zones. Thanks to its extreme versatility, it can be grown as a pluriannual crop (three to five years) in temperate to warm climates and, as an annual crop, in colder regions [5–7]. Unlike other traditional sweetener crops, such as sugar beet and sugar cane, stevia cultivation reduces agronomic inputs, including nutrients, water, and energy [8].

Besides these agronomic benefits, stevia has an exceptional phytocomplex composition that can be exploited in several applications, from the pharmaceutical to cosmetic and nutraceutical industry. Metabolic disorders such as type-II diabetes and obesity, associated with an excessive sugar consumption, are becoming more prevalent and stevia is the perfect sugar substitute in foods and drinks, given that its leaves contain non-calorie high-intensive sweetener compounds, namely steviol glycosides (SVglys) [9–11]. Its leaves also contain a complex mixture of triterpenoids, sterols, essential oils, phenols, and flavonoids, with functional and health-promoting properties [12–17]. Thanks to its positive agronomic and phytochemical characteristics, stevia could thus offer new solutions in terms of cropping system diversification and marketing opportunities.

Stevia cultivation is widespread all over the world and it has been introduced as a commercial crop in several countries [8,18]. Experimental cultivations are increasing in Europe aimed at producing higher-performing and yielding stevia genotypes [19–23]. Access to genotypes that can adapt to different environmental conditions is important for the selection of those characterized by both high levels of secondary metabolites and biomass yield.

To improve the competitiveness of stevia production in the Mediterranean region, it is important to identify higher-performing genotypes in terms of yield and quality, resource use efficiency, and resistance/tolerance to a wide range of biotic/abiotic stress combinations. Therefore, the aim of this study was to determine the best performing genotypes suitable for developing site-specific recommendations for stevia cultivation under the Mediterranean climate of central Italy. Nine stevia genotypes and two harvest times were compared for two consecutive years, exploring differences in plant growth traits (growth cycle, biometric characteristics, leaf yield) and in desired compounds (SVglys content and profile, total phenol and flavonoid accumulation, and antioxidant activity).

2. Materials and Methods

2.1. Experimental Design and Plant Materials

A field-plot trial was carried out during two growing seasons (2018 and 2019) at the experimental Centre of the Department of Agriculture, Food and Environment (DAFE, University of Pisa), located in San Piero a Grado, Pisa, central Italy (43°40' N; 10°19' E, 5 m above sea level). A completely randomized block experimental design, with four replications, was adopted, with the plot as the experimental unit and selected plants within the plot as the observational unit. The plot size was 3 m × 1.8 m (width × length), and each plot consisted of six rows of six plants (36 plant plot^{-1}), with a plant density of 6.67 plants m^{-2} by adopting an inter-row and intra-row spacing of 0.5 × 0.3 m. Seedbed preparation included moldboard plowing (30 cm depth), disk harrowing, and use of a cultivator. Pre-planting phosphorus and potassium fertilizations were performed at a rate of 100 kg ha^{-1} of P_2O_5 by triple superphosphate and 80 kg ha^{-1} of K_2O by potassium sulfate. Nitrogen (as ammonium nitrate) was applied at a rate of 40 kg N ha^{-1} after transplanting. Nine genotypes of different origins, belonging to DAFE's germplasm collection were used for the field-plot trial (Table 1). Plants for each genotype were first reproduced by micropropagation and then multiplied by stem cuttings to avoid plant genetic variability in terms of both morphological and phytochemical characteristics. Subsequently, they were kept under controlled conditions at the DAFE greenhouse until open field transplanting between mid-May and early June 2018.

Table 1. Origin of *S. rebaudiana* genotypes.

Genotype ID	Origin
PL	Israel
BR16	Brazil
RG	Italy
SL	Israel
NU	Italy
CO	Israel
BR5	Brazil
SW30	Italy
BR1	Brazil

Weather parameters (daily minimum, T_{min}, maximum, T_{max}, and mean temperatures, T_{mean}, and cumulative rainfall) were recorded by an automated weather station nearby the experimental site, from the beginning of vegetative plant development (June) to the full plant flowering (October) in each year of cultivation. Throughout the experiment, total rainfalls were 164.2 and 233.2 mm, in 2018 and 2019, respectively, with a mean average temperature of 19.0 °C and 16.4 °C in 2018 and 2019, respectively. Mean average T_{max} and T_{min} temperatures did not differ notably between 2018 and 2019 (Table 2).

Table 2. Monthly cumulative (mm) rainfall and mean temperatures (T_{mean}, T_{max}, and T_{min}) throughout the entire experimental period (May 2018–September 2019).

Year	Month	T_{max} (°C)	T_{min} (°C)	T_{mean} (°C)	Rainfall (mm)
2018	May	21.7	15.0	18.4	87.4
	June	25.6	17.6	21.6	5.0
	July	28.0	20.3	24.1	49.6
	August	29.3	20.4	24.8	25.2
	September	26.1	17.2	21.6	19.4
	October	22.7	14.1	18.4	65.0
	November	16.8	9.9	13.4	111.6
	December	13.3	5.5	9.4	51.0
2019	January	11.4	1.9	6.6	41.6
	February	13.9	4.1	9.0	58.6
	March	15.6	6.6	11.1	3.0
	April	17.6	9.0	13.3	112.2
	May	18.6	11.6	15.1	87.6
	June	26.3	17.8	22.1	2.8
	July	28.6	20.2	24.4	88.4
	August	28.9	20.2	24.6	3.6
	September	25.5	17.3	21.4	72.6

Before beginning the experiment, soil samples were collected at 0 to 30 cm depth in order to evaluate the physical and chemical characteristics of the soil (Table 3). The soil was loam with a sub-alkaline reaction, a good content of total nitrogen, organic matter, and exchangeable potassium, and with a low level of available phosphorus and salinity.

Table 3. Physical and chemical characteristics of the soil at the experimental site.

Characteristics	Values
Clay (<0.002 mm, %)	14.5
Silt (0.05–0.002 mm, %)	38.9
Sand (2–0.05 mm, %)	46.5
pH (H_2O 1:2.5 soil:water suspension; McLean method)	7.6
N tot (Kjeldahl method, g kg^{-1})	1.4
S.O. (Walkley–Black method, g kg^{-1})	3.1
Available phosphorus (Olsen method, mg kg^{-1})	9.9
Exchangeable potassium (Thomas method, mg kg^{-1})	211.5
CE (mS cm^{-1})	0.4
CSC (Method $BaCl_2$, pH 8.1, meq 100 g^{-1})	20.1

2.2. Crop Sampling and Agronomic Measurements

Phenological observations, from the beginning of vegetative plant development until full plant flowering, were performed in the ratooning crop, starting from the regrowth of new shoots, per each genotype and in each plot. For every observation, six plants per genotype were randomly selected in the field, with 54 total plants observed. The length of both the vegetative and reproductive phases was evaluated using accumulated thermal time (°C/day) and the accumulated growing degree days (GDDs) were calculated daily according to the equation (1) presented by McMaster and Wilhelm [24],

$$GDD = \Sigma \left[(T_{max} + T_{min})/2 - T_b\right] \quad (1)$$

where T_{max} is the daily maximum air temperature, T_{min} is the daily minimum air temperature, and T_b is the base temperature, for which 10 °C was used [22].

For each growing season, two destructive samplings (namely HT1 and HT2) were carried out at the following phenological stages (according to the BBCH scale by Le Bihan et al. [25]): HT1–Stage 48 = about 80% of final leaf biomass is developed; and HT2–Stage 55 = 50% of apex leaves are differentiated and present inflorescence, but flower buds are still closed.

HT1 corresponded to the full vegetative growth which was reached between 27 July–2 August in the 1st year and between 15–17 July in the 2nd year, depending on the genotype. HT2 corresponded to the beginning of flowering, occurred in early September in the 1st year (10–13 September 2018) and between the end of August and the beginning of September in the 2nd year (from 20 August to 2–3 September 2019). Samplings were manually performed, collecting six plants per genotype (1 sample = 1 plant). After each harvest, plant height (cm), basal stem diameter (mm), branching (n. stem/plant), fresh and dry weight of leaves and stems (g plant^{-1}), harvest index (HI), and specific leaf weight (LSW) were measured. Leaves of each sample were air-dried in a ventilated oven from 30 to 40 °C until constant weight. Dry leaves were ground to a fine powder by a laboratory mill (Grindomix GM 200, Retsch, Pedrengo (BG), Italy) and stored until the subsequent analyses. The harvest index (HI), which represents the plant's efficiency at producing leaves, was calculated as the ratio between leaf dry yield (g plant^{-1}) and total aboveground biomass yield (leaves plus stems). Specific leaf weight (SLW) was calculated as the ratio between dry leaf weight and leaf area (mg cm^{-2}). The leaf area was measured by collecting and subjecting to a color scan, two apical leaves, two middle leaves, and two basal leaves. The leaf images were then analyzed with ImageJ (Fiji Particle Analysis plug-in).

2.3. Phytochemical Analysis
2.3.1. Chemicals

Common Steviol Glycosides Standards Kit (Rubusoside, Dulcoside A, Stevioside, Rebaudiosides A, C, D, E and M) was purchased from Chromadex (LGC Standards S.r.L., Milan, Italy). Ferrous sulphate, 2,2-diphenyl-1-picrylhydrazyl (DPPH), gallic acid monohydrate (3,4,5-trihydroxybenzoic acid), 2,4,6-tri(2-pyridyl)-triazine (TPTZ), trizma acetate, Folin-Ciocalteu reagent, Trolox (6-hydroxy-2,5,7,8-tetramethylchroman-2-carboxylic acid), sodium carbonate, and ferric chloride were obtained from Sigma-Aldrich Chemical Co. (Milan, Italy). All chemicals used in the present study, including solvents, were of analytical grade.

2.3.2. Sample Extraction

A total of 0.1 g of leaf powder per sample was extracted with 10 mL of 70% (v/v) EtOH, and sonicated for 30 min at 60 °C. At the end of the sonication, the extracts were centrifuged (3500 rpm for 10 min) and filtered with a syringe filter (Ø 0.45 µm) to remove any suspended material. The extracts obtained were stored at 4 °C until subsequent analyses.

2.3.3. Steviol Glycosides Determination

The extraction procedure and the determination of steviol glycosides (SVglys) were carried out following Zimmerman et al. [26]. Steviol glycosides were analyzed using a Jasco PU980 HPLC system (JASCO Benelux B.V., Utrecht, Netherlands) coupled with a UV-visible wavelength detector. A hydrophilic column (Luna HILIC 200A, Phenomenex Inc., Torrance, CA, USA), 5 µm, 250 mm × 4.6 mm (Phenomenex Inc., Torrance, CA, USA) in combination with the corresponding pre-column (4 × 3.0 mm) was used. UV detection was carried out at 205 nm at room temperature with a flow rate of 0.68 mL/min and a run time of 20 min. Separation was achieved in acetonitrile/water (80:20) as isocratic mobile phase at pH 3.6 regulated with acetic acid. Chromatograms were acquired online, and data were collected using a Jasco interface (Hercules 2000 Interface Chromatography). Steviol glycosides were quantified using authentic standards, through calibration curves (0.005–1.00 g L^{-1}), obtained from standard mixtures containing Rubusoside (Rub), Dulcoside A (Dulc A), Stevioside (Stev), and Rebaudiosides A, C, D, E, and M (Reb A, C, D, E, and M).

2.3.4. Analysis of Total Phenols and Flavonoids

Total phenols were determined using the Folin-Ciocalteu method according to Dewanto et al. [27] and expressed as gallic acid equivalents (mg GA g^{-1} dry leaf). This method involved the reduction of Folin-Ciocalteu reagent by phenolic compounds, with a blue complex formation determined at 765 nm by UV-Vis spectrophotometer (Varian Cary 1E, Palo Alto, CA, USA). Total flavonoids were determined using the aluminum trichloride method according to Jia et al. [28] and expressed as catechin equivalents (mg CE g^{-1} dry leaf). The flavonoids-aluminum reaction created a pink complex formation measured at 510 nm using a UV-Vis spectrophotometer (Varian Cary 1E, Palo Alto, CA, USA).

2.3.5. Ferric Reducing Antioxidant Power (FRAP) Assay and Free Radical-Scavenging Assay

The determination of ferric reducing antioxidant power and the free radical-scavenging activity (FRAP and DPPH assay) of stevia leaf extracts followed Tavarini et al. [29]. The FRAP method is based on the ability of the antioxidant compounds to reduce Fe^{3+} to Fe^{2+} which, in the presence of TPTZ (2,4,6-tris(2-pyridyl)-s-triazine) led to a blue complex formation (Fe^{2+}-TPTZ), measured at 593 nm using a UV-Vis spectrophotometer (Varian Cary 1E, Palo Alto, CA, USA). The DPPH assay is based on the reducing activity of the antioxidant molecules against the 1,1-diphenyl-2-picryl-hydrazil (DPPH) radical which was characterized by a purple red color. The extent of the disappearance of DPPH is directly proportional to the amount of antioxidant present in the reaction measured at 517 nm using a UV-Vis spectrophotometer (Varian Cary 1E, Palo Alto, CA, USA). Total antioxidant activity and free radical-scavenging capacity were expressed as trolox equivalents (mmol TE g^{-1} dry leaf).

2.4. Statistical Analyses

All data were subjected to analysis of variance (ANOVA) using GraphPad Prism v. 8.0.2 (GraphPad Software, Inc., La Jolla, CA, USA). A three-way ANOVA analysis was conducted to assess the effect of genotype (G), crop age (CA), harvest time (HT), and their reciprocal interactions, on the agronomic characteristics and on total phenols and flavonoids, total SVglys and antioxidant activities (FRAP and DPPH). Means were separated on the basis of the least significant difference (LSD) only when the ANOVA F test showed significance at 0.05 or 0.01 probability level.

Hierarchical cluster (HC) and principal component (PC) analyses were performed on (i) the total content of phenols, flavonoids, and steviol glycosides, and (ii) on the individual steviol glycoside (Dulcoside A; Rebaudiosides A, C, D, E and M; Rubusoside, Stevioside) concentrations with JMP® Pro 13.2.1 (SAS Institute Inc., Cary, NC, USA). As unsupervised methods, the groups of samples obtained with hierarchical cluster analysis (HCA) and PCA can be observed even when there are no reference samples that can be used as a training set to establish the model. For both observation groups, the hierarchical cluster analysis (HCA) was conducted on the normalized average values, with Ward's algorithm, using Euclidean distances as a measure of similarity among the samples. In addition, principal component analyses (PCA) were carried out in order to reduce the dimensionality of the multivariate data of the matrix, whilst preserving most of the variance [30]. For the PCA of the total content of phenols, flavonoids, and steviol glycosides, a 35 × 3 (35 samples, 3 metabolite contents, 105 total data) dimensional matrix was used. The score plot obtained was defined by a PC1 and a PC2 covering 78.3 and 21.3% of the variance, respectively, for a total explained variance of 99.6%. For the PCA of the total content of individual steviol glycoside (Dulcoside A; Rebaudiosides A, C, E, M, and D; Rubusoside, Stevioside) concentrations, a 35 × 8 (35 samples, 8 steviol glycosides, 280 total data) dimensional matrix was used. The score plot obtained was defined by a PC1 and a PC2 covering 68.2 and 19.3% of the variance, respectively, for a total explained variance of 87.5%.

3. Results

3.1. Growth, Biometric, and Productive Measurements

Figure 1 shows the accumulation of thermal time, expressed as growing degrees days (GDDs), required by each genotype to pass from the vegetative phase to reproductive one in the second year after transplanting. Significant differences among genotypes were observed: RG, SL, CO, BR1, and NU (1414.4 GDD °C d^{-1}) accumulated fewer GDDs to develop the first flower buds than the others (BR16 and SW30 = 1779.15 °C d^{-1}, PL, and BR5 = 1575.2 °C d^{-1}).

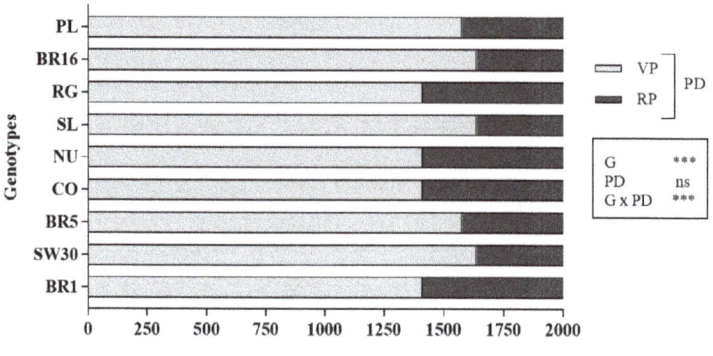

Figure 1. Effect of genotype (G) and plant development (PD), consisting of a vegetative (VP) and reproductive phase (RP), on the thermal time accumulation (°C d^{-1}). The significance of variability factors according to the F-test: ns, not significant; ***, significant at $p \leq 0.001$ level.

Genotype (G), crop age (CA), harvest time (HT) and their reciprocal interactions significantly affected biometric and productive traits, except for (i) branching, which did not vary depending on CA, (ii) basal stem diameter, in relation to G × HT interaction and (iii) plant height and SLW which were not affected by CA×HT (Table 4).

Taking into account the genetic effect (G), plant height ranged from 47.31 to 58.19 cm, reaching the highest values in RG and NU genotypes and the lowest values in SL, PL, and SW30. On the other hand, SW30 was characterized by the greatest stem basal diameter (8.80 mm), while BR16 showed the lowest value (5.06 mm), although along with RG, BR16 had the greatest branching than the other genotypes (PL, SL, NU, CO, BR5, SW30, and BR1) (Table 4).

Regarding leaf dry yield per plant, NU and BR1 were the most productive, while BR16 appeared to be the least productive genotype. The other genotypes showed medium leaf dry yields. Harvest index ranged from 73.51% (SL) to 52.79% (BR16), reflecting the plant architecture, in terms of height and branching, and leaf yield. PL and SL exhibited the highest specific leaf weight (SLW), while RG and NU showed the lowest weights.

In terms of crop age, all biometric and productive measurements showed significant increases in the second year of cultivation, except for the stem basal diameter, which was significantly smaller in 2-year-old plants, and branching, which did not vary in the two years of cultivation (Table 4).

Additionally, harvest time played a key role in defining the agronomic responses of stevia genotypes. In general, maximum plant height, stem basal diameter, branching and leaf dry yield increased from the point of full vegetative growth to the beginning of flowering (Table 4). On the other hand, the highest values of specific leaf weight and harvest index were recorded during the full vegetative growth, with a significant decrease in the subsequent harvest.

Table 4. Results of three-factorial analysis of variance (ANOVA) for agronomic characteristics of stevia genotypes tested in the open-field experimental trial.

Factor Main Effects	Factor Level	Height (cm)	Basal Stem Diameter (mm)	Branching (n. Stem Plant^{-1})	HI (%)	SLW (mg cm^{-2})	Leaf Dry Yield (g Plant^{-1})
Genotype (G)	PL	47.9 ± 3.1 [de]	7.5 ± 0.4 [bc]	18.0 ± 2.3 [b]	65.1 ± 2.5 [cd]	11.3 ± 0.3 [ab]	43.7 ± 3.6 [bc]
	BR16	48.5 ± 2.4 [d]	5.1 ± 0.2 [e]	27.0 ± 3.3 [a]	52.8 ± 2.7 [g]	9.4 ± 0.5 [e]	26.9 ± 3.2 [f]
	RG	55.3 ± 4.4 [ab]	7.6 ± 0.3 [b]	24.0 ± 2.8 [a]	56.0 ± 3.4 [f]	9.9 ± 0.4 [de]	36.9 ± 4.3 [cde]
	SL	44.5 ± 3.7 [e]	6.8 ± 0.4 [cd]	15.0 ± 1.7 [b]	73.5 ± 2.8 [a]	11.5 ± 0.4 [a]	29.1 ± 4.3 [ef]
	NU	58.2 ± 4.7 [a]	7.7 ± 0.3 [b]	19.0 ± 1.7 [b]	55.5 ± 3.4 [f]	9.5 ± 0.3 [e]	55.7 ± 5.9 [a]
	CO	56.6 ± 3.9 [a]	6.9 ± 0.4 [cd]	15.0 ± 1.3 [b]	64.0 ± 2.4 [d]	10.2 ± 0.3 [cd]	41.5 ± 3.0 [bcd]
	BR5	52.9 ± 2.7 [bc]	7.2 ± 0.4 [bcd]	17.0 ± 1.3 [b]	67.1 ± 2.8 [b]	10.6 ± 0.4 [c]	44.3 ± 3.8 [bc]
	SW30	47.3 ± 2.2 [de]	8.8 ± 0.2 [a]	15.0 ± 1.6 [b]	62.2 ± 2.7 [e]	10.8 ± 0.5 [bc]	33.9 ± 2.1 [def]
	BR1	50.0 ± 4.5 [cd]	6.6 ± 0.2 [d]	15.0 ± 1.3 [b]	65.8 ± 2.2 [bc]	10.6 ± 0.5 [c]	48.5 ± 5.4 [ab]
Crop age (CA)	2018	48.8 ± 1.6 [b]	7.5 ± 0.2 [a]	19.0 ± 1.1 [a]	56.0 ± 0.8 [b]	9.3 ± 0.1 [b]	35.4 ± 2.9 [b]
	2019	53.9 ± 1.8 [a]	6.8 ± 0.2 [b]	18.0 ± 1.1 [a]	67.9 ± 1.6 [a]	11.4 ± 0.2 [a]	44.6 ± 1.9 [a]
Harvest time (HT)	July	39.2 ± 0.8 [b]	6.6 ± 0.2 [b]	17.0 ± 1.0 [b]	69.3 ± 1.5 [a]	10.8 ± 0.2 [a]	34.3 ± 2.8 [b]
	September	63.1 ± 1.2 [a]	7.7 ± 0.2 [a]	19.0 ± 1.1 [a]	55.4 ± 0.9 [b]	10.0 ± 0.2 [b]	45.8 ± 2.0 [a]
Significance	Main effects						
	G	***	***	***	***	***	***
	CA	***	*	ns	***	***	***
	HT	***	***	***	***	***	***
	G × CA	***	***	***	***	***	***
	G × HT	***	ns	***	***	***	***
	CA × HT	ns	***	***	***	ns	***
	G × CA × HT	***	***	***	***	***	***

Values followed by identical letters are not significantly different for $p < 0.05$, according to the LSD post-hoc test. The significance of variability factors according to the F-test: ns, not significant; *, significant at $p \leq 0.05$; ***, significant at $p \leq 0.001$ level. HI, harvest index; SLW, specific leaf weight.

3.2. Phytochemical Evaluation and Antioxidant Activities

Table 5 shows data on the secondary metabolites and antioxidant activities, depending on genotype, crop age and harvest time. Total phenols and flavonoids, total SVglys, as well as in vitro antioxidant activities (measured by FRAP and DPPH assays) were significantly affected by all variability factors (G, CA, and HT) and their interactions. Regarding the effect of genotype, SL showed the highest values of both total phenols and DPPH. Conversely, the highest total flavonoids and FRAP capacity were observed for BR5 and BR1. Finally, SW30 showed the highest SVglys total content.

Interestingly, compared to the other genotypes, BR16, RG, and NU exhibited the lowest values of all secondary metabolites (total phenols, total flavonoids and total SVglys), as well as the lowest in vitro antioxidant activities (Table 5). Crop age (CA) positively influenced the phytochemical characteristics of stevia leaf extracts, with an increase in their values in the plants in the second year after transplanting. Only the total SVglys showed a significantly decrease passing from the first to the second year of cultivation. Taking into account harvest time, total phenols, total flavonoids, and the antioxidant activities decreased from the vegetative phase (July) to the beginning of the reproductive one (September). On the other hand, an opposite trend, with a significant increase from the 1st to the 2nd harvest time, was observed for total SVglys, confirming that the development of the first flower buds corresponds to the time of major accumulation of SVglys in stevia plants.

Table 6 reports the SVglys profile and related statistical significance. A significant effect of genotype was observed for almost all steviol glycosides, with the exception of Dulcoside A. Interestingly, the identified compounds were not present in all genotypes. This difference was evident for Reb M which was not detected in RG, SL, NU, and CO, and for Reb D which was not detected in NU and CO. The reduction in SVglys content observed in plants at the second year of age was also seen in the most represented compounds such as Stev, Reb C, Reb A, and Reb C. On the other hand, Rubusoside increased with crop age, while no effect was observed for Reb M and Reb D. With regard to harvest time, as already noted for the total content, each steviol glycoside significantly increased from the vegetative phase to the beginning of flowering, except for Rubusoside which remained stable in both harvest times.

HCA and PCA were carried out for total phenols, flavonoids, and SVglys content. The two-way dendrogram of the HCA is reported in Figure 2. The first macro-cluster (red) was grouped by itself, while the second comprised three sub-clusters (green, blue, and yellow). Among the metabolites analyzed, the total contents of phenols and flavonoids were clustered together, while the total SVglys concentration was grouped by itself, thus highlighting a higher degree of dissimilarity based on this parameter. Based on their total SVglys concentration, samples in the yellow cluster, all from 2018, shared the highest SVglys content; samples of the red cluster, all from 2019, were, instead, characterized by intermediate concentrations of SVglys; the lowest concentrations of these metabolites were, instead, common to samples of the blue and green clusters, especially those from 2019. As evidenced by their grouping in a common macro-cluster, the total phenols and total flavonoids showed a common quantitative distribution among the samples. In fact, higher concentrations of both these chemical classes were evidenced for samples of the red cluster, all from 2019; also, intermediate concentrations were evidenced for samples from 2019 in the green cluster; finally, samples belonging to the blue and yellow clusters exhibited the lowest contents of these compounds. With the exception of the green cluster, the samples were homogeneously distributed in the groups based on the harvest time.

Some genotypes showed very similar total abundances of the analyzed compounds between the two harvest times (July and September) within the same year: i.e., CO 2019, NU 2018 and 2019, RG 2018 and 2019, SW30 2018 and 2019, BR5 and BR1 2019 (Figure 2). The score plot of the PCA (Figure 3) confirmed the HCA grouping: with the exception of two samples (PL_2_2019 and BR5_2_2018), only the red samples were grouped in the upper quadrants (PC2 > 0), due to their higher phenols and flavonoids contents.

Among the other groups, the higher content of SVglys detected in the yellow samples meant that they were plotted in the right quadrants (PC1 > 0), whereas all the other samples were in the left quadrants (PC1 < 0). The same multivariate analyses were conducted on the SVglys profile of all the genotypes studied (Figures 4 and 5). The two-way dendrogram (Figure 4) of the HCA on the compounds showed two main groups driving the dissimilarities: Rubusoside, Stevioside, Rebaudiosides A, M, and D for macro-cluster 1;

Rebaudioside E, Rebaudioside C, and Dulcoside A for macro-cluster 2. The samples in the dendrogram (Figure 4) were distributed in two macro-clusters: the first comprised the red and green sub-groups; the second comprised the blue and yellow sub-groups, of which the latter only contained one sample (SW30_2_2018). The second macro-cluster was mainly composed of 2018 samples, while the first macro-cluster comprised all the 2019 samples, as well as a few of the 2018 ones (Figure 4).

Compared to the HCA of the total compounds, the dendrogram of the SVglys profile suggested that the differences between the samples were less due to the year and more to the steviol glycosides pool produced by each genotype (i.e., all the PL samples are in the red cluster; all the BR16 samples are in the green cluster). In addition, the SVglys pool of samples RG and NU appeared very similar, thus suggesting the proximity of these two genotypes; the same seemed true for BR5, BR1, and BR16 (Figure 4). Regarding SVglys profile, the influence of the year on the sample distribution was also confirmed by the PCA. The score plot is reported in Figure 5. Samples belonging to the first HCA macro-cluster, thus almost all 2019 samples were mostly distributed in the bottom quadrants (PC2 < 0); in particular, green samples were almost all grouped in the bottom left quadrant (PC1 < 0), while red ones were all positioned in the bottom right quadrant (PC1 > 0). The blue and yellow HCA clusters, so all 2018 samples were instead almost all plotted in the upper quadrants (PC2 > 0); blue samples were chiefly plotted in the left quadrant (PC1 < 0), while the only yellow sample was plotted in the right one (PC1 > 0). The similarity evidenced for genotypes RG-NU and BR5-BR1-BR16 was confirmed by the PCA plot (Figure 5); the former (RG-NU), in particular, were grouped closely in the upper left quadrant in the 2018 genotypes, while the 2019 ones were plotted together in the bottom left one. Sample BR5_2_2018 is plotted quite separately from all other samples, in the upmost area of the upper right quadrant, due to its high Dulcoside A concentration.

Table 5. Results of three-way ANOVA for secondary metabolites (total phenols, flavonoids, and steviol glycosides) and antioxidant activities of stevia leaf extracts.

Factor		Total Phenols (mg GAE g^{-1} DW)	Total Flavonoids (mg CE g^{-1} DW)	FRAP (mmol TE g^{-1} DW)	DPPH (mmol TE g^{-1} DW)	Total SVglys (g 100 g^{-1} DW)
Main Effects	Factor Level					
Genotype (G)	PL	58.2 ± 5.4 e	54.5 ± 3.9 d	0.34 ± 0.02 e	0.26 ± 0.03 d	25.2 ± 2.1 b
	BR16	41.6 ± 4.7 f	41.3 ± 4.6 e	0.29 ± 0.03 f	0.22 ± 0.02 e	16.2 ± 1.5 d
	RG	37.3 ± 3.5 g	33.3 ± 3.2 f	0.28 ± 0.02 f	0.20 ± 0.02 f	15.5 ± 1.9 de
	SL	71.3 ± 7.0 a	74.1 ± 5.2 bc	0.47 ± 0.02 bc	0.36 ± 0.04 a	24.6 ± 1.4 b
	NU	38.9 ± 3.7 fg	39.9 ± 2.9 e	0.27 ± 0.03 f	0.23 ± 0.02 e	12.2 ± 0.5 e
	CO	63.7 ± 6.6 cd	74.9 ± 7.4 b	0.45 ± 0.03 c	0.32 ± 0.04 c	19.7 ± 1.4 c
	BR5	66.1 ± 5.5 bc	81.2 ± 8.4 a	0.49 ± 0.03 ab	0.34 ± 0.04 b	23.9 ± 2.6 b
	SW30	61.5 ± 5.9 d	70.9 ± 6.9 c	0.42 ± 0.04 d	0.31 ± 0.04 c	29.3 ± 2.7 a
	BR1	67.9 ± 6.5 ab	84.1 ± 7.3 a	0.51 ± 0.02 a	0.31 ± 0.04 c	19.5 ± 1.6 c
Crop age (CA)	2018	32.2 ± 1.3 b	37.3 ± 1.8 b	0.32 ± 0.01 b	0.17 ± 0.003 b	24.3 ± 1.3 a
	2019	78.3 ± 2.1 a	83.8 ± 3.2 a	0.46 ± 0.02 a	0.39 ± 0.01 a	17.1 ± 0.6 b
Harvest time (HT)	July	58.7 ± 2.8 a	64.2 ± 3.6 a	0.47 ± 0.02 a	0.31 ± 0.02 a	18.5 ± 0.6 b
	September	53.2 ± 3.7 b	58.3 ± 4.0 b	0.32 ± 0.01 b	0.26 ± 0.01 b	22.5 ± 1.3 a
Significance	Main effects					
	G	***	***	***	***	***
	CA	***	***	***	***	***
	HT	***	***	***	***	***
	G × CA	***	***	***	***	***
	G × HT	***	***	***	***	**
	CA × HT	***	***	***	***	***
	G × CA × HT	***	***	***	***	***

Values followed by identical letters are not significantly different for $p < 0.05$, according to the LSD post-hoc test. The significance of variability factors according to the F-test: ns, not significant; **, significant at $p \leq 0.01$; ***, significant at $p \leq 0.001$ level. FRAP, ferric reducing antioxidant power; DPPH, 1,1-diphenyl-2-picrylhydrazyl; Total SVglys, total steviol glycosides.

Table 6. Results of three-way ANOVA for steviol glycoside profile.

Figure Main Effects	Factor Level	Rub	Dulc A	Stev	Reb C	Reb A	Reb E	Reb M	Reb D
Genotype (G)	PL	0.59 ± 0.12 c	1.01 ± 0.04 b	12.05 ± 0.66 b	1.29 ± 0.05 b	7.47 ± 0.53 a	0.93 ± 0.14 bc	0.09 ± 0.001 b	1.36 ± 0.05 bc
	BR16	1.53 ± 0.12 b	1.71 ± 0.22 ab	3.86 ± 0.23 de	1.59 ± 0.19 b	5.15 ± 0.34 b	0.87 ± 0.06 c	0.08 ± 0.001 b	1.43 ± 0.03 ab
	RG	2.77 ± 0.38 a	1.93 ± 0.19 ab	1.19 ± 0.14 ef	5.31 ± 0.33 a	0.88 ± 0.08 c	1.19 ± 0.16 ab	nd	1.37 ± 0.01 bc
	SL	2.47 ± 0.57 a	2.22 ± 0.65 ab	8.83 ± 0.23 c	1.72 ± 0.10 b	7.55 ± 0.20 a	0.89 ± 0.01 c	nd	1.25 ± 0.02 bcd
	NU	2.19 ± 0.53 a	2.08 ± 0.22 ab	0.94 ± 0.09 f	5.61 ± 0.34 a	1.01 ± 0.19 c	1.35 ± 0.27 a	nd	nd
	CO	0.92 ± 0.02 bc	2.36 ± 0.25 a	4.51 ± 0.17 cd	5.70 ± 0.29 a	4.49 ± 0.31 b	0.94 ± 0.15 bc	nd	nd
	BR5	0.92 ± 0.04 bc	2.83 ± 0.52 a	5.71 ± 0.55 cd	7.79 ± 1.19 a	6.10 ± 0.81 ab	1.12 ± 0.25 ab	0.27 ± 0.01 ab	0.74 ± 0.01 d
	SW30	0.84 ± 0.04 c	1.76 ± 0.20 ab	16.02 ± 1.13 a	2.07 ± 0.30 b	4.39 ± 0.46 b	1.38 ± 0.26 a	0.54 ± 0.23 a	1.97 ± 0.19 a
	BR1	0.86 ± 0.14 c	2.31 ± 0.31 a	4.50 ± 0.29 cd	5.42 ± 0.38 a	4.68 ± 0.50 b	0.88 ± 0.13 c	nd	1.60 ± 0.01 a
Crop age (CA)	2018	0.40 ± 0.10 b	2.86 ± 0.28 a	7.17 ± 1.49 a	4.87 ± 0.87 a	5.30 ± 0.75 a	1.40 ± 0.13 a	0.81 ± 0.05 a	1.63 ± 0.18 a
	2019	2.03 ± 0.21 a	1.23 ± 0.06 b	5.54 ± 0.96 b	3.42 ± 0.40 b	3.84 ± 0.53 b	0.61 ± 0.04 b	0.11 ± 0.02 a	1.32 ± 0.10 a
Harvest time (HT)	July	1.50 ± 0.32 a	1.75 ± 0.17 b	5.97 ± 1.13 a	3.59 ± 0.48 b	3.88 ± 0.53 b	0.88 ± 0.11 b	0.11 ± 0.001 a	1.35 ± 0.08 a
	September	1.20 ± 0.24 a	2.19 ± 0.35 a	6.74 ± 1.43 a	4.48 ± 0.85 a	5.05 ± 0.76 a	1.26 ± 0.18 a	0.39 ± 0.02 a	1.69 ± 0.25 a
Significance	Main effects								
	G	**	ns	***	*	**	**	***	***
	CA	***	***	*	**	*	***	**	ns
	HT	ns	*	ns	*	**	*	ns	ns
	G × CA	**	ns	**	**	*	ns	***	ns
	G × HT	ns	ns	ns	**	**	ns	ns	ns
	CA × HT	ns	*	**	**	**	**	**	**
	G × CA × HT	***	ns	***	***	**	**	***	ns

Values followed by identical letters are not significantly different for $p < 0.05$, according to the LSD post-hoc test. The significance of variability factors according to the F-test: ns, not significant; *, significant at $p \leq 0.05$; **, significant at $p \leq 0.01$; ***, significant at $p \leq 0.001$ level. Rub, Rubusoside; Dulc A, Dulcoside A; Stev, Stevioside; and Reb A, C, D, E, and M, Rebaudiosides A, C, D, E, and M.

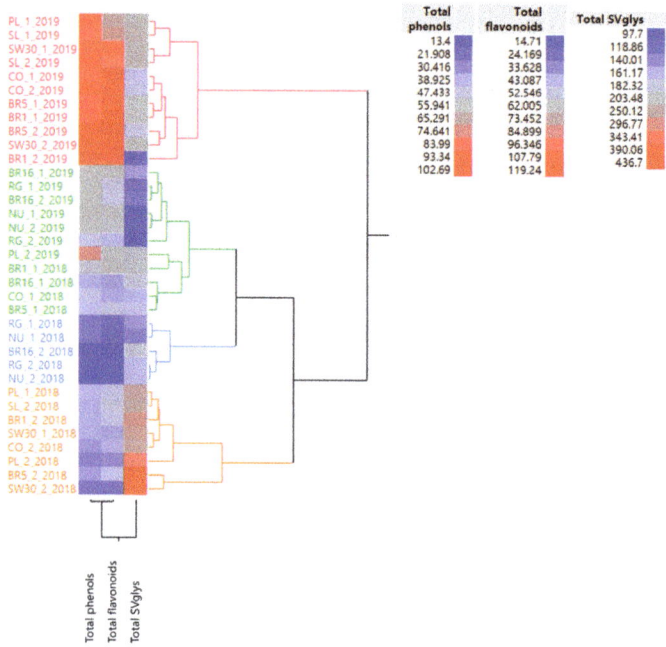

Figure 2. Hierarchical cluster analysis (HCA) on total phenols, flavonoids and SVglys. Each genotype is followed by the indication of harvest time (1 = first harvest; 2 = second harvest) and crop age (2018 and 2019).

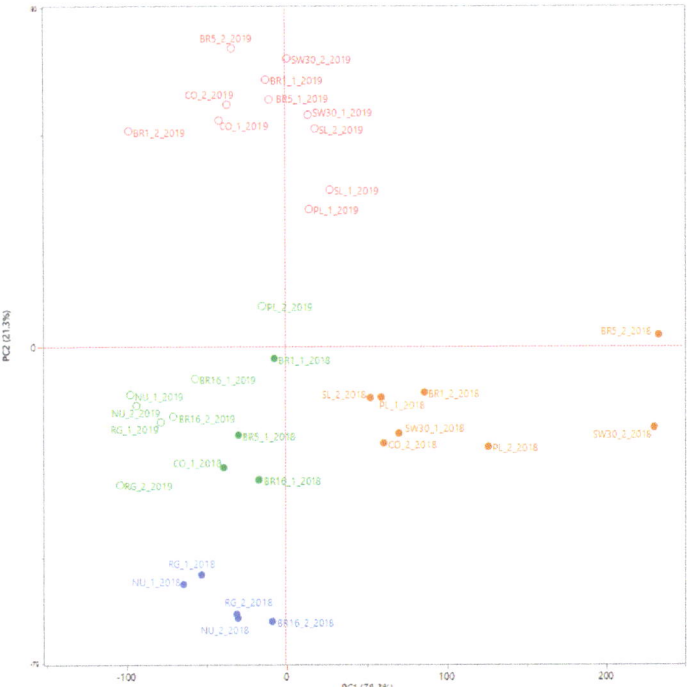

Figure 3. Principal component analysis (PCA) on total phenols, flavonoids, and SVglys. Each genotype is followed by the indication of harvest time (1 = first harvest; 2 = second harvest) and crop age (2018 and 2019).

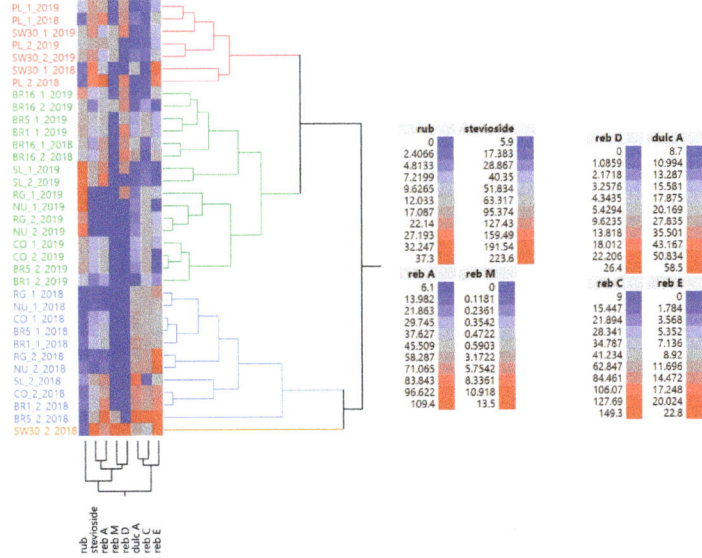

Figure 4. Hierarchical cluster analysis (HCA) on SVglys profile (Rub, Dulc A, Stev, Reb A, C, D, E, and M). Each genotype is followed by the indication of harvest time (1 = first harvest; 2 = second harvest) and crop age (2018 and 2019).

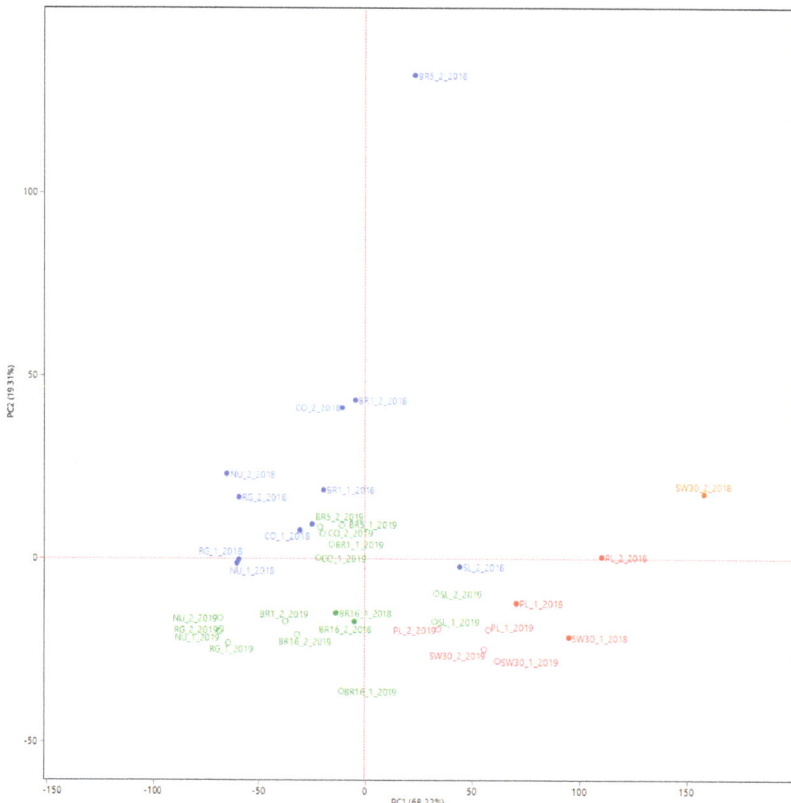

Figure 5. Principal component analysis (PCA) on SVglys profile (Rub, Dulc A, Stev, Reb A, C, D, E and M). Each genotype is followed by the indication of harvest time (1 = first harvest; 2 = second harvest) and crop age (2018 and 2019).

4. Discussion

We assessed the agronomic and phytochemical performances of nine *S. rebaudiana* (Bertoni) genotypes for two consecutive growing seasons. The aim was to select the best productive genetic resources to develop new and improved cultivars for the Mediterranean area. To improve the competitiveness of stevia production as a semi-perennial crop in this environment, it is important to produce higher-performing crops in terms of yield and quality. Therefore, the identification of new varieties/cultivars of *S. rebaudiana* with long stand duration, consistently high and stable leaf and SVglys yields, as well as a high level of other beneficial substances, is the top priority for the successful introduction of stevia into Mediterranean cropping systems.

Our results showed that some of the genotypes tested were characterized by a longer vegetative period before flowering, as demonstrated by the higher thermal requirements (Figure 1). This behavior promotes both higher leaf yield and steviol glycoside concentration, which in turn determine higher SVglys yield. This is important given the time required for the plant to synthesize and accumulate the steviol glycosides in the leaves. In fact, the maximum accumulation of these compounds, which depends on several environmental, agronomic, and physiological factors, is reached when the plant moves from the budding phase to an initial flowering stage with less than 10% flowers [31]. Therefore, the greater the vegetative period, the greater the accumulation of steviol glycosides in the leaves. When the plant starts to flower, nutrients accumulate in the reproductive organs and, as a result, vegetative growth declines.

Given that the leaves are the commercially important part of stevia, a delay in flowering can enhance vegetative growth and economic yield. In this regard, Ceunen and Geuns [31] observed large amounts of SVglys within the upper leaves during the budding phase and, as more and more buds become flowers, leaves contained lower amounts of SVglys. This is important for the choice of best harvesting time, since the maximum SVglys concentration is reached in the physiological stage of flowering-bud formation [32,33]. Flowering and, in general, plant growth are also affected by the photoperiod and temperature. Stevia is, in fact, a short-day plant with a critical photoperiod of between 12–13 h of day length [34,35]. Regarding air temperature, in our environment vegetative growth was lower when the maximum day temperature was below 10 °C or over 35 °C.

Overall, our results showed that genotype, crop age and harvest time represented key pre-harvest factors for defining the morphological, phenological, and quanti-qualitative traits in stevia.

Regarding differences in plant morphology and canopy architecture, we confirmed previous findings [21,36,37], which reported a very high variability, due to genetic characteristics, has been reported for plant height, basal stem diameter, branching, specific leaf weight, and leaf yield. In order to investigate this great variability among stevia genotypes and populations, some authors [21,22] have observed that increased yields were correlated to a high leaf area index (LAI), which is responsible for a greater light interception and, in turn, for a higher leaf photosynthesis.

We assessed the specific leaf weight (SLW), a leaf thickness index that is positively correlated with leaf photosynthesis: leaves with high SLW values are thicker and generally have a higher chlorophyll density per area unit ($\mu g\ cm^{-2}$) and, therefore, greater photosynthetic capacities than thinner leaves [38]. SLW significantly changed depending on genotype, crop age and harvest time. In particular, in 2-year-old plants, an increase in SLW was accompanied by increased leaf dry yields, suggesting that, thanks to a greater SLW leaf photosynthesis may have been enhanced with a consequent increase in crop yield [39].

Plants with a high SLW have leaves with a low surface/volume ratio, which is more efficient in terms of water use [40]. This ratio plays an important role in leaf functioning and is related to the strategies for acquiring and using the energy resources [41], as well as a tool to evaluate plant productivity [42] with a view to sustainable agriculture. For a better comprehension of the relationships between morphotype and light interception and photosynthetic activity, future studies are needed in order to improve leaf yield in *S. rebaudiana*.

Regarding stevia productivity, we found that the least productive genotypes are those with more branching, as reported by Tateo et al. [43]. In addition, our findings suggested a positive relation between plant height and leaf dry yield in agreement with previous studies [36,37]. Finally, the aerial biomass variability among genotypes exhibited high heritability, and the differences can also be partly attributed to crop age, pedo-climatic conditions of the cultivation site and plant development stage, as already observed in previous studies [19,44,45].

In our environment, stevia productivity significantly increased from the first to the second year of cultivation, with a very high winter survival rate. This suggests that stevia could grow as a semi-perennial crop in central Italy. This could bring positive agronomic advantages, such as, reductions in soil erosion and nutrient leaching, C soil sequestration, which in turn have important implications in multifunctional and sustainable agriculture.

Our findings confirmed previous results [5,6] in which in the temperate area of central Italy, the cultivation of stevia is long-term (5–7 years), with a vegetative period from April–May to September-October. In the tested climatic conditions, during winter, above-ground parts of the plant became dry, and there was regrowth through new shoots in the buried rhizome the next spring, thus producing a new crop without replanting. On the other hand, at higher latitudes as in central Europe (such as Belgium and Germany) and Canada [46–48], with cooler winters, stevia behaves like an annual crop with annual replanting.

We thus believe that stevia can be cultivated as a promising new crop in the Mediterranean climate of central Italy, though the SVgly content and composition still need to be optimized.

We found that the content and composition in bioactive compounds varied greatly depending on the genotype, crop age and harvest time. Two-year-old plants generally exhibited the highest content of total phenols and flavonoids, as well as the antioxidant activities; conversely, total SVglys content peaked in the establishment year. The sampling carried out at the beginning of the reproductive phase only improved total SVglys, again confirming that, in stevia, plant development is a primary factor in influencing the biosynthetic pathway.

Harvesting when plants were in full vegetative development (July harvest) maximized the content of polyphenol compounds and improved the antioxidant activities. Similar results have been obtained in a previous work [49] highlighting how the identification of the optimal harvest time was able to maximize the bioactive compounds of interest and, consequently, the health-promoting properties of stevia leaves. However, since the biosynthesis of secondary metabolites is a plant defense mechanism against biotic and abiotic stresses, through many physiological, biochemical, and molecular changes in plant metabolism, it is not always easy to identify the optimal time in which the different metabolites are maximally expressed.

The accumulation of polyphenols during the beginning of flowering stage could be related to the ecological roles of these compounds, such as intensifying antifungal defenses and attracting pollinators [32]. In addition, since different interactions among environmental and agronomic factors may occur, it is difficult to select individual stimuli that can influence a single metabolic pathway.

Of the secondary metabolites that have been synthesized and accumulated in stevia leaves, the presence of phenolic compounds is the subject of increasing interest because of their significant practical use for nutritional and medicinal applications. In fact, there are important implications for the growing market of natural stevia products [50], where they are employed as natural preservatives, thanks to their capacity to delay the oxidative degradation of lipids and to improve the shelf life of foods and beverages. They are also involved in the prevention of oxidative stress in humans, thanks to the hydroxyl groups in their molecules which have antioxidant anti-inflammatory properties [51,52].

Our findings provide new knowledge about the dissimilarity among genotypes, which was above all generated by the total SVglys content with respect to total phenols and flavonoids. Previous studies have investigated the effect of farming practices and genotype × environment interaction on stevia crop productivity, Reb A and Stev accumulation and on their reciprocal ratio (Reb A/Stev) [19,21,22,53–55]. Our results showed that the differences in the steviol glycoside profile are less due to crop age and more to genetics, as suggested by the dendrogram of the SVglys profile. It is well-known that, while the SVglys content in the leaves can vary depending on environmental and agronomic factors [56,57], the SVglys qualitative profile remains quite stable, indicating the high genotypic determinism of this trait [19].

We thus believe that our findings are a good starting point to screen the best genotypes in terms of SVglys profile, taking into account that breeding programs for stevia are increasingly aimed at varieties with an optimal Reb A/Stev ratio, which is considered a good qualitative measure of sweetness, or with a high content of Reb M and Reb D, characterized by a very sweet taste with no liquorice-like taste. We found that genotypes, such as SL, BR5, PL, and SW30 accumulated significantly high amounts of Reb A, Reb M, and Reb D. Among these genotypes, SL and BR5 were characterized by relatively low content of Stevioside. Conversely, PL and SW30 exhibited high contents of Stevioside, which negatively affect the Reb A/Stev ratio, lowering it below 1.

In our study, we try to identify the stevia ideotype, deriving from the combination of morphological, productive, and phytochemical traits. Consequently, considering not only the SVglys profile, but also leaf yield, growth crop cycle and the composition of their whole phytocomplex (polyphenols, total SVglys and their related antioxidant activities), SL, BR5, PL, and SW30 seemed to be the best performing genotypes, in the given environment, and thus suitable as starting point for future breeding programs.

5. Conclusions

Our findings highlighted that stevia, in the tested environment, represents a promising semi-perennial crop, which can contribute to the diversification of traditional cropping systems, thus increasing their sustainability and, at the same time, generating functional products with high added value.

We revealed a strong dependence of growth, crop yield and quality on genotype, harvest time, and crop age. The high variability among genotypes highlighted the importance of morpho-productive and qualitative traits for the identification of the best stevia ideotype, for our environment and for targeted end uses. The best combination of morphological, productive, and phytochemical traits was observed for SL, BR5, PL, and SW30 genotypes.

In conclusion, identifying genotypes characterized by specific SVglys profile (high Reb A, Reb D, Reb M), together with high content of all bioactive compounds and satisfactory yields, is crucial to establish future breeding programs for this crop.

Author Contributions: Conceptualization, L.G.A.; methodology, S.T. and C.C.; software, R.A.; formal analysis, C.C. and S.T.; investigation, L.G.A., C.C. and S.T.; data curation, C.C. and R.A.; writing—original draft preparation, C.C., R.A. and S.T.; writing—review and editing, L.G.A. and S.T.; supervision, L.G.A. All authors have read and agreed to the published version of the manuscript.

Funding: This research received no external funding.

Institutional Review Board Statement: Not applicable.

Informed Consent Statement: Not applicable.

Data Availability Statement: Data sharing not applicable.

Conflicts of Interest: The authors declare no conflict of interest.

References

1. Vandermeer, J.; Lawrence, D.; Symstad, A.; Hobbie, S. Effect of biodiversity on ecosystem functioning in managed ecosystems. In *Biodiversity and Ecosystem Functioning: Synthesis and Perspectives*; Oxford University Press: Oxford, UK, 2002; pp. 221–233.
2. Freibauer, A.; Rounsevell, M.D.A.; Smith, P.; Verhagen, J. Carbon sequestration in the agricultural soils of Europe. *Geoderma* **2004**, *122*, 1–23. [CrossRef]
3. Martini, A.; Tavarini, S.; Macchia, M.; Benelli, G.; Canale, A.; Romano, D.; Angelini, L.G. Influence of insect pollinators and harvesting time on the quality of *Stevia rebaudiana* (Bert.) Bertoni seeds. *Plant Biosyst.* **2017**, *151*, 341–351. [CrossRef]
4. Zhang, Y.; Li, Y.; Jiangb, L.; Tian, C.; Li, J.; Xiao, Z. Potential of Perennial Crop on Environmental Sustainability of Agriculture. *Procedia Environ. Sci.* **2011**, *10*, 1141–1147. [CrossRef]
5. Andolfi, L.; Macchia, M.; Ceccarini, L. Agronomic-productive characteristics of two genotype of *Stevia rebaudiana* in central Italy. *Ital. J. Agron.* **2006**, *2*, 257–262. [CrossRef]
6. Lavini, A.; Riccardi, M.; Pulvento, C.; De Luca, S.; Scamosci, M.; D'Andria, R. Yield, Quality and Water Consumption of *Stevia rebaudiana* Bertoni Grown under Different Irrigation Regimes in Southern Italy. *Ital. J. Agron. Riv. Agron.* **2008**, *2*, 135–143. [CrossRef]
7. Angelini, L.; Tavarini, S. Crop productivity, steviol glycoside yield, nutrient concentration and uptake of *Stevia rebaudiana* Bert. under Mediterranean field conditions. *Commun. Soil Sci. Plant. Anal.* **2014**, *45*, 2577–2592. [CrossRef]
8. Angelini, L.G.; Martini, A.; Passera, B.; Tavarini, S. Cultivation of *Stevia rebaudiana* Bertoni and Associated Challenges. In *Sweeteners*; Mérillon, J.M., Ramawat, K., Eds.; Reference Series in Phytochemistry; Springer: Cham, Switzerland, 2018. [CrossRef]
9. Lemus-Mondaca, R.; Vega-Gálvez, A.; Zura-Bravo, L.; Ah-Hen, K. *Stevia rebaudiana* Bertoni, source of a high-potency natural sweetener: A comprehensive review on the biochemical, nutritional and functional aspects. *Food Chem.* **2012**, *132*, 1121–1131. [CrossRef] [PubMed]
10. Singh, D.P.; Kumari, M.; Prakash, H.G.; Rao, G.P.; Solomon, S. Phytochemical and Pharmacological Importance of Stevia: A Calorie-Free Natural Sweetener. *Sugar Tech.* **2019**, *21*, 227–234. [CrossRef]
11. Masoumi, S.J.; Ranjbar, S.; Keshavarz, V. The Effectiveness of Stevia in Diabetes Mellitus: A Review. *Int. J. Food Sci. Nutr.* **2020**, *5*, 49–53.
12. Kennelly, E.J. Sweet and non-sweet constituents of *Stevia rebaudiana* (Bertoni). In *Stevia, the Genus Stevia Medicinal and Aromatic Plants-Industrial Profiles*; Kinghorn, A.D., Ed.; Taylor and Francis: London, UK, 2002; pp. 68–85.

13. Geuns, J.M.C. *Stevia and Steviol Glycosides: Properties, Techniques, Uses, Exposure, Toxicology, Pharmacological Effects*; Euprint: Heverlee, Belgium, 2010.
14. Wölwer-Rieck, U. The Leaves of *Stevia rebaudiana* (Bertoni), Their Constituents and the Analyses Thereof: A Review. *J. Agric. Food Chem.* **2012**, *60*, 886–895. [CrossRef]
15. Gaweł-Bęben, K.; Bujak, T.; Nizioł-Łukaszewska, Z.; Antosiewicz, B.; Jakubczyk, A.; Karaś, M.; Rybczyńska, K. *Stevia rebaudiana* Bert. leaf extracts as a multifunctional source of natural antioxidants. *Molecules* **2015**, *20*, 5468–5486. [CrossRef]
16. Buniowska, M.; Carbonell-Capella, J.M.; Znamirowska, A.; Zulueta, A.; Frígola, A.; Esteve, M.J. Steviol glycosides and bioactive compounds of a beverage with exotic fruits and *Stevia rebaudiana* Bert as affected by thermal treatment. *Int. J. Food Prop.* **2020**, *23*, 255–268. [CrossRef]
17. Yilmaz, F.M.; Görgüç, A.; Uygun, Ö.; Bircan, Y. Steviol glycosides and polyphenols extraction from *Stevia rebaudiana* Bertoni leaves using maceration, microwave, and ultrasound-assisted techniques. *Sep. Sci. Technol.* **2020**, 1–15. [CrossRef]
18. Debnath, M.; Ashwath, N.; Midmore, D.J. Physiological and morphological responses to abiotic stresses in two cultivars of *Stevia rebaudiana* (Bert.) Bertoni. *S. Afr. J.* **2019**, *123*, 124–132. [CrossRef]
19. Barbet-Massin, C.; Giuliano, S.; Alletto, L.; Daydé, J.; Berger, M. Towards a semi-perennial culture of *Stevia rebaudiana* (Bertoni) under temperate climate: Effects of genotype, environment and plant age on steviol glycoside content and composition. *Genet. resour. Crop. Evol.* **2015**, *63*, 1–10. [CrossRef]
20. Grevsen, K.; Sørensen, J.N. 2nd year cultivation results of the Danish "Green Stevia" project—A natural sweetener for organic food products. In Proceedings of the 9th Stevia Symposium 2016, From Field to Fork, Gothenburg, Sweden, 15 September 2015; pp. 115–126.
21. Hastoy, C.; Cossona, P.; Cavaignac, S.; Boutié, P.; Waffo-Teguo, P.; Rolin, D.; Schurdi-Levrauda, V. Deciphering performances of fifteen genotypes of *Stevia rebaudiana* in southwestern France through dry biomass and steviol glycoside evaluation. *Ind. Crops Prod.* **2019**, *128*, 607–619. [CrossRef]
22. Munz, S.; Präger, A.; Merkt, N.; Claupeina, W.; Graeff-Hönninger, S. Leaf area index, light interception, growth and steviol glycoside formation of *Stevia rebaudiana* Bertoni under field conditions in southwestern Germany. *Ind. Crops Prod.* **2018**, *111*, 520–528. [CrossRef]
23. Dyduch-Siemińska, M.; Najda, A.; Gawroński, J.; Balant, S.; Świca, K.; Żaba, A. *Stevia Rebaudiana* Bertoni, a Source of High-Potency Natural Sweetener-Biochemical and Genetic Characterization. *Molecules* **2020**, *25*, 767. [CrossRef] [PubMed]
24. McMaster, G.G.; Wilhelm, W. Growing degree-days: One equation, two interpretations. *Agric. Forest Meteorol.* **1997**, *87*, 291–300. [CrossRef]
25. Le Bihan, Z.; Cosson, P.; Rolin, D.; Schurdi-Levraud, V. Phenological growth stages of stevia (*Stevia rebaudiana* Bertoni) according to the Biologische Bundesanstalt Bundessortenamt and Chemical Industry (BBCH) scale. *Ann. Appl. Biol.* **2020**, 1–13. [CrossRef]
26. Zimmermann, B.F.; Woelwer-Rieck, U.; Papagiannopoulos, M. Separation of Steviol Glycosides by Hydrophilic Liquid Interaction Chromatography. *Food Anal. Methods* **2012**, *5*, 266–271. [CrossRef]
27. Dewanto, V.; Wu, X.; Liu, R. Processed Sweet Corn Has Higher Antioxidant Activity. *J. Agric. Food Chem.* **2002**, *50*, 4959–4964. [CrossRef]
28. Jia, Z.; Mengcheng, T.; Jianming, W. The determination of flavonoid contents in mulberry and their scavenging effects on superoxide radicals. *Food Chem.* **1999**, *64*, 555–559.
29. Tavarini, S.; Sgherri, C.; Ranieri, A.M.; Angelini, L.G. Effect of nitrogen fertilization and harvest time on steviol glycosides, flavonoid composition, and antioxidant properties in *Stevia rebaudiana* Bertoni. *J. Agric. Food Chem.* **2015**, *63*, 7041–7050. [CrossRef]
30. Ascrizzi, R.; Flamini, G.; Giusiani, M.; Stefanelli, F.; Deriu, V.; Chericoni, S. VOCs as fingerprints for the chemical profiling of hashish samples analyzed by HS-SPME/GC–MS and multivariate statistical tools. *Forensic Toxicol.* **2018**, *36*, 243–260. [CrossRef]
31. Ceunen, S.; Geuns, J. Influence of photoperiodism on the spatio-temporal accumulation of steviol glycosides in *Stevia rebaudiana* (Bertoni). *Plant Sci.* **2013**, *198*, 72–82. [CrossRef] [PubMed]
32. Ramesh, K.; Singh, V.; Megeji, N.W. Cultivation of stevia [*Stevia rebaudiana* (Bert.) Bertoni]: Comprehensive review. *Adv. Agron.* **2006**, *89*, 137–177.
33. Serfaty, M.; Ibdah, M.; Fischer, R.; Chaimovitsh, D.; Saranga, Y.; Dudai, N. Dynamics of yield components and stevioside production in *Stevia rebaudiana* grown under different planting times, plant stands and harvest regime. *Ind. Crops Prod.* **2013**, *50*, 731–736. [CrossRef]
34. Midmore, D.J.; Rank, A.H. *A New Rural Industry—Stevia—To Replace Imported Chemical Sweeteners*; Technical Report No. 02/022; Report for the Rural Industries Research and Development Corporation; Rural Industries Research and Development Corporation: Wagga Wagga, Australia, 2002.
35. Valio, I.F.M.; Rocha, R.F. Effect of photoperiod and growth regulators on growth and flowering of *Stevia rebaudiana* Bertoni. *Jap. J. Crop. Sci.* **1977**, *46*, 243–248. [CrossRef]
36. Gaurav, S.; Singh, Y.; Sirohi, S. Genetic variability for yield and quality traits in *Stevia rebaudiana* (Bertoni). *Progress. Res.* **2008**, *3*, 95–96.
37. Abdullateef, R.A.; Osman, M. Influence of Genetic Variation on Morphological Diversity in Accessions of *Stevia Rebaudiana* Bertoni. *Int. J. Biol.* **2011**, *3*, 66–72. [CrossRef]
38. Thompson, J.A.; Nelson, R.L.; Schweitzer, L.E. Relationships among specific leaf weight, photosynthesis rate, and seed yield in soybean. *Crop. Sci.* **1995**, *35*, 1575–1581. [CrossRef]
39. Shamim, M.J.; Tanaka, Y.; Sakoda, K.; Shiraiwa, T.; Nelson, R.L. Physiological analysis of leaf photosynthesis of backcross-derived progenies from soybean (*Glycine max* (L.) Merrill) and *G. tomentella* Hayata. *Plant. Prod. Sci.* **2020**, 1–9. [CrossRef]

40. Khan, H.R.; Link, W.; Hocking, T.J.; Stoddard, F.L. Evaluation of physiological traits for improving drought tolerance in faba bean (*Vicia faba* L.). *Plant Soil* **2007**, *292*, 205–217. [CrossRef]
41. Amanullah, I. Specific Leaf Area and Specific Leaf Weight in Small Grain Crops Wheat, Rye, Barley, and Oats Differ at Various Growth Stages and NPK Source. *J. Plant Nutr.* **2015**, *38*, 1694–1708.
42. White, J.W.; Montes Rojas, C. Variation in parameters related to leaf thickness in common bean (*Phaseolus vulgaris* L.). *Field Crops Res.* **2005**, *91*, 7–21. [CrossRef]
43. Tateo, E.; Mariotti, M.; Bononi, M.; Lubian, E. Stevioside content and morphological variabiity in a population of *Stevia rebaudiana* (Bertoni) from Paraguay. *Ital. J. Food Sci.* **1998**, *10*, 261–267.
44. Parris, C.A.; Shock, C.C.; Qian, M. Dry Leaf and Steviol Glycoside Productivity of *Stevia rebaudiana* in the Western United States. *J. Am. Soc. Hortic. Sci.* **2016**, *51*, 1220–1227. [CrossRef]
45. Tavarini, S.; Angelini, L.G. *Stevia rebaudiana* Bertoni as a source of bioactive compounds: The effect of harvest time, experimental site and crop age on steviol glycoside content and antioxidant properties. *J. Sci. Food Agric.* **2013**, *93*, 2121–2129. [CrossRef] [PubMed]
46. Brandle, J.E.; Starratt, A.N.; Gijzen, M. *Stevia rebaudiana*: Its agricultural, biological and chemical properties. *Can. J. Plant Sci.* **1998**, *78*, 527–536. [CrossRef]
47. Lankes, C.; Pude, R. Possibilities for growth of Stevia in European temperate zones. In *Steviol Glycosides: Technical and Pharmacological Aspects, Proceedings of the 2nd Stevia symposium, Leuven, Belgium, 27 June 2008*; Geuns, J.M.C., Ed.; Euprint: Leuven, Belgium, 2008; pp. 103–116.
48. Woelwer-Rieck, U.; Lankes, C.; Wawrzun, A.; Wüst, M. Improved HPLC method for the evaluation of the major steviol glycosides in leaves of *Stevia rebaudiana*. *Eur. Food Res. Technol.* **2010**, *231*, 581–588. [CrossRef]
49. Tavarini, S.; Clemente, C.; Bender, C.; Angelini, L.G. Health-promoting compounds in stevia: The effect of mycorrhizal symbiosis, phosphorus supply and harvest time. *Molecules* **2020**, *25*, 5399. [CrossRef] [PubMed]
50. Christaki, E.; Bonos, E.; Giannenas, I.; Florou-Paneri, P. Aromatic Plants as a Source of Bioactive Compounds. *Agriculture* **2012**, *2*, 228–243. [CrossRef]
51. Rivera, T.; Oesterling, J. Naturally Sweetened Juice Beverage Products. U.S. Patent 9788562B2, 17 October 2017.
52. Letenneur, L.; Proust-Lima, C.; Le Gouge, A.; Dartigues, J.F.; Barberger-Gateau, P. Flavonoid Intake and Cognitive Decline over a 10-Year Period. *Am. J. Epidemiol.* **2007**, *165*, 1364–1371. [CrossRef] [PubMed]
53. Kumar, H.; Kaul, K.; Bajpai-Gupta, S.; Kumar, V.; Kumar, S. A comprehensive analysis of fifteen genes of steviol glycosides biosynthesis pathway in *Stevia rebaudiana* (Bertoni). *Gene* **2012**, *492*, 276–284. [CrossRef]
54. Pal, P.K.; Mahajana, M.; Prasad, R.; Pathania, V.; Singh, B.; Ahuj, P.S. Harvesting regimes to optimize yield and quality in annual and perennial *Stevia rebaudiana* under sub-temperate conditions. *Ind. Crops Prod.* **2015**, *65*, 556–564. [CrossRef]
55. Moraes, R.M.; Donegac, M.A.; Cantrelld, C.L.; Melloc, S.C. McChesneyea, J.D. Effect of harvest timing on leaf production and yield of diterpene glycosides in *Stevia rebaudiana* Bert: A specialty perennial crop for Mississippi. *Ind. Crops Prod.* **2013**, *51*, 385–389. [CrossRef]
56. Pal, P.K.; Kumar, R.; Guleria, V.; Mahajan, M.; Prasad, R.; Pathania, V.; Gill, B.S.; Singh, D.; Chand, G.; Singh, B.; et al. Crop-ecology and nutritional variability influence growth and secondary metabolites of *Stevia rebaudiana* Bertoni. *BMC Plant. Biology* **2015**, *15*, 1–16. [CrossRef]
57. Geuns, J. Stevioside. *Phytochemistry* **2003**, *64*, 913–921. [CrossRef]

Article

Flavouring Extra-Virgin Olive Oil with Aromatic and Medicinal Plants Essential Oils Stabilizes Oleic Acid Composition during Photo-Oxidative Stress

Salvatore Barreca [1,†], Salvatore La Bella [2,†], Antonella Maggio [3,*], Mario Licata [2,*], Silvestre Buscemi [3], Claudio Leto [4], Andrea Pace [3] and Teresa Tuttolomondo [2]

1. Department of Pharmaceutical Sciences, Università degli Studi di Milano, Via L. Mangiagalli 25, 20131 Milan, Italy; salvatore.barreca@unimi.it
2. Department of Agricultural, Food and Forest Sciences, Università degli Studi di Palermo, Viale delle Scienze, Building 4, 90128 Palermo, Italy; salvatore.labella@unipa.it (S.L.B.); teresa.tuttolomondo@unipa.it (T.T.)
3. Department of Biological, Chemical and Pharmaceutical Sciences and Technologies, Laboratory of Chemistry, Università degli Studi di Palermo, Viale delle Scienze, Building 16, 90128 Palermo, Italy; silvestre.buscemi@unipa.it (S.B.); andrea.pace@unipa.it (A.P.)
4. Research Consortium for the Development of Innovative Agro-Environmental Systems (CoRiSSIA), Via della Libertà 203, 90143 Palermo, Italy; claudio.leto@unipa.it
* Correspondence: antonella.maggio@unipa.it (A.M.); mario.licata@unipa.it (M.L.)
† These authors contributed equally to the work.

Abstract: Essential oils (EOs) from medicinal and aromatic plants (MAPs) are well-known as natural antioxidants. Their addition to extra-virgin olive oil (EVOO) can contribute to reducing fat oxidation. The main aim of this study was to improve both food shelf-life and aromatic flavour of EVOO, adding different EOs of Sicilian accessions of common sage, oregano, rosemary and thyme. The morphological and production characteristics of 40 accessions of MAPs were preliminarily assessed. EOs from the most promising accessions of MAPs were analysed by gas-chromatography and mass spectrometry. Photo-oxidative studies of the EOs were carried out and the determination of the EVOO fatty acids obtained from 4 Italian olive varieties was also made. EO content was on average 1.45% (v/w) for common sage, 3.97% for oregano, 1.42% for rosemary and 5.90% for thyme accessions. The highest average EO yield was found in thyme (172.70 kg ha^{-1}) whilst the lowest (9.30 kg ha^{-1}) in rosemary accessions. The chemical composition of EOs was very different in the four MAPs in the study. No significant change of oleic acid percentage was detected in the mixture of EVOO with EO samples. The results seem to highlight the presence of an antioxidant effect of EOs on EVOO.

Keywords: extra-virgin olive oil; aromatic and medicinal plants; essential oil; gas-chromatography and mass spectrometry analyses; antioxidant; oleic acid

1. Introduction

Extra-virgin olive oil (EVOO), an appreciated food especially in the countries of the Mediterranean area, in recent years has gained interest among the consumers in North America and Northern Europe, especially flavoured with spices, herbs or fruits. The production of flavoured olive oils is a traditional practice in Mediterranean area, in order to enhance the sensorial characteristics of the original olive oils and to improve the sensory properties of the foods [1,2]. The success of these innovative products is associated with various health benefits as well as with the taste and flavour. Flavoured olive oil can represent an extremely interesting product for oil and aromatic plant producers to diversify their offer.

The numerous health benefits of EVOO depend on its particular composition [3,4]. It is composed primarily of triacylglycerols (around 97.00–98.00%), minor amounts of free fatty acids and glyceridic compounds, phospholipids and oxidized triacylglycerols and around 1% of unsaponifiable constituents of varied structure and polarity. There are a high proportion of monounsaturated fatty acids, mainly oleic acid, and a modest presence of polyunsaturated fatty acids. In addition, some natural antioxidants such as tocopherols, carotenoids, sterols and phenolic compounds are present. Some health properties of EVOO depend on content of oleic acid that serves to slow down penetration of fatty acids into arterial walls. The EVOO stability rate depends on both antioxidant compounds and storage conditions. During storage nutritional and health properties of extra-virgin oil may be changed by oxidative phenomena. The addition of naturally occurring antioxidants causes an improvement on the shelf life and nutritional value of oil.

Recently a study has showed that the initial quality of EVOO enriched and not enriched with lycopene commercially available prevails over the slight antioxidant activity that lycopene could exert. The EVOO composition in the main and minor components is the key factor that determines its performance in gastrointestinal conditions [5]. The methods commonly used to flavour olive oils are different [6]. The "gourmet oils" are prepared left herbs, spices and fruits mixed with oil at room temperature for a defined time. The mixture is, then, filtered to remove turbidity and solid parts. Infusion is the oldest method of oil aromatization and the most considered from the producers [7]. On the other hand, comilling the olives with herbs, spices or fruits such as lemons and bergamots during the oil productive process is a new approach for preparing clear and safe flavoured olive oils. Another method is the ultrasound-assisted maceration.

The flavouring technique significantly influences the chemical and sensory quality of the olive oils. In particular, the infusion of oils with spices caused a greater oxidative degradation due to lower content of total phenols. On the other hand, the olive oils obtained by combined malaxation of olives and spices were less bitter. The aromatic quality was not significantly affected by the method of flavouring, except for sulphur compounds that were greater in oils obtained by malaxation [8].

Rarely, the addition of essential oils (EOs) of medicinal and aromatic plants (MAPs) is used to obtain flavoured olive oils. These plants have drawn more attention due to the antimicrobial, antifungal, insecticidal and antioxidant effects [9–11]. Application of antioxidants is one of the technically simplest ways of reducing fat oxidation. Natural antioxidants from edible aromatic plants have many advantages such as (a) to be accepted by consumers, (b) to be considered safe, (c) to come from natural resources and (d) to have functional and sensory properties. Considering the beneficial effects of fatty acid contained in olive oils, addition of the MAPs EOs could represent a useful technique to preserve fatty acid profile in the EVOO composition. The results of preliminary studies showed that the presence of oregano EO, preserve sensory quality of extra virgin olive oil prolonging the shelf life of this product [12].

In Sicily (Italy), MAPs are widely present in the native flora. The great diversity in climate, soil and habitat of this Mediterranean region has contributed and provided good plant genetic resources. As reported in numerous studies [13–17], the exploitation of MAPs biodiversity and the cultivation of native accessions under open field conditions highlighted the effects of genetic and environmental factors on the qualitative and quantitative production of the EOs [18]. Furthermore, when analysing the EO profile of various aromatic species, specific Sicilian chemotypes were found. Following these findings, a number of Sicilian MAPs have now been distinguished from the same species growing in other Mediterranean regions, thus making their EOs of particular interest due to specific compounds and aroma.

In this study, to keep the biological properties of EVOO unaltered and improve them in long-term period, it was then decided to investigate the effect of adding EOs of MAPs to prevent oxidative processes. Native accessions of common sage, oregano, rosemary and thyme were used for obtaining EOs. Among different MAPs found in Sicilian areas, these species were chosen as they were widely used as a condiment in the Mediterranean diet and had already been used for the production of flavoured oils.

Thus, the aims of this study were: (i) to select the Sicilian accessions of common sage, oregano, rosemary and thyme, grown in collection fields, considering the best ratio between EO content and plant dry weight and (ii) to test the efficacy of the different EOs of these species as a natural antioxidant to improve both the shelf life and the quality of EVOO.

2. Materials and Methods

2.1. Experimental Site of MAPs

A number of Sicilian native accessions of common sage (*Salvia officinalis* L.), oregano (*Origanum vulgare* ssp. *hirtum* (Link) Ietswaart), rosemary (*Rosmarinus officinalis* L.) and thyme (*Thymbra capitata* (L.) Cav.) were grown in 4 separate collection fields (one for species) at the "Orleans" experimental farm (Palermo, 31 m a.s.l., 38°06'26.2" N, 13°20'56.0" E) belonging to the University of Palermo, north-west Sicily. These accessions had been previously collected in various areas of Sicily and subsequently subjected to taxonomic characterization using analytical keys and comparing them to exsiccata stored at the Botanical Gardens of the University of Palermo.

Soil at "Orleans" experimental area was sandy clay loam (56% sand, 23% clay, 21% silt) with a pH of 7.91, 19 g kg^{-1} organic carbon, 58 g kg^{-1} total carbonates, 37 g kg^{-1} active carbonates, 13.2 g kg^{-1} total nitrogen, 18.11 mg kg^{-1} assimilable phosphorus and 320 mg kg^{-1} exchangeable potassium.

The climate of the area is Mediterranean with mild, humid winters and hot, dry summers [19]. The average annual temperature is 18.40 °C, with average minimum and maximum temperatures of 14.80 and 21.70 °C, respectively. Annual average rainfall is approx. 600 mm.

2.2. Collection Fields of MAPs and Main Cultivation Practices

In 2018, a total of 40 accessions of Sicilian MAPs (5 for common sage, 15 for oregano, 10 for rosemary and 10 for thyme) were assessed at the 4 collection fields (Figure 1).

Figure 1. Accessions of medicinal and aromatic plants (MAPs) grown in the collection fields. (**a**) refers to common sage, (**b**) referes to oregano, (**c**) refers to rosemary and (**d**) refers to thyme plants.

Common sage and rosemary were planted at 2 m × 1 m spacing while oregano and thyme at 1.5 m × 1 m spacing. The accessions had previously been collected in various Sicilian areas which were different for soil and climate conditions. Plants were 4 years old. Each collection field was equipped with a drip irrigation system. However, plants were grown in dry conditions, this being a traditional practice used for cultivation of MAPs in Sicily. Plants exploited the residual soil fertility and no chemical fertilization application was given during the year. Weed control was manual and no pesticides were used. Harvest was carried out when most plants of the 4 species were at the full blooming stage.

2.3. Morphological and Production Parameters

Morphological and production measurements were made on a sample plot of 10 plants per accession, excluding the border rows.

The main morphological parameters (data not shown) of each species, such as plant height, number of branches, number of stems, floral spike length, flowers length and width were recorded. The plant fresh weight of above and below-ground plant parts was determined by harvesting the plants. The harvested plant material was, subsequently, dried in an oven at 65 °C for 48 h. The plant dry weight was, then, calculated. Dry matter yield per hectare was also estimated.

Samples of leaves and flowers were collected from the accessions of each species when 80% of the plants were in full flowering stage. These samples were, subsequently, hydrodistilled for 2 h in a Clevenger-type apparatus with a separated extraction chamber. EOs obtained were dried over anhydrous sodium sulphate and stored in dark flask at -18 °C in freezer until the EO samples were analysed by gas–liquid chromatography (GC) and mass spectrometry (MS) or used in the studies. EO yield was calculated by multiplying dry matter yield by oil content, by 0.90 (approximate specific gravity of oil) [20–22].

Finally, for each accession within the 4 species, the EO content was related to dry weight in order to select the most promising accessions.

2.4. Chemicals

Potassium hydroxide and solvents were purchased from Sigma Aldrich (Steinhem, Germany) (analytical grade) hexanal. (E)-2-hexenal and 2-methyl-pentanol were purchased from Sigma (Steinhem, Germany). Activated charcoal (0.50–1.00 mm; 18–35 mesh ASTM) was purchased from E. Merck (Schuchardt, Germany). The charcoal was cleaned by treatment in a Soxhlet apparatus with diethyl ether and was tested in order to verify the absence of any absorbed substances.

2.5. GC-MS Analyses of EOs

Analyses of EOs were performed by GC-MS Shimadzu QP 2010 plus equipped with an AOC-20i auto-injector (Shimadzu, Kyoto, Japan), a split/split-less injector (t = 280 °C) and a capillary column (30 m, 0.25 mm i.d. 0.25 mL film thickness) coated with DB WAX (polyethylene glycol. JW).

The temperature program was 40 °C for 5 min and from 40 to 250 with a rate of 2 °C min^{-1} and from 250 to 270 with a rate of 10 °C min^{-1}. The temperature of injector was maintained at 250 °C while the temperature of detector was maintained at 280 °C. The compounds were identified by comparing their retention time and mass spectra with published data, Adams, NIST 11, Wiley 9 and FFNSC 2 mass spectral database.

The samples were injected in splitless mode. The quantitative composition was obtained by peak area normalization, and the response factor for each component was considered to equal 1.

2.6. Photo-Oxidative Studies

100 µL of olive oil was transmethylated with 10 mL of KOH (2M) methanol solution according to the European Standard NF EN ISO 5509 (2000). Fatty acid methyl esters (FAME) obtained were transferred in hexane and analysed according to the European

Standard NF EN ISO 5508 (1995). Analyses were performed by using a Shimadzu gas chromatograph (GC) equipped with a split/split-less injector (t = 280 °C) and flame ionization detector (FID) (t = 250 °C). A capillary column (30 m, 0.25 mm i.d., 0.25 mm film thickness) coated with DB WAX (polyethylene glycol, JW) was used. The inlet pressure of the hydrogen as carrier gas was 154 kPa with a ratio of 70. The oven temperature program was as follows: 40 °C for 5 min and from 40 to 250 with a rate of 2 °C min^{-1} and from 250 to 270 with a rate of 10 °C min^{-1}. The temperature of injector was maintained at 250 °C while the temperature of detector was maintained at 280 °C. The compounds were identified by comparing their retention time and mass spectra with published data, Adams, NIST 11, Wiley 9 and FFNSC 2 mass spectral database.

In this study, EVOOs of four varieties of olive (*Olea europaea* L.) were considered. In Table 1, the initials of the varieties both in the preveraison (green fruit) and veraison (brown fruit) stages are reported.

Table 1. List of the varieties of olive in two diverse stages with relative initials.

Variety of Olive	Preveraison Stage (Green Fruit)	Veraison Stage (Brown Fruit)
Arbequina	AV	AI
Biancolilla	BV	BI
Cerasuola	CV	CI
Nocellara del Belice	NV	NI

2.7. Fatty Acid Determination before and after Photo-Oxidative Studies

Olive oil in n-hexane was transmethylated with a solution of KOH (2M) according to the European Standard NF EN ISO 5509 (2000). Fatty acid methyl esters (FAME) were analysed according to the European Standard NF EN ISO 5508 (1995). Analyses were performer by using a GC-MS Shimadzu QP 2010 plus equipped with an AOC-20i autoinjector (Shimadzu, Kyoto, Japan), a split/split-less injector (t = 280 °C) and a capillary column (30 m, 0.25 mm i.d. 0.25 mL film thickness) coated with DB WAX (polyethylene glycol. JW) was used. Helium was the carrier gas (1 mL min^{-1}); ionization voltage 70 eV. The temperature was initially kept at 40 °C for 5 min. Then gradually increased to 250 °C at 2 °C min^{-1} rate. Held for 15 min and finally raised to 270 °C at 10 °C min^{-1}. One µL of samples was injected at 250 °C automatically and in the splitless mode; transfer line temperature, 295 °C.

2.8. Statistical Analysis

Statistical analyses were performed using the package MINITAB 17 (State College, PA, USA) for Windows. Data of all production parameters of the MAPs accessions were processed using analysis of variance. The difference between means was carried out using Tukey's test. Concerning percentage content of oleic acid in the EVOO samples and in the mixture of EVOO with EO, all the representative values were shown using mean ± standard error calculation.

3. Results and Discussion

3.1. Agronomic Assessment of Common Sage, Oregano, Rosemary and Thyme Accessions

Data regarding the production parameters of the Sicilian MAPs are shown in Table 2. In the table, the initials CS indicate accessions of common sage, OR of oregano, RSM of rosemary and THY of thyme.

Table 2. Production characteristics of the four MAPs during the study period.

Species	Dry Weight (g plant^{-1})	Dry Matter Yield (kg ha^{-1})	EO Content (% v/w)	EO Yield (kg ha^{-1})
Common sage				
CS1	134.06 d	671.33 d	1.08 d	6.35 e
CS2	329.53 b	1645.10 b	1.10 cd	16.32 b
CS3	105.53 e	529.07 e	2.04 a	9.47 d
CS4	284.27 b	1423.60 c	1.20 c	15.43 c
CS5	361.30 a	1810.73 a	1.80 b	29.55 a
Oregano				
OR1	770.13 b	5139.44 b	5.82 b	268.51 a
OR2	399.60 k	2666.30 j	6.15 a	147.12 e
OR3	221.73 n	1478.53 m	3.01 k	36.74 n
OR4	504.03 g	3360.72 f	5.11 e	154.72 d
OR5	698.12 c	4652.69 c	5.12 d	218.52 b
OR6	503.93 g	3358.79 f	5.55 c	167.85 c
OR7	468.80 h	3125.88 g	4.02 f	113.12 h
OR8	441.77 i	2952.40 h	3.45 h	91.68 j
OR9	379.93 m	2529.14 l	3.52 g	80.14 k
OR10	413.20 j	2754.71 i	4.01 f	99.41 i
OR11	905.80 a	6039.59 a	2.26 n	122.33 g
OR12	684.12 d	4562.01 d	3.15 i	129.18 f
OR13	521.50 e	3476.67 e	3.11 j	97.30 i
OR14	390.70 l	2604.79 k	2.82 l	65.92 m
OR15	503.07 f	3364.21 f	2.32 m	70.51 l
Rosemary				
RSM1	42.20 h	211.33 h	1.51 b	2.62 h
RSM2	79.23 f	396.51 f	1.21 d	4.20 g
RSM3	274.53 c	1376.77 c	0.78 g	9.65 c
RSM4	339.20 b	1709.11 b	1.01 e	15.37 b
RSM5	534.76 a	2684.33 a	1.52 b	36.61 a
RSM6	47.67 g	240.07 g	2.32 a	5.04 f
RSM7	18.53 j	93.52 j	2.31 a	2.03 i
RSM8	205.63 d	1034.81 d	0.91 f	8.42 d
RSM9	26.20 i	131.61 i	1.42 c	1.66 i
RSM 10	117.23 e	586.43 e	1.21 d	6.29 e
Thyme				
THY1	337.67 i	2252.70 i	4.51 i	91.28 h
THY2	624.50 c	4174.77 c	5.96 d	223.94 c
THY3	416.67 g	2777.27 g	5.20 h	130.26 g
THY4	494.60 f	3294.80 f	6.02 c	178.45 e
THY5	656.17 a	4374.20 a	6.52 b	256.57 a
THY6	626.17 b	4177.20 b	6.80 a	256.51 a
THY7	75.17 j	501.73 j	5.52 g	25.58 i
THY8	381.30 h	2541.93 h	5.78 f	133.24 f
THY9	522.60 e	3486.93 e	5.90 e	185.94 d
THY10	607.20 d	4048.20 d	6.80 a	249.13 b

Means followed by the same letter in the same column are not significantly different for $p \leq 0.05$ according to test of Tukey.

Results of one-way ANOVA revealed significant differences between the accessions within each species for all the production parameters tested.

For common sage, CS5 and CS2 were of considerable interest regarding dry weight, dry matter yield and EO yield. The highest average value of EO content was obtained by CS3 (2.04%), while the lowest value was found in CS1 (1.08%).

In the case of oregano, EO content percentage ranged from 2.26% (OR11) to 6.15% (OR2). Average EO content of the 15 accessions was 4.01%. OR1 (268.51 kg ha^{-1}) and OR5

(218.52 kg ha^{-1}) obtained the highest average EOs yields. Concerning other parameters in the study, OR11 and OR1 performed best among the oregano accessions.

High variability in production parameters was observed also for rosemary accessions. In particular, RSM5 obtained the highest average dry weight (534.36 g), dry matter yield (2684.33 kg ha^{-1}) and EO yield (36.61 kg ha^{-1}); in contrast, lowest average values were found in RSM7. EO percentage content ranged between 2.32% and 0.78%; the highest average values of EO content were recorded by RSM6 and RSM7. It is worth noting that RSM7, in contrast, recorded the lowest values for dry weight, dry matter yield and EO yield.

Observing the data of thyme accessions, THY5 and THY6 performed the best and produced values higher than the average for the field. Among the accessions, THY10 also performed the best for EO content percentage (6.80%).

When considering the ratio between EO content and plant dry weight, it was found to be highly different among accessions of each species (Figure 2).

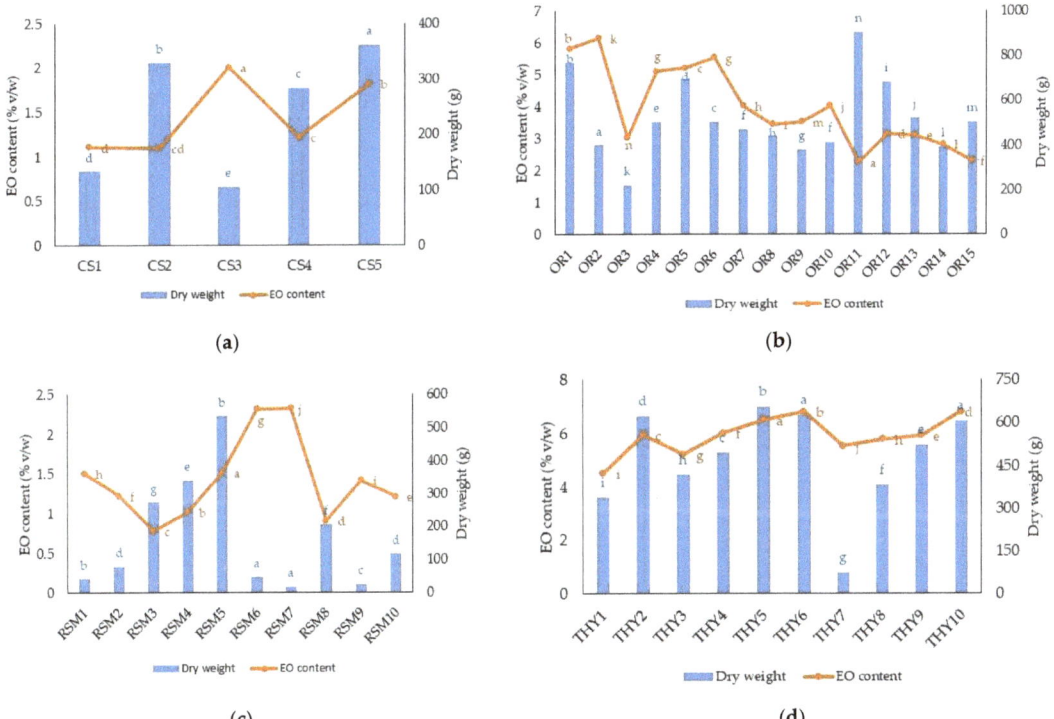

Figure 2. Ratio between essential oils (EO) content and plant dry weight. Means followed by the same letter in the same column are not significantly different for $p \leq 0.05$ according to test of Tukey. Graph (**a**) refers to common sage, graph (**b**) refers to oregano, graph (**c**) refers to rosemary and graph (**d**) refers to thyme.

For common sage, CS3 obtained the best performance while CS2 and CS4 recorded a high average dry weight compared to those of EO content. In the case of oregano, the best ratio between EO content and plant dry weight were found in OR2. On the other hand, OR11 and OR12 showed low average EO content compared to dry weight. For rosemary, RSM7 performed better than other accessions. However, RSM6 also showed high average EO content compared to dry weight. Finally, THY7 was the most promising accession of the thyme accessions. THY2 and THY5, however, were not remarkable in this respect.

In this study, significant differences in terms of yield parameters were found between accessions of each MAP during the test period. These differences mainly depend on the effects of genetic and environmental factors. Various authors [23–27] have reported that genetic variation occurring in native germplasm significantly affects EO content and yield of MAPs. Others [17,26–32] have found that climate factors can greatly influence the performance of MAPs in terms of dry weight, dry matter and essential oil yields. In our study, all accessions of common sage, oregano, rosemary and thyme were assessed under the same climate, soil and growth conditions, thus, the differences in production parameters were due to genotype response to environmental conditions. In the Mediterranean area, literature [33–37] shows different mean values of EO content and yield mainly depending on the climate and soil characteristics of the area and cultivation practices of MAPs. When comparing our findings with those reported in these studies, differences and similarities were found which can be explained by the study of genetic and environmental factors. On the basis of that, it is possible to say that changes in these factors can produce differences in the chemical composition of the EOs in plants of the same species in different environments.

In this study, the ratio between EO content and dry weight yield was also assessed in order to select the most promising accessions within each MAP. High production of EO in relation to dry weight yield is crucial in the cultivation of MAPs. If the EO content is found to be too low compared to plant dry weight, the production process would not be viable for farmers from an economic point of view. On the contrary, when EO content is higher, the production process is, instead, profitable. Therefore, the creation of mixtures of EVOO with MAP EOs requires a preliminary assessment of the EO content and its availability in the long term. This could allow for an estimation of the real cost of this condiment in the market.

3.2. Chemical Composition of the MAPs EOs

The chemical composition of EOs of common sage, oregano, rosemary and thyme accessions are shown in Table 3.

Table 3. Chemical composition of EOs of the common sage, oregano, rosemary and thyme accessions.

RT [a]	KI [b]	Compound	Common Sage	Oregano	Rosemary	Thyme
		Monoterpene Hydrocarbons	27.73	31.31	31.47	28.86
10.22	938	α thujene	0.37	0.79	0.11	0.91
10.60	944	α pinene	4.93	0.90	11.25	1.04
11.59	957	camphene	6.88	t	6.88	0.41
11.92	961	thuja-2.4(10)-diene			0.32	
13.10	974	verbenene			0.16	
13.23	975	sabinene	0.39	0.16	0.05	
13.46	978	β pinene	8.39	0.01	3.53	0.16
14.61	990	β-myrcene	3.83	2.24	1.93	2.36
15.39	997	Mentha-1(7),8-diene			0.04	
15.56	999	phellandrene α	0.04	0.36	0.28	0.38
15.66	1000	δ-3-carene		0.11	2.64	
16.34	1012	terpinene α	0.23	3.42	0.35	2.87
16.60	1017	p-cymene		0.01	0.03	
16.89	1022	o-cymene	0.22	9.70	3.09	8.73
17.24	1028	sylvestrene	1.65	0.76		0.75
17.99	1042	(Z)-β-ocimene	0.15	0.02		
18.70	1053	(E)-β-ocimene	0.02	0.10		0.05
19.37	1064	terpinene γ	0.43	12.59	0.32	10.98
21.32	1094	Mentha-2.4(8)-diene	0.20	0.13	0.40	0.22
21.70	1099	Cymenene			0.09	

Table 3. Cont.

RT [a]	KI [b]	Compound	Common Sage	Oregano	Rosemary	Thyme
		Oxygenated Monoterpenes	51.20	64.06	61.87	65.00
17.37	1031	eucalyptol	20.58	0.06	39.38	
20.21	1077	sabinene hydrate cis	0.14	0.64		0.43
22.43	1109	Sabinene hydrate <trans->	0.10	0.12		
22.51	1110	linalool				0.10
22.57	1111	Pinene oxide<α->	0.09	0.36	2.27	0.22
22.79	1114	thujone<cis->	8.25	0.02		
23.62	1125	thujone<trans->	5.78			
23.68	1126	fenchol <exo>			0.05	
24.29	1134	Campholenal α			0.03	
25.14	1144	Pinocarveol trans			0.30	
25.52	1149	camphor	13.77		4.81	
26.29	1158	Eucarvone	0.09			
26.55	1161	pinocamphone trans			0.16	
26.69	1163	pinocarvone			0.60	
27.29	1170	isocitral<(Z)->			0.10	
27.42	1171	Borneol	0.80	0.07	10.40	1.00
27.65	1174	pinocamphone cis			0.50	
28.08	1178	terpinen-4-ol	0.27	0.32	0.46	0.61
29.18	1190	α terpineol	0.09	0.02	0.78	
29.31	1192	dihydro carvone cis				0.06
29.91	1198	verbenone			0.26	
31.68	1228	thymol methyl ether		0.24		
32.27	1238	carvacrol methyl ether		3.70	0.17	
32.68	1245	carvone			0.17	
35.29	1287	Isobornyl acetate	0.04		1.43	
35.54	1291	thymol	1.09	58.40		0.23
36.56	1307	carvacrol		0.11		62.35
39.42	1348	thymol acetate	0.03			
		Sesquiterpene Hydrocarbons	18.17	1.17	5.30	4.13
38.42	1334	elemene<δ>	0.03			
40.80	1367	ylangene α	0.24	0.05		
41.15	1372	copaene α	0.06			
41.64	1379	Bourbonene β	0.04			
43.16	1398	Longipinene<β>	1.40			
43.34	1401	Longifolene	0.09			
43.83	1409	caryophyllene (Z)	6.08	0.72	4.91	3.93
44.47	1419	caryophyllene (E)	0.31		0.02	
44.93	1427	Aromadendrene	3.98			
45.52	1436	Barbatene<β>	0.35			
46.05	1444	caryophyllene α	2.88	0.02	0.37	0.07
46.38	1449	Muurola-3-5-diene cis	0.30			
47.21	1462	Unknown	0.03			
47.42	1465	Cadina-1(6).4-diene cis	0.09	0.08		
47.68	1469	Muurola-4(14).5-diene	0.07			
48.16	1476	Aristolochene<4.5-di-epi->	0.20			
48.39	1480	Muurolene (γ)	0.84			
48.55	1482	bicyclogermacrene	1.06			
49.70	1499	amorphene γ	0.03	0.15		0.13
50.02	1503	amorphene δ	0.09	0.15		
		Other	0.26	0.00	0.67	0.65
14.14	985	octen-3-ol	0.20	t	0.52	0.58
14.40	988	octanone	0.06	t	0.15	
15.36	997	octan-3-ol				0.07
		Total	97.36	96.54	99.31	98.64

[a] Retention time. [b] Kováts retention index.

The chemical composition of EOs was very different in the four MAPs in the study, highlighting high diversity among the species.

The average percentage content of monoterpene hydrocarbons was found, however, to be similar for the various species. It was, in fact, 27.73% in common sage EO, 31.31% in oregano EO, 31.47% in rosemary EO and 28.86% in thyme EO, respectively. Among the monoterpene hydrocarbons, α pinene was found to be one third of the total EO content in rosemary, while γ terpinene represented almost half of the total EOs content in oregano and thyme plants.

Oxygenated monoterpenes were the prevailing group of compounds in EOs of the four species. The average content percentage of this group ranged between 51.20% (common sage EO) and 65.00% (thyme EO). In particular, eucalyptol was found to be the main compound in common sage (20.58%) and rosemary (39.38%) EOs. Analyses of EOs revealed that borneol (10.40%) was more abundant in rosemary EO while its oxidation product, camphor (13.77%), was more represented in common sage EO. In the case of oregano and thyme plants, the main compounds among the oxygenated monoterpenes were thymol (58.40%) and carvacrol (62.35%), respectively.

Concerning the average percentage content of sesquiterpene hydrocarbons, it was found to be different in EOs of the four MAPs. It was 18.17% in common sage EO, 4.13% in oregano EO, 5.30% in rosemary EO and 1.17% in thyme EO, respectively.

3.3. Photo-Oxidative Studies

In order to study the possible employment of EOs as antioxidant to EVOO, the concentration 0.15% v/v of the MAPs EO/EVOO mixtures was chosen. This percentage was the minimum that guaranteed the presence of plant volatile organic compounds (VOCs) in the headspace. An amount of 10 mL of different mixtures were placed in a pyrex tube and irradiated at 360 nm at 40 °C for different time by using an UV test. Oleic acid was the most abundant fatty acid considered in the EVOO samples.

The percentage variation of oleic acid in the oil samples of Cerasuola "green" (CV) after photo-oxidative at different irradiation time is showed in Figure 3.

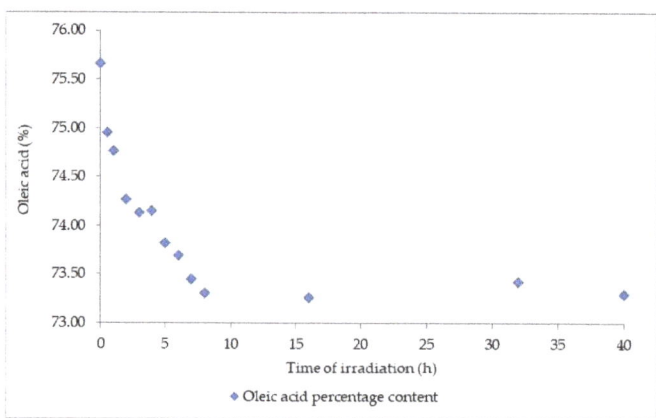

Figure 3. Percentage content of oleic acid in Cerasuola "green" (CV); sample irradiated at different time.

The higher percentage variation was detected during 8 h of irradiation, while, after 16 h, no significant percentage variation was detected.

Accordingly, the photo-oxidative experiments were carried out by irradiating the different EVOOs and mixture samples for 32 h in the same experimental conditions. The main results are showed in Figure 4.

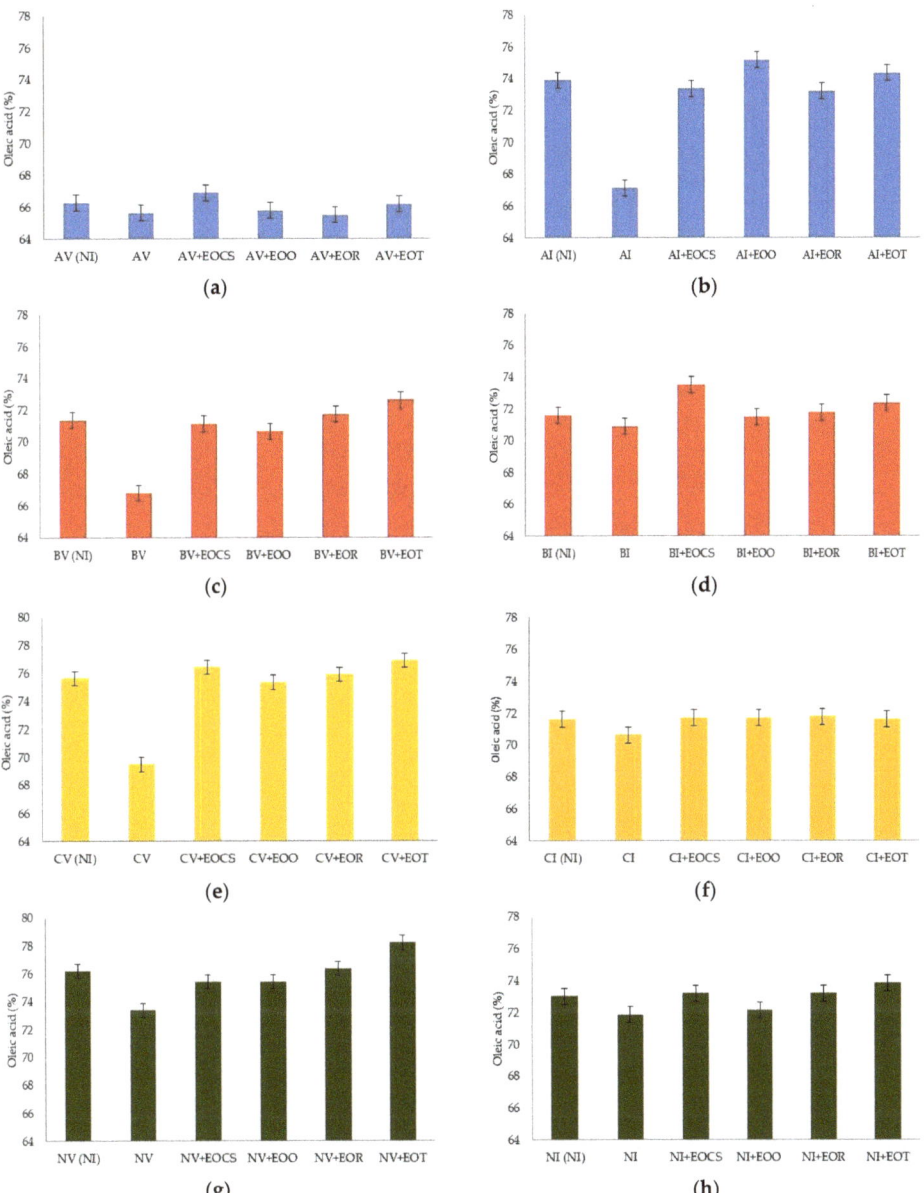

Figure 4. Percentage content of oleic acid in the extra-virgin olive oil (EVOO) samples not irradiated, irradiated and in the mixture of EVOO with EO. Average values (±standard error) are shown. The initials NI in brackets in all graphs stand for not irradiated; EOCS stands for essential oil of common sage; EOO stands for essential oil of oregano; EOR stands for essential oil of rosemary; EOT stands for essential oil of thyme. (**a**) refers to Arbequina "green" (AV), (**b**) refers to Arbequina "brown" (AI), (**c**) refers to Biancolilla "green" (BV), (**d**) refers to Biancolilla "brown" (BI), (**e**) refers to Cerasuola "green" (CV), (**f**) refers to Cerasuola "brown" (CI), (**g**) refers to Nocellara del Belice "green" (NV) and (**h**) refers to Nocellara del Belice "brown" (NI).

In all EVOOs samples exposed at irradiation performed at 360 nm, variations in oleic acid composition were detected. Furthermore, in three different EVOOs samples

(CV, BV and AI), a relevant variation in the percentage content of oleic acid (8.00%) was recorded. This variation can be ascribed at photo-oxidative reaction. In fact, as reported in literature [38], the fatty acid alkyl chain is susceptible to oxidation both at double bonds and adjacent allylic carbons. In fact, free-radical and photo-oxidative reactions at allylic carbons are responsible for deterioration of unsaturated oils and fats, resulting in rancid flavours and reduced nutritional quality. In this context, light and oxygen, promotes oxidation of unsaturated fatty acids. Ultraviolet radiation decomposes existing hydroperoxides, peroxides and carbonyl and other oxygen-containing compounds, producing radicals that initiate autoxidation. Moreover, naturally present pigments such as chlorophyll, hematoporphyrins and riboflavin act as sensitizers. Light excites these sensitizers to the triplet state that promotes oxidation processes.

No significant change of oleic acid percentage was detected in the mixture of EVOO with EO samples. These results seem to bring out the presence of an antioxidant effect of EOs on EVOO. This result is independent from the presence of phenolic compounds, such as in the oregano and thyme EOs, which have a high percentage of phenolic compounds. It would seem, therefore, a synergistic effect of all the components of EOs. Further studies are needed to highlight a correlation between the chemical composition of the MAPs EOs and antioxidant effects.

4. Conclusions

This work is an innovative study to assess the quality of flavoured EVOO prepared by adding MAPs EO. The proposed preparation method is valid to obtain flavoured EVOO and can represent a useful alternative to traditional methods. The addition of EO allows, in fact, to maintain the biological properties of EVOO and prevent any oxidation process which can occur due to light and oxygen. Furthermore, being that EO is rich in antioxidants, its addition causes an improvement of the shelf life of EVOO. The study carried out through the GC analysis can be also a support to the sensory analysis for the evaluation of the oil quality. From a health point of view, it is possible to sustain that the use of flavoured EVOO by adding MAPs EO can have several benefits for humans such as to come from natural resources, to be considered safe and to have functional and sensory properties. Furthermore, the addition of natural antioxidants from EOs, instead of synthetic ones, could be of high interest to food industry whose primary objective is to offer good, healthy and safe food products, with a balanced nutritional profile and economically accessible to all consumers. The creation of natural mixtures and "new" condiments requires, however, an evaluation of the production and availability of MAPs EOs. A high production of EO in relation to plant dry weight seems to be considered crucial in the cultivation of MAPs. Further studies are required to confirm these findings and assess the oxidative stability of the mixtures in the long term, being that the literature in this field is very minimal.

Author Contributions: Conceptualization, S.LB., A.M., A.P. and T.T.; methodology, S.B. (Salvatore Barreca), S.L.B., A.M., A.P. and T.T.; software, S.B. (Salvatore Barreca) and M.L.; validation, S.B. (Silvestre Buscemi) and C.L.; formal analysis, A.M. and T.T.; investigation, S.B. (Salvatore Barreca) and S.L.B.; resources, S.L.B., A.P. and T.T.; data curation, S.B. (Salvatore Barreca) and M.L.; writing—original draft preparation, S.L.B., A.M. and M.L.; writing—review and editing, S.L.B., A.M., M.L. and A.P.; visualization, S.B. (Silvestre Buscemi) and T.T.; supervision, S.B. (Silvestre Buscemi) and C.L.; project administration, C.L.; funding acquisition, C.L. All authors have read and agreed to the published version of the manuscript.

Funding: This research was funded by the European Union, Italian Ministry of Education, University and Research and Italian Ministry of Economic Development, grant number: PON02_00667—PON02_00451_3361785.

Institutional Review Board Statement: Not applicable.

Informed Consent Statement: Not applicable.

Data Availability Statement: Data are available by contacting the authors.

Acknowledgments: The authors would like to thank the European Union, Italian Ministry of Education, University and Research and Italian Ministry of Economic Development, funding the "DI.ME.SA.—Valorisation of typical products of the Mediterranean diet and their use for health and nutraceutical purposes", research project.

Conflicts of Interest: The authors declare no conflict of interest. The funders had no role in the design of the study; in the collection, analyses, or interpretation of data; in the writing of the manuscript, or in the decision to publish the results.

References

1. Caporaso, N.; Padano, A.; Nicoletti, G.; Sacchi, R. Capsaicinoids, antioxidant activity, and volatile compounds in olive oil flavoured with dried chilli pepper (*Capsicum annuum*). *Eur. J. Lipid Sci. Technol.* **2013**, *10*, 1434–1442. [CrossRef]
2. Baiano, A.; Gambacorta, G.; La Notte, E. *Aromatization of Olive Oil*, 1st ed.; Transworld Research Network: Kerala, India, 2010; pp. 1–29.
3. Harwood, J.L.; Yaqoob, P. Nutritional and health aspects of olive oil. *Eur. J. Lipid Sci. Tech.* **2002**, *104*, 685–697. [CrossRef]
4. Vitaglione, P.; Savarese, M.; Paduano, A.; Scalfi, L.; Fogliano, V.; Sacchi, R. Healthy virgin olive oil: A matter of bitterness. *Crit. Rev. Food Sci. Nutr.* **2015**, *55*, 1808–1818. [CrossRef] [PubMed]
5. Nieva-Echevarría, B.; Goicoechea, E.; Guillén, M.D. Oxidative stability of extra-virgin olive oil enriched or not with lycopene. Importance of the initial quality of the oil for its performance during in vitro gastrointestinal digestion. *Food Res. Int.* **2020**, *130*, 108987. [CrossRef] [PubMed]
6. Clodoveo, M.L.; Dipalmo, T.; Crupi, P.; Durante, V.; Pesce, V.; Maiellaro, I.; Lovece, A.; Mercurio, A.; Laghezza, A.; Corbo, F.; et al. Comparison between different flavored olive oil production techniques: Healthy value and process efficiency. *Plant Food Hum. Nutr.* **2016**, *71*, 81–87. [CrossRef]
7. Issaouia, M.; Flamini, G.; Souid, S.; Bendini, A.; Barbieri, S.; Gharbi, I.; Gallina Toschi, T.; Cioni, P.L.; Hammami, M. How the addition of spices and herbs to virgin olive oil to produce flavored oils affects consumer acceptance. *Nat. Prod. Comm.* **2016**, *11*, 775–780. [CrossRef]
8. Caponio, F.; Durante, V.; Varva, G.; Silletti, R.; Previtali, M.a.; Viggiani, I.; Squeo, G.; Summo, C.; Pasqualone, A.; Gomes, T.; et al. Effect of infusion of spices into the oil vs. combined malaxation of olive paste and spices on quality of naturally flavoured virgin olive oils. *Food Chem.* **2016**, *202*, 221–228. [CrossRef]
9. Miguel, M.G. Antioxidant activity of medicinal and aromatic plants. A review. *Flavour Fragr. J.* **2010**, *25*, 291–312. [CrossRef]
10. Parham, S.; Khazari, A.Z.; Bakhsheshi-Rad, H.R.; Nur, H.; Ismail, A.F.; Sharif, S.; RamaKrishna, S.; Berto, F. Antioxidant, antimicrobial and antiviral properties of herbal materials. *Antioxidants* **2020**, *9*, 1309. [CrossRef]
11. Mutlu-Ingok, A.; Devecioglu, D.; Nur Dikmetas, D.; Karbancioglu-Guler, F.; Capanoglu, E. Antibacterial, antifungal, antimycotoxigenic, and antioxidant activities of essential oils: An updated review. *Molecules* **2020**, *25*, 4711. [CrossRef]
12. Asensio, C.M.; Nepote, V.; Grosso, N.R. Sensory attribute preservation in extra virgin olive oil with addition of oregano essential oil as natural antioxidant. *J. Food Sci.* **2012**, *77*, S294–S301. [CrossRef]
13. Tuttolomondo, T.; Iapichino, G.; Licata, M.; Virga, G.; Leto, C.; La Bella, S. Agronomic evaluation and chemical characterization of Sicilian *Salvia sclarea* L. accessions. *Agronomy* **2020**, *10*, 1114. [CrossRef]
14. Licata, M.; Tuttolomondo, T.; Dugo, G.; Ruberto, G.; Leto, C.; Napoli, E.M.; Rando, R.; Fede, M.R.; Virga, G.; Leone, R.; et al. Study of quantitative and qualitative variations in essential oils of Sicilian oregano biotypes. *J. Essent. Oil Res.* **2015**, *27*, 293–306. [CrossRef]
15. Tuttolomondo, T.; Dugo, G.; Ruberto, G.; Leto, C.; Napoli, E.M.; Cicero, N.; Gervasi, T.; Virga, G.; Leone, R.; Licata, M.; et al. Study of quantitative and qualitative variations in essential oils of Sicilian *Rosmarinus officinalis* L. *Nat. Prod. Res.* **2015**, *29*, 1928–1934. [CrossRef] [PubMed]
16. Tuttolomondo, T.; Dugo, G.; Leto, C.; Cicero, N.; Tropea, A.; Virga, G.; Leone, R.; Licata, M.; La Bella, S. Agronomical and chemical characterisation of *Thymbra capitata* (L.) Cav. biotypes from Sicily, Italy. *Nat. Prod. Res.* **2015**, *29*, 1289–1299. [CrossRef]
17. La Bella, S.; Tuttolomondo, T.; Dugo, G.; Ruberto, G.; Leto, C.; Napoli, E.M.; Potortì, A.G.; Fede, M.R.; Virga, G.; Leone, R.; et al. Composition and variability of the essential oil of the flowers of *Lavandula stoechas* from various geographical sources. *Nat. Prod. Comm.* **2015**, *10*, 2001–2004. [CrossRef]
18. Maggio, A.; Rosselli, S.; Bruno, M. Essential oils and pure volatile compounds as potential drugs in Alzheimer's disease therapy: An updated review of the literature. *Curr. Pharm. Design* **2016**, *22*, 4011–4027. [CrossRef] [PubMed]
19. Kottek, M.; Grieser, J.; Beck, C.; Rudolf, B.; Rubel, F. World map of the Köppen-Geiger climate classification updated. *Meteorol. Z.* **2006**, *15*, 259–263. [CrossRef]
20. Yaseen, M.; Singh, M.; Ram, D.; Singh, K. Production potential, nitrogen use efficiency and economics of clary sage (*Salvia sclarea* L.) varieties as influenced by nitrogen levels under different locations. *Ind. Crop. Prod.* **2014**, *54*, 86–91. [CrossRef]
21. Rostro-Alanis, M.J.; Báez-González, J.; Torres-Alvarez, C.; Parra-Saldívar, R.; Rodriguez-Rodriguez, J.; Castillo, S. Chemical composition and biological activities of oregano essential oil and its fractions obtained by vacuum distillation. *Molecules* **2019**, *24*, 1904. [CrossRef]
22. Atti-Santos, A.C.; Rossato, M.; Pauletti, G.P.; Duarte Rota, L.; Rech, J.C.; Pansera, M.R.; Agostini, F.; Atti Serafini, L.; Moyna, P. Physico-chemical evaluation of *Rosmarinus officinalis* L. essential oils. *Braz. Arch. Biol. Technol.* **2005**, *48*, 1035–1039. [CrossRef]

23. Barra, A. Factors affecting chemical variability of essential oils: A review of recent developments. *Nat. Prod. Commun.* **2009**, *4*, 1147–1154. [CrossRef] [PubMed]
24. Tsusaka, T.; Makino, B.; Ohsawa, R.; Ezura, H. Genetic and environmental factors influencing the contents of essential oil compounds in *Atractylodes lancea*. *PLoS ONE* **2019**, *14*, e0217522. [CrossRef]
25. Bakha, M.; El Mtili, N.; Machon, N.; Aboukhalid, K.; Amchra, F.Z.; Khiraoui, A.; Gibernau, M.; Tomi, F.; Al Faiz, C. Intraspecific chemical variability of the essential oils of Moroccan endemic *Origanum elongatum* L. (Lamiaceae) from its whole natural habitats. *Arab. J. Chem.* **2020**, *13*, 3070–3081. [CrossRef]
26. Moghaddam, M.; Farhadi, N. Influence of environmental and genetic factors on resin yield, essential oil content and chemical composition of *Ferula assa-foetida* L. populations. *J. Appl. Res. Med. Aromat. Plants* **2015**, *2*, 69–76. [CrossRef]
27. Tuttolomondo, T.; Dugo, G.; Ruberto, G.; Leto, C.; Napoli, E.M.; Potortì, A.G.; Fede, M.R.; Virga, G.; Leone, R.; D'Anna, E.; et al. Agronomical evaluation of Sicilian biotypes of *Lavandula stoechas* L. spp. *stoechas* and analysis of the essential oils. *J. Essent. Oil Res.* **2015**, *27*, 115–124. [CrossRef]
28. Yeddes, W.; Wannes, W.A.; Hammami, M.; Smida, M.; Chebbi, A.; Marzouk, B.; Tounsi, M.S. Effect of environmental conditions on the chemical composition and antioxidant activity of essential oils from *Rosmarinus officinalis* L. growing wild in Tunisia. *J. Essent. Oil. Bear. Pl.* **2018**, *21*, 972–986. [CrossRef]
29. Chrysargyris, A.; Mikallou, M.; Petropoulos, S.; Tzortzakis, N. Profiling of essential oils components and polyphenols for their antioxidant activity of medicinal and aromatic plants grown in different environmental conditions. *Agronomy* **2020**, *10*, 727. [CrossRef]
30. Yavari, A.; Nazeri, V.; Sefidkon, F.; Hassani, M.E. Influence of some environmental factors on the essential oil variability of *Thymus migricus*. *Nat. Prod. Commun.* **2010**, *5*, 943–948. [CrossRef]
31. Aboukhalid, K.; Al Faiz, C.; Douaik, A.; Bakha, M.; Kursa, K.; Agacka-Mołdoch, M.; Machon, N.; Tomi, F.; Lamiri, A. Influence of environmental factors on essential oil variability in *Origanum compactum* Benth. growing wild in Morocco. *Chem. Biodivers.* **2017**, *14*, e1700158. [CrossRef]
32. Stefanaki, A.; Cook, C.M.; Lanaras, T.; Kokkini, S. The Oregano plants of Chios Island (Greece): Essential oils of *Origanum onites* L. growing wild in different habitats. *Ind. Crop. Prod.* **2016**, *82*, 107–113. [CrossRef]
33. Kokkini, S.; Karousou, R.; Hanlidou, E.; Lanaras, T. Essential oil composition of Greek (*Origanum vulgare* ssp. *hirtum*) and Turkish (*O. onites*) oregano: A tool for their distinction. *J. Essent. Oil Res.* **2004**, *16*, 334–338.
34. Zaouali, Y.; Messaoud, C.; Salah, A.B.; Boussaid, M. Oil composition variability among populations in relationship with their ecological areas in Tunisian *Rosmarinus officinalis* L. *Flavour Frag. J.* **2005**, *20*, 512–520. [CrossRef]
35. Bakhy, K.; Benlhabib, O.; Al Faiz, C.; Bighelli, A.; Casanova, J.; Tomi, F. Wild *Thymbra capitata* from Western Rif (Morocco): Essential oil composition, chemical homogeneity and yield variability. *Nat. Prod. Commun.* **2013**, *8*, 1155–1158. [CrossRef] [PubMed]
36. Couladis, M.; Tzakou, O.; Mimica-Dukíc, N.; Jančić, R.; Stojanović, D. Essential oil of *Salvia officinalis* L. from Serbia and Montenegro. *Flavour Frag. J.* **2002**, *17*, 119–126. [CrossRef]
37. Arraiza, M.P.; Arrabal, C.; Lopez, J.V. Seasonal variation of essential oil yield and composition of sage (*Salvia officinalis* L.) grown in Castilla—La Mancha (Central Spain). *Not. Bot. Horti. Agrobo.* **2012**, *40*, 106–108. [CrossRef]
38. Angerosa, F. Influence of volatile compounds on virgin oli.ve oil quality evaluated by analytical approaches and sensor panels. *Eur. J. Lipid Sci. Tech.* **2002**, *104*, 639–660. [CrossRef]

Article

Four-Year Study on the Bio-Agronomic Response of Biotypes of *Capparis spinosa* L. on the Island of Linosa (Italy)

Salvatore La Bella [1,†], Francesco Rossini [2,†], Mario Licata [1], Giuseppe Virga [3,*], Roberto Ruggeri [2,*], Nicolò Iacuzzi [1], Claudio Leto [1,2] and Teresa Tuttolomondo [1]

1. Department of Agricultural, Food and Forest Sciences, Università degli Studi di Palermo, Viale delle Scienze 13, Building 4, 90128 Palermo, Italy; salvatore.labella@unipa.it (S.L.B.); mario.licata@unipa.it (M.L.); nicolo.iacuzzi@unipa.it (N.I.); claudio.leto@unipa.it (C.L.); teresa.tuttolomondo@unipa.it (T.T.)
2. Department of Agricultural and Forest Sciences, Università degli Studi della Tuscia, 01100 Viterbo, Italy; rossini@unitus.it
3. Research Consortium for the Development of Innovative Agro-Environmental Systems (Corissia), Via della Libertà 203, 90143 Palermo, Italy
* Correspondence: giuseppe.virga@corissia.it (G.V.); r.ruggeri@unitus.it (R.R.)
† These authors are equally contributed.

Abstract: The caper plant is widespread in Sicily (Italy) both wild in natural habitats and as specialized crops, showing considerable morphological variation. However, although contributing to a thriving market, innovation in caper cropping is low. The aim of the study was to evaluate agronomic and production behavior of some biotypes of *Capparis spinosa* L. subsp. *rupestris*, identified on the Island of Linosa (Italy) for growing purposes. Two years and seven biotypes of the species were tested in a randomized complete block design. The main morphological and production parameters were determined. Phenological stages were also observed. Analysis of variance showed high variability between the biotypes. Principal component analysis and cluster analysis highlighted a clear distinction between biotypes based on biometric and production characteristics. Production data collected in the two-year period 2007–2008 showed the greatest production levels in the third year following planting in 2005. In particular, biotype SCP1 had the highest average value (975.47 g) of flower bud consistency. Our results permitted the identification of biotypes of interest for the introduction into new caper fields. Further research is needed in order to characterize caper biotypes in terms of the chemical composition of the flower buds and fruits.

Keywords: caper plant; island of Linosa; morphological and productive characteristics; growing

1. Introduction

The caper plant belongs to the *Capparaceae* family, which includes approximately 40–60 tropical, subtropical and temperate genera, 700–900 species of which belong to tree, shrub and herbaceous plants [1]. The genus *Capparis* L. includes approximately 250 species distributed in tropical and subtropical regions [2,3]. It is a minor crop but since the origins of civilization man has taken an interest in this species due to its healing and nutritional properties [4–7]. Populations of *Capparis* L. on the continent of Europe include *Capparis spinosa* L. with two subspecies: subsp. *spinosa* and subsp. *rupestris* (Sibth and Sm) Nyman [8].

In Sicily (Italy) and the surrounding islands, *C. spinosa*, with the two intraspecific taxa, subsp. *spinosa* and subsp. *rupestris*, is widespread both wild in natural habitats and as specialized crops [9], showing considerable morphological variation due to a number of factors, such as phenotypic plasticity, eco-geographical differentiation and hybridization processes, which promote the presence of intermediate phenotypes [3,10].

Caper buds, harvested from both wild and cultivated plants, are mainly used for food and medicinal purposes [5]. Immature flower buds, called "capers", the fruits, known as "cucunci" or "capperone" and the tender leaves, preserved in salt or vinegar, are popular in cuisine, enjoying good levels of global trade [11–15]. The consistency of the caper berries is central to the quality of the berry. Consistency is important in the creation maturation indices, for the handling and preservation of the product and in customer sensory perception. The size of the bud is also fundamental for commercialization purposes, with a customer preference for small buds [16,17].

It is known for its medicinal use due to the marked therapeutic effects of its extracts. Ethnobotanical research carried out in Sicily [18] shows how extracts of the stem of the caper plant have been used in traditional Sicilian medicine to treat toothache for many years. Various pharmacological properties are attributed to the extracts of leaves, stems, flowers, fruits and roots, such as anti-hypertensive [19], anti-hepatic [20], anti-diabetic [21,22], anti-obesity [23], anti-allergic [24], anti-inflammatory [25] and antibiotic [26] properties. It is to be noted that the biochemical compounds of the caper are influenced by geographical and environmental conditions, by the harvesting period of the immature flower bud and its size, by storage methods, by genotype and by method of extraction and processing [12,27] as evidenced in other Mediterranean species of the same Country [28]. Phenolic and flavonoid compounds are amongst those bioactive compounds found in abundance in the various parts of the caper plant [29–35]. According to various authors [36], in particular, rutin is the most abundant phenolic compound in fresh berries, whereas quercetin (produced by the hydrolysis of rutin and which has not been found in fresh caper berries), is the most abundant phenolic compound in fermented berries. Recent studies [37–39] on quercetin have shown that this flavonoid would interfere with the SARS-COV-2 virus by reducing or eliminating the possibility of replication.

In addition to food and medicinal uses, the aesthetic properties of *Capparis spinosa* also make it popular as an ornamental plant for gardens, walls and terraces [40–42]. Furthermore, due to its xerophilic nature highly extensive root system, extremely high root/stem ratio and moderate water consumption [43,44], the caper is highly suited as a crop to regions with harsh climatic conditions, such as those in the Mediterranean area. The root system architecture and aerial biomass help limit erosion and protect the soil from high temperatures, even in the presence of extreme climate change, thus creating conditions suitable for microbiota and ensuring the agroecosystems are sustainable [10]. Therefore, it is a species of agronomic interest, able to reduce erosion and slow down the desertification process [45]. It is also widely used in re-forestation and re-naturalization in Sicily [46].

However, although contributing to a thriving market, innovation in caper cropping is low. Caper crop specialization is limited by the absence of improved cultivars and the lack of studies on the characterization and valorization of Sicilian caper germplasm. Current knowledge on *Capparis spinosa* does not allow us to define the characteristics of the genetic material being cultivated with any degree of certainty. Therefore, we cannot say that there are any caper cultivars. Individual plants used in production are frequently of uncertain origin, coming either from seedlings or from cuttings of plants harvested from the wild. They are often selected by farmers based on certain highly appreciated characteristics from within local populations [47].

The aim of the study was to evaluate agronomic and production behavior of seven biotypes of *Capparis spinosa* L. subsp. *rupestris*, identified on the Island of Linosa (Italy), over a four-year test period and to identify the most promising biotypes for cultivation.

2. Materials and Methods

2.1. Experimental Site, Cropping Techniques and Plant Material

The test was carried out over the four-year period 2005–2008, on the Island of Linosa, Sicily, Italy, (35°51'43" N 12°52'37" E: Google Earth) at a local farm located in the village Calcarella, between Monte Vulcano and Montagna Rossa, at an altitude of 32 m a.s.l. The test site lies on North–West facing rolling terrain. The soil is typic xerorthents; volcanic, shallow, loose and with scarce organic matter [48] and vegetation is synanthropic, typical of abandoned cropland on these soils (*Euphorbia* spp., *Brassicaceae*, etc.).

Prior to planting, biotypes identified in a previous study [49], classified as *Capparis spinosa* L. subsp. *rupestris* and marked with the abbreviation SCP1-7 (Table 1), underwent virological investigation for caper latent virus (CapLV) at the Rome Experimental Institute of Plant Pathology (Italy) in order to ensure only "healthy material" was used.

Table 1. Main morphological characteristics of *Capparis spinosa* L. subsp. *rupestris* biotypes.

Biotype	Leaf Color (Code *n*.)	Leaf Morphology	Spiny Stipulates	Flower Bud Color (Code Number) *	Bud Morphology
SCP1	brown-green (371)	obovate leaves with retuse apices	absent	deep-green (412)	rounded
SCP2	brown-green (371)	obovate leaves with retuse apices	absent	deep-green (412)	rounded
SCP3	brown-green (371)	ovate leaves with marked retuse apices	absent	deep-green (412)	rounded/pyramidal
SCP4	deep-green (421)	ovate leaves	absent	deep-green (411)	rounded/pyramidal
SCP5	deep-green (421)	ovate leaves with marked retuse apices	absent	deep-green (412)	rounded/pyramidal
SCP6	deep-green (426)	ovate leaves with marked retuse apices	absent	deep-green with dark spots (422)	rounded/pyramidal
SCP7	deep-green (426)	Ovate leaves with marked retuse apices	absent	deep-green with dark spots (423)	rounded

* Seguy E.: Code universel des couleurs (Universal color code).

In December 2005, an experimental plot with a randomized block design with three replicates was created using the plants of the 7 biotypes under evaluation, with a planting spacing of 2.50 × 2.50 m. The photos of the experimental field and caper biotypes are presented in Supplementary Figures S1–S4.

Local cultivation practices were used for the planting: rooted cuttings were placed in holes 30 cm deep and 300 g of blond peat was placed at the bottom of each hole in order to increase soil water holding capacity.

Subsequently, 3 to 4 lava stones were placed around the plantlings to protect them from the wind and to limit water loss from evaporation (Figure 1).

Figure 1. Mitigating effect on evaporation of the lava stones.

During the first year of growth, five rescue irrigation were carried out in summer to encourage establishment of the young plantlings. Pruning was carried out at the end of each year during the autumn-winter period (November–December) by cutting branches to approximately 6–10 cm from the base (long pruning), (Figure 2). Crop care included manual weeding 5 times and hoeing 3 times.

Figure 2. Pruning carried out in December 2007.

2.2. Plant Measurement

During the first year (2006), 6 months after planting, production was considered negligible and no measurements were taken. In the two-year period (2007–2008), however, weekly measurements of the main phenological stages were carried out according to extended BBCH scale [50]: start of plant growth, flower bud formation, flowering, fruit formation and plant dormancy. Each phenological phase was identified when each parcel showed 70–80% of the plants in the considered phase. The following parameters were also determined for each caper biotype: flower bud fresh weight (FW); flower bud dry weight (DW); weight of 100 flower buds; percentage of flower bud dry matter; flower bud diameter (Figure S5); flower bud consistency (Figure S6); average length of primary branches; number of nodes per cm on primary branch; number of secondary branches on primary branch; number of flower buds per primary branch; number of flower buds per secondary branch. Data of all parameters showed normal distribution.

A penetrometer test (FT02, 0–1 kg) with a 2 mm ferrule was used to determine bud consistency; values are expressed in grams.

2.3. Statistical Analysis

All biometric and production parameter data were subjected to analysis of variance. The difference between means was carried out using the Tukey test.

In order to assess the correlation between the biometric and production parameters, Pearson's correlation coefficient was calculated for each year, prior to standardization of data. By grouping the data from the two years, principal component analysis (PCA) was carried out to evaluate the relationship between the different characteristics and how the accessions behaved along the component axes. In addition, cluster analysis (UPGMA) was performed and shown graphically on the principal components plot. Before conducting principal component analysis (PCA) and cluster analysis (UPGMA), the data was standardized. Data analysis was performed using Minitab 19 software for Windows. Principal Component Analysis (PCA) score plots and cluster analysis (UPGMA) were performed with Past 4.03 software for Windows.

3. Results

3.1. Analysis of Rainfall and Air Temperature Trends at the Test Site

Rainfall and air temperature trends during the test period are shown in Figure 3.

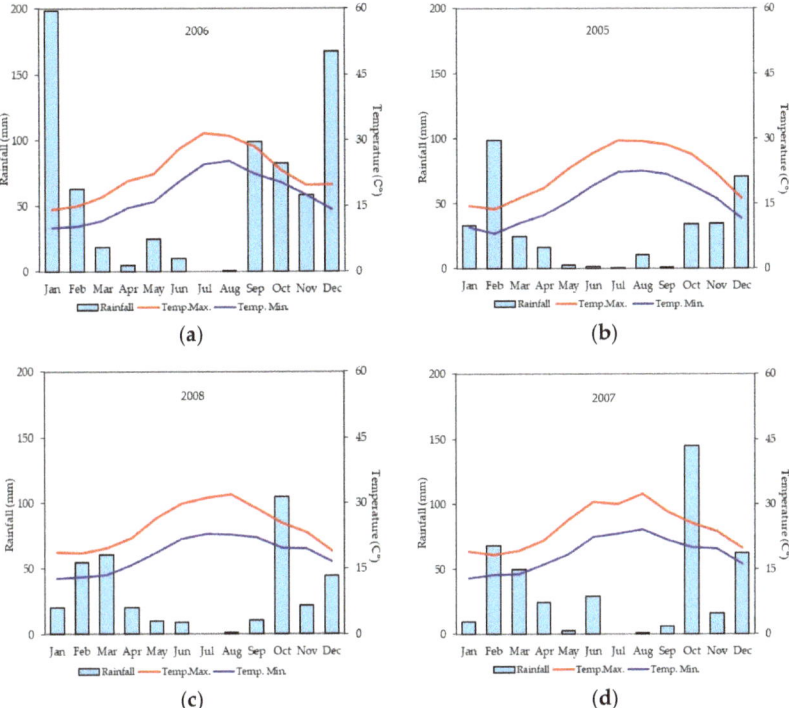

Figure 3. Rainfall and air temperature trends during the test period in the experimental area. Graph (**a**) refers to 2005, graph (**b**) refers to 2006, graph (**c**) refers to 2007 and graph (**d**) refers to 2008.

Rainfall levels during the 4 test years were not always typical of the test environment. In 2005, the year the caper plants were planted in the test field, precipitation depth was consistent with the test environment, whereas in 2006, it was high at 725 mm of rain. In 2007, when phenological and production measuring began, rainfall levels were approximately 100 mm greater than typical of the test environment (414.40 mm). In 2008, the final test year, rainfall was consistent with the test environment (359.00 mm). Rainfall events have always been concentrated mostly in January and in the months between September and December. Average minimum air temperatures (2005:15.80 °C—2006:17.30 °C—2007:18.60 °C—2008:18.60 °C) and average maximum air temperatures (2005:22.10 °C—2006:22.60 °C—2007:24.60 °C—2008:24.60 °C) were found to be consistent with the test environment.

3.2. Analysis of Biometric and Production Parameters

The biotype and year factors and biotype-by-year interaction determined significant differences for almost all parameters in the study. Differences found in parameters during the test years highlight the influence of plant age on biometric and production characteristics. Only for variables: weight of 100 flower buds, flower bud diameter, flower bud consistency, nodes/cm primary branch and ratio flower bud/secondary branches were no statistical differences found (Table 2).

Table 2. Effects of biotype, year and interaction biotype-by-year on biometric and production parameters. Average values are shown.

Factor	Flower Bud Fresh Weight (g)	Flower Bud Dry Weight (g)	Weight 100 Flower Bud (g)	Flower Bud Dry Matter (%)	Flower Bud Diameter (mm)	Flower Bud Consistency (g)	Primary Branch Average Length (cm)	Primary Branch Nodes cm^{-1} (n)	Secondary Branch/ Primary Branch (n)	Flower Buds/ Primary Branch (n)	Flower Buds/ Secondary Branch (n)
Biotype (B)											
SCP1	1031.89 ab	187.24 ab	20.63 c	17.79 bc	7.57 c	975.47 a	104.21 d	0.66 b	5.77 c	36.08 b	15.13 f
SCP2	1167.61 a	220.83 a	22.84 b	18.64 ab	8.19 ab	886.67 d	139.19 a	0.70 a	6.37 c	46.58 a	25.51 e
SCP3	1056.70 ab	184.57 ab	22.08 b	17.23 c	7.65 c	855.97 e	127.24 b	0.64 bc	19.07 a	25.43 c	47.01 a
SCP4	1114.43 ab	160.70 b	23.40 b	14.23 d	8.15 ab	842.16 f	117.78 c	0.61 c	18.58 a	24.17 c	43.72 b
SCP5	1099.20 ab	212.70 a	23.04 b	19.11 a	8.49 a	856.59 e	117.92 c	0.65 bc	18.94 a	25.00 c	37.68 d
SCP6	919.16 bc	177.44 ab	28.12 a	19.02 a	7.58 c	922.85 c	99.74 e	0.64 bc	17.37 b	24.46 c	41.41 c
SCP7	745.17 c	147.97 b	27.80 a	19.48 a	7.97 bc	942.10 b	114.77 c	0.66 b	6.62 c	35.94 b	14.16 f
Year (Y)											
2007	446.00 b	77.53 b	24.11 a	17.46 b	8.01 a	898.82 a	114.13 b	0.66 a	12.87 b	29.67 b	31.77 a
2008	1592.33 a	291.44 a	23.86 a	18.39 a	7.87 a	895.99 a	120.40 a	0.65 a	13.62 a	32.52 a	32.41 a
Y × B	*	*	*	*	*	**	*	**	*	*	**

Means followed by the same letter in the same column are not significantly different according to Tukey's test ($p \leq 0.05$). * significant at $p \leq 0.05$; ** significant at $p \leq 0.01$.

During the test years, fresh weight, dry weight and flower bud dry matter were greater in 2008 despite the fact that rainfall was approximately 50 mm lower than in 2007. The same trend was found when evaluating the morphological characteristics, such as average length of primary branch, number of secondary branches/primary branch and number of flower buds/primary branch.

The highest average values of fresh and dry weight of flower bud were found, in accessions SCP2, SCP3, SCP4, SCP5 and SCP1 (FW: 1167.61–1031.89 g; DW: 220.83–160.70 g), while the lowest averages were recorded in SCP7 (FW:745.17 g; DW: 147.97 g) which were also distinguished by the greatest weight of 100 flower bud (Table 2). The percentage of bud dry weight varied from 19.84% (SCP7) to 14.23% (SCP4). The diameter of the largest flower bud (8 mm) was recorded in SCP5, SCP2 and SCP4, while that of the smallest flower bud (7 mm) was observed in SCP1, SCP7, SCP6 and SCP3. The highest flower bud consistency (957.47 g) was determined in SCP1, while the lowest (842.16 g) in SCP4.

As regards the biometric parameters of the caper accessions in the study, the greatest average length of the primary branch, the greatest number of nodes/cm of primary branch and the highest number of flower buds/primary branch ratio were observed in SCP2. SCP3, SCP5 and SCP4 had the highest number of secondary branches/primary branch ratio while SCP7, together with accessions SCP1 and SCP5 showed the lowest. The highest number of flower buds/secondary branch ratio (47.10) was recorded in SCP3 while the lowest in SCPI (15.13) and SCP7 (14.16) for which no significant differences were found.

The main results for production characteristics of the caper accessions (Table S1) in two year-study highlight that SCP2 and SCP5 obtained the best performance while SCP7 was the least productive accession. Evaluation of results for the first year of biometric and production characteristics showed that both flower bud fresh weight and flower bud dry weight were greater in SCP2 (FW: 533.76 g—DW: 97.07 g), whilst SCP7 was found to have lower flower bud fresh. SCP6 and SCP7 were found to have greater 100 flower bud weight, while SCP1 and SCP3 recorded the lowest in this weight. The greatest percentages for flower bud dry matter varied from 18.82% (SCP5 and SCP7) to 13.88% (SCP4), while while the flower bud diameter varied from 8.53 mm (SCP5) to 7.60 mm (SCP6). Greatest flower bud consistency was found in SCP1 (980.44 g), while the lighest consistency was found in SCP4 (840.40 g), SCP3 (851.08) and SCP5 (856.69).

The greatest average length of the primary branch and the greatest number of nodes/cm of primary branch were found in SCP2. The greatest number of secondary branches/primary branch was recorded in SCP5 and SCP3, while the lowest values for this ratio were found in SCP7. The greatest number of flower buds/primary branch was found in SCP2 while the lowest number in SCP5, SCP4, SCP6 and SCP3. The greatest number of flower buds/secondary branch was determined in SCP3 and SCP4 whilst the lowest in SCP7.

In the second year, the greatest fresh weight of flower buds were found in SCP2, SCP4, SCP5, SCP3 and SCP1 while the lowest in SCP7. The greatest dry weight of flower buds were found in SCP2 and SCP5 while the lowest in SCP7.

By analyzing the results of the accession for each year, no variations were found either in the 100-flower bud weight or the flower bud dry matter %. Furthermore, the order of the accession classification remained unchanged for both of the parameters. The same trend was found for both of the parameters flower bud diameter and flower bud consistency. It is worth noting that, in 2008, the results were slightly lower above all regarding flower-bud diameter, and greater uniformity in characteristics was found between accessions. Flower bud diameter ranged, in 2008, between 8.44 mm (SCP5) and 7.39 mm (SCP1), and flower bud consistency ranged between 970.51 g (SCP1) and 843.92 g (SCP4). Accession SCP2 demonstrated the greatest production of longer primary branches, in the same way that SCP7, SCP4 and SCP5 produced the highest number of shorter primary branches.

The number of nodes cm^{-1} on the primary branch was again greater in SCP2, whilst the lower numbers were found in SCP7, SCP4, SCP6 and SCP3.

The number of secondary branches on the primary branch was greater in SCP3, SCP4, SCP5 and SCP6, whilst in the remaining accession, values were approximately one third of the former: SCP1, SCP2 and SCP7.

The greatest number of flower buds/primary branch was found in SCP2 followed by SCP1 and SCP7.

A similar trend as the previous year was also found for the number of flower buds/ secondary branches, with the greatest number of buds found for accession SCP3, whilst accessions SCP1 and SCP7 developed the fewest flower buds/secondary branches.

3.3. Correlation Matrix

Table 3 shows correlations between the various morphological and production parameters divided by year.

Many correlations were founds between the characteristics observed, albeit only a few were considered significant ($p < 0.05$; $p < 0.01$) and sometimes divergent.

In particular, worthy of note is the fact that the relationship between the fresh weight of the flower buds per plant (FWFB/P) and the dry weight of the flower buds per plant (DWFB/P) was found to be positive and significant only in 2007, whilst it remained medium high ($r = 0.70$) in 2008. Furthermore, in 2008, the parameter fresh weight of the flower buds per plant (FWFB/P), showed a significant but negative correlation with 100-flower bud weight (W100FB) whilst, in 2007, these two parameters were found to be always negatively correlated, but medium-high in value ($r = -0.61$).

The relationship between the number of nodes/primary branch (PBN) and the dry matter % of the flower buds (FBDM) was significant and positive for 2007 but somewhat absent ($r = 0.16$) in 2008. The same relationship was found, albeit with a stronger relationship ($r = 0.66$) in 2008 regarding the number of flower buds/primary branch (FBPB) and the number of nodes/primary branch (PBN). In contrast, the positive correlation between the number of flower buds/secondary branch (FBSB) and the number of secondary branches/primary branch (SBPB) was considered highly significant for both years.

All of the negative and significant correlations number of flower bud/secondary branches (FBSB) and flower bud consistency (FBC); number of secondary branches/primary branch (SBPB) and number of nodes/primary branch (PBN); number of flower buds/secondary branches (FBSB) and number of nodes/cm/primary branch (PBN); number of flower buds/primary branch (FBPB) ad number of secondary branches/primary branch (SBPB); number of flower buds/secondary branches (FBSB) and number of flower buds/primary branch (FBPB), found in 2007 corresponded to those found in 2008, with the exception of number of secondary branches/primary branch (SBPB) and number of nodes/primary branch (PBN), and of number of flower buds/secondary branches (FBSB) and number of nodes/primary branch (PBN), which were negligible in 2008.

3.4. PCA Analysis

PCA analysis, carried out not only to assess relationships between the variables and their importance, but also to reveal the behavior of the accessions along the component axes, showed that the 3 principal components accounted for over 77.00 % of total variability (Table 4).

For analytical purposes, however, only the first three were considered to be of interest.

In Table 5, it is clear that the largest principal component (PC1), at 36.44%, is strongly correlated with as many as 6 out of 11 characteristics.

In particular, it is positively correlated with the percentage of flower bud dry matter, flower bud consistency, number of nodes/cm on primary branch and number of flower buds/primary branch, and negatively correlated with the number of secondary branches/primary branch and the number of flower buds/secondary branches.

The second component, which accounts for 23.82% of the total variance, is positively linked to the flower bud fresh weight/plant, flower bud dry weight/plant and the number of primary branch average length and negatively to the 100-flower bud weight.

Table 3. Correlation matrix of biometric and production parameters.

	Characters	FBFW/P	FBDW/P	W100FB	FBDM	FBD	FBC	PBAL	PBN	SBPB	FBPB	FBSB
	FBFW/P		0.7002	−0.8101 *	−0.5108	0.3024	−0.5619	0.4757	0.6516	0.2236	0.0232	0.3796
	FBDW/P	0.7830 *		−0.564	0.2515	0.2249	−0.2139	0.4032	0.8684 *	−0.0813	0.3368	0.0534
	W100FB	−0.6078	−0.3776		0.4343	−0.019	0.2082	−0.3884	−0.5368	0.0527	−0.1581	−0.0324
	FBDM	−0.3115	0.3465	0.3651		−0.0941	0.4765	−0.1324	0.1596	−0.3932	0.3782	−0.4446
	FBD	0.5340	0.5728	−0.1761	0.0887		−0.6163	0.4530	0.0717	0.2176	0.0069	0.1378
2007	FBC	−0.6513	−0.3213	0.1985	0.4819	−0.5068		−0.5733	−0.0409	0.7607 *	0.4574	−0.821 *
	PBAL	0.6478	0.5935	−0.3399	−0.059	0.5296	−0.5468		0.5314	−0.028	0.3941	0.1760
	PBN	−0.1514	0.3560	0.1526	0.7813 *	0.2248	0.5360	0.2839		−0.4022	0.6623	−0.1507
	SBPB	0.2767	0.0482	0.0363	−0.3481	0.0432	−0.7105	−0.1299	−0.8002 *		−0.9116 **	0.9351 **
	FBPB	0.1124	0.3145	−0.1459	0.3193	0.1362	0.4115	0.5069	0.8235 *	−0.8903 **		−0.7551 *
	FBSB	0.4199	0.1261	−0.016	−0.4437	−0.0361	−0.7621 *	0.0469	−0.7885 *	0.9491 **	−0.7396 *	

FBFW = flower bud fresh weight; FBDW = flower bud dry weight; W100FB = weight 100 flower bud; FBDM = flower bud dry matter; FBD = flower bud diameter; FBT = flower bud consistency; PBAL = primary branch average length; PBN = primary branch nodes cm^{-1}; SBPB = secondary branch/ primary branch; FBPB = flower buds/primary branch; FBSB = flower buds/secondary branch. * correlation is significant at the 0.05 level. ** correlation is significant at the 0.01 level.

Table 4. Variance in principal components and cumulative contribution to total variance.

	PC1	PC2	PC3
Eigenvalues	4.01	2.62	1.88
% variance	36.44	23.82	17.04
% cumulative variance	36.44	60.26	77.30

Table 5. Factor weights of properties on the three principal components.

	PC1	PC2	PC3
Flower bud fresh weight/plant (g)	−0.0170	0.7726	0.5975
Flower bud dry weight/plant (g)	0.0781	0.7551	0.6347
Weight 100 flower buds (g)	0.0483	−0.4720	0.2717
Flower bud dry matter (%)	0.5838	0.0633	0.3437
Flower bud diameter (mm)	−0.1576	0.3101	−0.6735
Flower bud consistency (g)	0.7826	−0.4748	0.3420
Primary branch average length (cm)	0.0313	0.8279	−0.4013
Primary branch nodes/cm^{-1} (n/cm)	0.7238	0.2111	−0.3868
Second. branches/primary branch (n)	−0.9526	0.0376	0.1472
Flower buds/primary branch (n)	0.8570	0.3944	−0.1708
Flower buds/secondary branches (n)	−0.9240	0.1302	0.0663

The third component explains a lower percentage of variance (17.04%) compared to PC1 and PC2 and is negatively correlated with the flower bud diameter however, it was able to separate the accessions more distinctly compared to the second component, confirming the diversity of the accessions.

Figure 4 shows a loading plot of factor weights relating to the two main principal components.

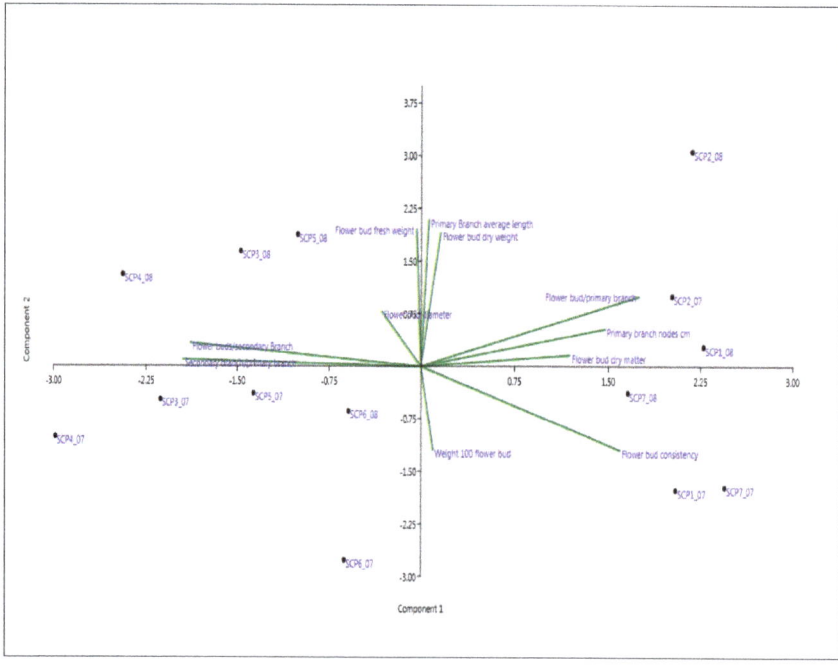

Figure 4. Factor weight and grouping of *Capparis spinosa* subsp. *rupestris* accessions.

Statistical data can be extracted from Figures 4 and 5, which projects the distribution of the accessions on the plot for the two principal components.

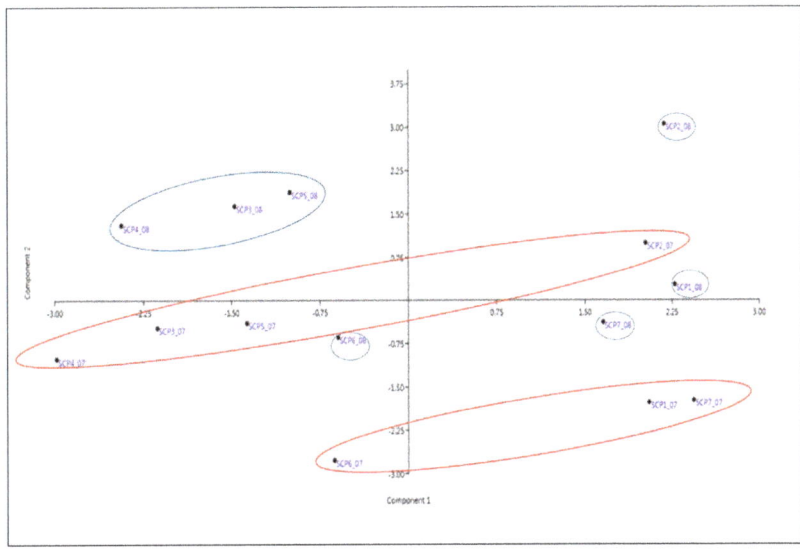

Figure 5. Distribution of the accessions on the score plot for the two principal components. In the graph, the dots refer to accessions of *Capparis spinosa* subsp. *rupestris* grown in the first year while the squares refer to accessions grown in the second year, the same color refers to the same accession.

Representation on the plots of relationships between the accessions showed a relatively wide variability. Cluster analysis lead to the identification of two main groups; the first group containing all of the accessions grown in 2007 (shown in red) and all those grown in 2008 in the second group (shown in blue) (Figure 5).

The first main group (2007) can be subdivided into two further subgroups, one which encompasses 4 accessions (SCP2, SCP5, SCP4, SCP3) and the other 3 accessions (SCP7, SCP1 SCP6). The second main group (2008) is formed by 5 subgroups. With the exception of one of these subgroups, which includes 3 accessions (SCP4, SCP5, SCP3), each of the other subgroups is formed by one accession only (SCP1, SCP2, SCP6, SCP7).

Apart from the conformity in behavior shown by the species in both years and made clear by the cluster analysis which formed two macro-groups, a number of subgroups also emerged based on expressions of the most significant morphological and production characteristics.

Accessions SCP2 and SCP1 from 2008, each of which form a group on their own, located in the top right quadrant, showed the best characteristics associated with PC1 and PC2 (Figure 4). It is worth noting, however, that for SCP1 (2008), component 2 had little weight whilst PC3 assumed greater significance (Supplementary Figure S7). Accession SCP1(2008) can be considered a good compromise of all the characteristics being examined, as it performed well regarding production and produced the best the biometric and quality parameters like SCP2 (2008), which performed the best for all of the characteristics. However, SCP2 (2008) differed from SCP1 (2008) as it produced larger flower buds. SCP2 (2008), in fact, is located in the lower right quadrant, as can be seen on the plot between component 1 and 3, similar to SCP2 in 2007 (Supplementary Figure S7).

SCP2 (2007) is located in this same quadrant (lower right). Although it presented characteristics favorably linked to PC1 and PC2 (Figure 5), unlike accessions SCP3, SCP4 and SCP5 (lower left quadrant), with which it shares a subgroup, all of the accessions are defined by PC3 (Supplementary Figure S7). The abovementioned accessions are located in the quadrants along which PC3 assumes a negative value and, as component 3 is negatively correlated to flower bud diameter, all the accessions produced larger flower buds (Supplementary Figure S7).

The second subgroup (2007) included 2 of the 3 accessions (SCP1, SCP7) associated with those parameters with greatest values for component 1, unlike the other accession SCP6, which is located to the left of the origin. All 3 accessions, however, are located in quadrants with negative values for PC2 (Figure 5). Regarding the characteristic linked to PC3, the 3 accessions (SCP1, SCP6, SCP7), as they are positioned close to the origin, they all have medium-sized buds. Furthermore, it is important to highlight that SCP6 produced the smallest flower buds of the 3 accessions, as located in the top left quadrant, whilst the other 2 accessions are located in the lower right quadrant (Supplementary Figure S7).

In addition, the flower buds produced by the accessions in the third subgroup (SCP3, SCP4, SCP5) in the main 2008 cluster are near in size to the buds of accession SCP6 in 2007 (Supplementary Figure S7). However, the abovementioned accessions are positively characterized by characteristics linked to PC2 and negatively for the parameters linked to PC1 (Figure 5) SCP6 and SCP7, each of which form a subgroup on their own in the main 2008 cluster, although having certain production characteristics which are similar, differ regarding the biometric characteristics linked to PC1 and PC3. In particular, SCP6 had a lighter flower bud consistency, shorter average primary branch length, lower number of flower buds/primary branch and a smaller flower bud diameter compared to SCP7.

3.5. Phenology

Table 6 shows average days, in the two test years, for the four phenological stages considered.

Table 6. Average days per year corresponding to phenological stages.

Year	Plant Dormancy (Day)	Plant Growth (Day)	Flower Bud Emergence (Day)	Fruiting (Day)
2007	90.14 b	267.00 a	212.57 b	175.71 b
2008	91.14 a	265.58 b	215.14 a	175.85 a
Significance	**	**	**	**

Means followed by the same letter in the same column are not significantly different according to Tukey's test ($p \leq 0.05$). ** significant at $p \leq 0.01$.

Statistically significant differences regarding number of days (for the phenological stages in consideration and in the two test years) were found for all the parameters under study.

Accessions (Table 7) presented statistically significant differences for all phenological stages measured.

Table 7. Average length of phenological stages based on accessions of C. spinosa subsp. rupestris.

Biotype	Plant Dormancy (Day)	Plant Growth (Day)	Flower Bud Emergence (Day)	Fruiting (Day)
SCP1	94.00 b	264.01 e	217.00 b	178.02 d
SCP2	80.01 g	297.02 a	233.02 a	181.51 a
SCP3	85.51 f	270.51 b	209.51 f	177.52 e
SCP4	98.50 a	260.01 f	212.02 e	178.02 c
SCP5	92.01 d	267.02 5	215.01 c	180.02 b
SCP6	91.01 e	266.52 d	208.02 g	170.03 f
SCP7	93.50 c	258.01 g	212.52 d	165.53 g
Significance	**	**	**	**

Means followed by the same letter in the same column are not significantly different according to Tukey's test ($p \leq 0.05$). ** significant at $p \leq 0.01$.

The duration of plant dormancy in particular was greater in accession SCP4 (98.50 days), whilst shorter in accession SCP2 (80.0 days) by 18 days.

Plant growth stage (297.02 days), flower bud emergence (233.02 days) and fruiting (181.51 days) were also longer in accession SCP2. Plant growth stage was shortest in accession SCP7 (258.01 days); this accession also recorded the shortest fruiting stage (165.52 days). However, shortest flower bud emergence stage was shortest for SCP6 (208.02).

In Figure 6, the flower and flower bud of the species are shown.

(a)

(b)

Figure 6. Flower (**a**) and flower buds (**b**) of caper plant.

4. Discussion

Increased demand for buds and caper fruits has prompted farmers to switch from wild plant harvesting to specialized crops of caper plants [51]. Of fundamental importance for the creation of new caper plants is undoubtedly the genetic material used for propagation purposes. Therefore, the identification of biotypes in the wild and characterized by high agronomic performance, which can be recommended to farmers or included in genetic improvement programs, is considered an excellent strategy [52–54] Previous studies carried out by Barbera [47] have led to specific characteristics to be identified which are deemed of interest in crop development; for example, high productivity, long stems, short internodes and high node fertility, spherical, dark green buds with closely-placed, non-pubescent and late opening bracts, oval fruits with a light green pericarp and few seeds, absence of stipular spines, easy separation of stems to simplify harvesting and post-harvest operations, suitability for agamic reproduction and resistance to biotic and abiotic stresses. Bud consistency is, without doubt, extremely important in the definition of quality. Amongst those characteristics most sought-after is the diameter. The Boletín Oficial del Estado [55] distinguishes seven classes of increasing diameter, from the smallest of 7 mm to the greatest of 13 mm, highlighting the fact that those most highly appreciated by consumers are actually smaller than 7 mm.

The characterization of the germplasm on the island of Linosa led to the evaluation of 7 biotypes with at least one characteristic not in common, previously identified on the island by Tuttolomondo et al. [49]. Biotypes included in the agronomic evaluation belong to the species C. *spinosa* subsp. *rupestris* (Sm) Nyman which exhibits a narrower range than species C. *spinosa* subsp. *spinosa* and is found in areas of the Mediterranean and North Africa [56]. It is a spineless chamaephyte with few or no ramifications of the primary branches and with uniform morphological traits [57].

Rainfall trends in the four test years were consistent with the test environment except for rainfall depth in 2006. Such high levels (725 mm) undoubtedly contributed to the successful establishment of the caper field and no failures were recorded (data not shown).

Production data collected in the two-year period 2007–2008 showed the greatest production levels in the fourth year following planting in 2005. This behavior is consistent with the characteristics of the species. The caper plant begins production, although in insignificant quantities, in the first year of planting. Full production is recorded as of the fourth year and can reach an average yield of 4–5 kg plant^{-1} and over. This level of production is thought to last up to 35–40 years and to be influenced not only by biotype, age and cropping techniques (fertilizing, irrigation, etc.) but also by the growth environment [10,47]. In our case, production in the year following planting was considered negligible and no measurements were taken.

In our study, in fact, already from 2008, all biotypes showed a significant increase in yields corresponding to approximately three/four times those recorded in 2007. A comparison of the test accessions showed that all seven biotypes differed significantly for all biometric and production parameters. More specifically, in both years, biotype SCP2 demonstrated greater production characteristics, both in terms of greater flower bud fresh weight and dry weight and in morphometric terms. Furthermore, a greater number of flower buds on the main branches were recorded for SCP2, in accordance with previous studies [12,58] which found that a longer primary branch determined a greater number of nodes, allowing greater differentiation of flower buds and, therefore, increased productivity. Aytac et al. [58] demonstrates how the length of the primary branches of caper plants increases by increasing the slope of the caper crop field.

In our study, conducted on a flat field, the length of the primary branches in the test accessions, both during the first and second year of the test, was considerably longer than the length obtained under similar environmental conditions but in older caper plants and using different agronomic management by Tuttolomondo et al. [49]. These differences are presumably due to genetic and non-environmental factors.

In 2008, yields expressed in grams of flower buds per plant, obtained from the remaining accessions are consistent with previous tests conducted by Barbera et al. [59]. Yields in these tests, albeit under different agronomic conditions and in different environments, ranged from 1 to 1.5 kg plant^{-1} (Island of Pantelleria) and from 2 to 3 kg plant^{-1} (Island of Salina). Biotype SCP2 also obtained the lowest number of secondary branches and relative flower buds. This characteristic is valued by farmers as it is seen to facilitate harvesting operations with lower production costs, as reported by Barbera [46]. Regarding flower bud size, a valuable characteristic from a commercial point of view (the smaller they are, the more they are valued), previous studies conducted by Aytac et al. [56] showed how a harvest interval of 5 days was found to produce the highest number of flower buds with a diameter of less than 7 mm—a diameter highly valued on a commercial level [53]— highlighting how reducing harvest intervals determines smaller flower buds. In our study, the smallest size of flower buds with a harvest interval of 8 days was found for accession SCP1 with 7.57 mm and SCP6 with 7.58 mm.

Another useful element in defining the quality of buds and valued by consumers is the consistency. This was measured using a penetrometer. Biotype SCP1 obtained the best result at 975.47 g; A difference of a little over 100 g from the least substantial in consistency (SCP4). Consistency determination, not previously noted by other authors, allowed caper bud quality indexes to be expanded. This characteristic has been studied for other species and is considered strategic as it seems that consumers are more sensitive to differences in consistency than in taste [60]. The measurement of fruit consistency using a penetrometer has long been used in apricots, peaches nectarines, peaches and plums as an index of ripeness [61]. The use of penetrometric analysis in order to identify best bud and caper fruit consistency not only adds value to the product in terms of consumer demand, but also helps innovate mechanical processing and develop the caper supply chain. In order to facilitate the design of machines to be used in the marketing of caper fruits, Lorestani [17] studied the physical and chemical characteristics of unripe buds and caper fruits through elasticity testing (Young Modulus) and the ZwickRoell universal testine machine.

Phenological analysis allowed biotypes to be differentiated according to the duration of the single phenological stage. Shorter plant dormancy and longer plant growth, and, therefore, longer flower bud emission and fruiting stage (which presumably contributed to the increased yield) was found for biotype SCP2. An earlier production stage inevitably led to better use of soil water resources, built up during the autumn-winter period. It is during this period that greatest rainfall levels were recorded for both years, thereby creating an agronomic benefit for the crop. The length of the phenological stages observed were consistent for all of the accessions in the test with those found in previous studies carried out in Sicily by Fici [3]. On average, all of the biotypes began growth stage during March and began to emit flower buds during April right up until November, when the biotypes stopped growth. The phenological trends of the various accessions is a further factor for accession characterization and is of great interest for the species and for cropping technique development. Furthermore, studies carried out by Melgarejo et al. [62] show that phenological behavior is fundamental for the improvement of cropping techniques of this species as various edible parts are included in the term 'yield', (flower buds, young sprouts and fruits) spanning over the entire annual growth cycle.

Characterization of biometric and production parameters (based on statistical methods such as correlation matrix, PCA, and cluster analysis) is the first step towards successful description and understanding of the variability of caper biotypes. It is well known that biometric and production parameters are strongly influenced by genetic and environmental factors [63]. Morphological and production variations found in plant populations can demonstrate adaptation strategies to various selection pressures from phenotypic plasticity or genetic differentiation due to natural selection or other evolutionary forces [64]. PCA and cluster analyses showed a clear distinction between biotypes based on biometric and production characteristics.

5. Conclusions

Agronomic characteristics linked to drought-resistance and tolerance to high temperatures together with the use of accessions with good production results, makes this species a good candidate for use in marginal lands from an environmental point of view. These lands are increasingly more fragile due to climate change, which has caused not only a reduction in rainfall levels but also anomalous intensity and irregular distribution.

The results of this study contribute to further knowledge on caper germplasm found on the Island of Linosa. The biotypes which were analyzed showed good adaptability of the test environment and good yield results. Although the best results in terms of flower buds, length of primary branch, number of nodes/primary branch and precocity were obtained with biotype SCP2, it is also worth noting that results for biotypes SCP1 and SCP5 were also satisfactory. Regarding quality parameters, such as average flower bud diameter and consistency, the best results for both years were obtained with SCP1.

This first 4 years of tests on caper germplasm characterization is the first test in the Mediterranean area to focus on the identification of accessions of interest for the introduction of innovation into new caper fields. This work can contribute to ex situ conservation of the species, since the best biotypes can be propagated and grown.

Further research is needed, however, in order to characterize caper accessions in terms of the chemical composition of the flower buds, fruits and other parts of the plant with application in the food, cosmetics, pharmaceutical and medicinal sectors.

Supplementary Materials: The following are available online at https://www.mdpi.com/article/10.3390/agriculture11040327/s1, Figure S1: A view of the experimental field. Figure S2: Caper plantlings after 5 months from the transplanting in open field. Figure S3: Flowering stage of caper biotypes. Figure S4: Growth stage of caper plants. Figure S5: Determination of flower bud diameter. Figure S6: Determination of flower bud consistency using a penetrometer. Figure S7: PC3. Table S1: Average values of the biometric and pro-duction parameters of accessions of *Capparis spinosa* subs. *rupestris* in 2007 and 2008.

Author Contributions: Conceptualization, S.L.B. and F.R.; methodology, S.L.B. and F.R.; software, N.I. and G.V.; validation, M.L. and C.L.; formal analysis, N.I., R.R. and T.T.; investigation, M.L., G.V. and R.R.; resources, S.L.B., F.R. and T.T.; data curation, M.L., G.V., R.R. and N.I.; writing—original draft preparation, S.L.B., F.R., N.I. and T.T.; writing—review and editing, M.L. and G.V.; visualization, R.R. and C.L.; supervision, S.L.B. and F.R.; project administration, C.L.; funding acquisition, C.L. All authors have read and agreed to the published version of the manuscript.

Funding: This research was funded by the Sicilian Regional Ministry of Agriculture and Forestry (Italy), grant number: DDS N. 509/2005.

Institutional Review Board Statement: Not applicable.

Informed Consent Statement: Not applicable.

Data Availability Statement: Data are available by contacting the authors.

Acknowledgments: The authors would like to thank the Sicilian Regional Ministry of Agriculture and Forestry, funding the "Characterization, genetic breeding and safeguard of capper plant (*Capparis spinosa* L.) of minor islands of Sicily", research project. Special thanks go to Lucie Branwen Hornsby for her linguistic assistance.

Conflicts of Interest: The authors declare no conflict of interest. The funders had no role in the design of the study; in the collection, analyses, or interpretation of data; in the writing of the manuscript, or in the decision to publish the results.

References

1. Hall, J.C.; Sytsma, K.J.; Iltis, H.H. Phylogeny of *Capparaceae* and *Brassicaceae* based on chloroplast sequence data. *Am. J. Bot.* **2002**, *89*, 1826–1842. [CrossRef]
2. Jacobs, M. The genus *Capparis* (*Capparaceae*) from the Indus to the Pacific. *Blumea* **1965**, *12*, 385–541.
3. Fici, S. Intraspecific variation and evolutionary trends in *Capparis spinosa* L. (*Capparaceae*). *Plant Syst. Evol.* **2001**, *228*, 123–141. [CrossRef]
4. Sozzi, O.G. Caper bush: Botany and horticulture. *Hortic. Rev.* **2001**, *27*, 125–188.
5. Rivera, D.; Inocencio, C.; Obón, C.; Alcaraz, F. Review of food and medicinal uses of *Capparis* L. subgenus *Capparis* (*Capparidaceae*). *Econ. Bot.* **2003**, *57*, 515–534. [CrossRef]
6. Bhoyar, M.S.; Mishra, P.G.; Naik, K.P.; Murkute, A.A.; Srivastavar, B.R. Genetic variability studies among natural populations of *Capparis spinosa* from cold arid desert of trans-Himalayas using DNA markers. *Natl. Acad. Sci. Lett.* **2012**, *35*, 505–515. [CrossRef]
7. Tlili, N.; El-Fallah, W.; Saadadoui, E.; Khaldi, A.H.; Triki, S.; Nasri, N. The caper (*Capparis* L.): Ethnopharmacology, phyto-chemical and pharmacological properties. *Fitoterapia* **2011**, *82*, 93–101. [CrossRef] [PubMed]
8. Heywood, V.H. *Capparis* L. In *Flora Europaea*; Tutin, T.G., Heywood, V.H., Burges, N., Valentine, D.H., Walters, S.M., Webb, D.A., Eds.; Cambridge University Press: Cambridge, UK, 1993; p. 312.
9. Gristina, A.S.; Fici, S.; Siragusa, M.; Fontana, I.; Garfi, G.; Carimi, F. Hybridization in *Capparis spinosa* L.: Molecular and morphological evidence from a Mediterranean island complex. *Flora Morphol. Distrib. Funct. Ecol. Plants* **2014**, *209*, 733–741. [CrossRef]
10. Chedraoui, S.; Abi-Rizk, A.; El-Beyrouthy, M.; Chalak, L.; Ouaini, N.; Rajjou, L. *Capparis spinosa* L. in a systematic review: A xerophilous species of multi values and promising potentialities for agrosystems under the threat of global warming. *Front. Plant Sci.* **2017**, *8*, 1845. [CrossRef]
11. Saadaoui, E.; Khaldi, A.; Khouja, M.L.; Mohamed, E.G. Intraspecific variation of *Capparis spinosa* L. in Tunisia. *J. Herbs Spices Med. Plants* **2009**, *15*, 9–15. [CrossRef]
12. Sozzi, O.G.; Vicente, A.R. Capers and caperberries. In *Handbook of Herbs and Spices*; Peter, K.V., Ed.; Woodhead Publishing Limited and CRC Press: Boca Raton, FL, USA, 2006; pp. 230–256.
13. Aliyazicioglu, R.; Tosun, G.; Eyupoglu, E. Characterisation of volatile compounds by spme and gc-fid/ms of capers (*Capparis spinosa* L.). *Afr. J. Agric. Res.* **2015**, *10*, 2213–2217.
14. Legua, P.; Martínez, J.; Melgarejo, P.; Hernández, F. Phenological growth stages of caper plant (*Capparis spinosa* L.) according to the Biologische Bundesanstalt, Bundessortenamt and CHemical scale. *Ann. Appl. Biol.* **2013**, *163*, 135–141. [CrossRef]
15. Romeo, V.; Ziino, M.; Giuffrida, D.; Condurso, C.; Verzera, A. Flavor profile of Capers (*Capparis spinosa* L.) from the Eolian Archipelago by HS-SPME/GC-MS. *Food Chem.* **2007**, *101*, 1272–1278. [CrossRef]
16. Barbera, G. *Il Cappero*; Edagricole: Bologna, Italy, 1993.
17. Lorestani, A.N. Some physical and mechanical properties of caper. *J. Agric. Technol.* **2012**, *8*, 1199–1206.
18. Tuttolomondo, T.; Licata, M.; Leto, C.; Gargano, M.L.; Venturella, G.; La Bella, S. Plant genetic resources and traditional knowledge on medicinal use of wild shrub and herbaceous plant species in the Etna Regional Park (Eastern Sicily, Italy). *J. Ethnopharmacol.* **2014**, *155*, 1362–1381. [CrossRef] [PubMed]

19. Ali, Z.N.; Eddouks, M.; Michel, J.B.; Sulpice, T.; Hajji, L. Cardiovascular effect of capparis spinosa aqueous extract. Part III: Antihypertensive effect in spontaneously hypertensive rats. *Am. J. Pharmacol. Toxicol.* **2007**, *2*, 111–115. [CrossRef]
20. Gadgoli, C.; Mishra, S.H. Antihepatotoxic activity of p-methoxy benzoic acid from *Capparis spinosa*. *J. Ethnopharmacol.* **1999**, *66*, 187–192. [CrossRef]
21. Kazemian, M.; Abad, M.; Haeri, M.R.; Ebrahimi, M.; Heidari, R. Anti-diabetic effect of *Capparis spinosa* L. root extract in diabetic rats. *Avicenna J. Phytomed.* **2015**, *5*, 325–332.
22. Mollica, A.; Zengin, G.; Locatelli, M.; Stefanucci, A.; Mocan, A.; Macedonio, G.; Carradori, S.; Onaolapo, O.; Onaolapo, A.; Adegoke, J.; et al. Anti-diabetic and anti-hyperlipidemic properties of *Capparis spinosa* L.: In vivo and in vitro evaluation of its nutraceutical potential. *J. Funct. Foods* **2017**, *35*, 32–42. [CrossRef]
23. Lemhadri, A.; Eddouks, M.; Sulpice, T.; Burcelin, R. Anti-hyperglycaemic and anti-obesity effects of *Capparis spinosa* and *Chamaemelum nobile* aqueous extracts in HFD mice. *Am. J. Pharmacol. Toxicol.* **2007**, *2*, 106–110. [CrossRef]
24. Trombetta, D.; Occhiuto, F.; Perri, D.; Puglia, C.; Santagati, N.A.; Pasquale, A.D.; Saija, A.; Bonina, F. Antiallergic and anti-histaminic effect of two extracts of *Capparis spinosa* L. flowering buds. *Phytother. Res.* **2005**, *19*, 29–33. [CrossRef] [PubMed]
25. Zhou, H.; Jian, R.; Kang, J.; Huang, X.; Li, Y.; Zhuang, C.; Yang, F.; Zhang, L.; Fan, X.; Wu, T.; et al. Anti-inflammatory effects of caper (*Capparis spinosa* L.) fruit aqueous extract and the isolation of main phytochemicals. *J. Agric. Food Chem.* **2010**, *58*, 12717–12721. [CrossRef] [PubMed]
26. Mahboubi, M.; Mahboubi, A. Antimicrobial activity of *Capparis spinosa* as its usages in traditional medicine. *Herba Pol.* **2014**, *60*, 39–48. [CrossRef]
27. Tlili, N.; Khaldi, A.; Triki, S.; Munné-Bosch, S. Phenolic compounds and vitamin antioxidants of caper (*Capparis spinosa*). *Plant Foods Hum. Nutr.* **2010**, *65*, 260–265. [CrossRef] [PubMed]
28. Perrino, E.V.; Wagensommer, R.P. Crop Wild Relatives (CWR) priority in Italy: Distribution, ecology, in situ and ex situ conservation and expected actions. *Sustainability* **2021**, *13*, 1682. [CrossRef]
29. Yang, T.; Wang, C.; Liu, H.; Chou, G.; Cheng, X.; Wang, Z. A new antioxidant compound from *Capparis spinosa*. *Pharm. Biol.* **2010**, *48*, 589–594. [CrossRef]
30. Yang, T.; Wang, C.-H.; Chou, G.-X.; Wu, T.; Cheng, X.-M.; Wang, Z.-T. New alkaloids from *Capparis spinosa*: Structure and X-ray crystallographic analysis. *Food Chem.* **2010**, *123*, 705–710. [CrossRef]
31. Zhang, S.; Hu, D.-B.; He, J.-B.; Guan, K.-Y.; Zhu, H.-J. A novel tetrahydroquinoline acid and a new racemic benzofuranone from *Capparis spinosa* L., a case study of absolute configuration determination using quantum methods. *Tetrahedron* **2014**, *70*, 869–873. [CrossRef]
32. Sharaf, M.; El-Ansari, M.A.; Saleh, N.A. Flavonoids of four *Cleome* and three *Capparis* species. *Biochem. Syst. Ecol.* **1997**, *25*, 161–166. [CrossRef]
33. Kulisic-Bilusic, T.; Blažević, I.; Dejanović, B.; Miloš, M.; Pifat, G. Evaluation of the antioxidant activity of essential oils from caper (*Capparis spinosa*) and sea fennel (*Crithmum maritimum*) by different methods. *J. Food Biochem.* **2010**, *34*, 286–302. [CrossRef]
34. Fu, X.P.; Wu, T.; Abdurahim, M.; Su, Z.; Hou, X.L.; Aisa, H.A.; Wu, H. New spermidine alkaloids from *Capparis spinosa* roots. *Phytochem. Lett.* **2008**, *1*, 59–62. [CrossRef]
35. Zhang, H.; Ma, Z.F. Phytochemical and pharmacological properties of *Capparis spinosa* as a medicinal plant. *Nutrients* **2018**, *10*, 116. [CrossRef]
36. Francesca, N.; Barbera, M.; Martorana, A.; Saiano, F.; Gaglio, R.; Aponte, M.; Moschetti, G.; Settanni, L. Optimised method for the analysis of phenolic compounds from caper (*Capparis spinosa* L.) berries and monitoring of their changes during fermen-tation. *Food Chem.* **2016**, *196*, 1172–1179. [CrossRef]
37. Smith, M.; Smith, J.C. Repurposing therapeutics for COVID-19: Supercomputer-based docking to the SARS-CoV-2 viral spike protein and viral spike protein-human ACE2 interface. *ChemRxiv* **2020**, in press.
38. Derosa, G.; Maffioli, P.; D'Angelo, A.; Di Pierro, F. A role for quercetin in coronavirus disease 2019 (COVID-19). *Phytother. Res.* **2021**, *35*, 1230–1236. [CrossRef] [PubMed]
39. Haslberger, A.G.; Jacob, U.; Hippe, B.; Karlic, H. Mechanisms of selected functional foods against viral infections with a view on COVID-19: Mini review. *Funct. Food Health Dis.* **2020**, *5*, 195–209. [CrossRef]
40. Bailey, L. *The Standard Cyclopedia of Horticulture*; Macmillan: New York, NY, USA, 1927; Volume 1, p. 658.
41. Baccaro, G. *Il Cappero: Pianta da Reddito*; Universale Edagricole, Edizioni Edagricole: Bologna, Italy, 1987.
42. Faran, M. Capparis spinosa—The plant on the wall. In *Medicinal and Aromatic Plants of the Middle-East (Medicinal and Aromatic Plants of the World)*; Yaniv, Z., Dudai, N., Eds.; Springer: Dordrecht, The Netherlands, 2014; pp. 59–65.
43. Zuo, W.; Ma, M.; Ma, Z.; Gao, R.; Guo, Y.; Jiang, W.; Liu, J.; Tian, L. Study of photosynthetic physiological characteristics of desert plant *Capparis spinosa* L. *J. Shihezi Univ.* **2012**, *3*, 006.
44. Gan, L.; Zhang, C.; Yin, Y.; Lin, Z.; Huang, Y.; Xiang, J.; Fu, C.; Li, M. Anatomical adaptations of the xerophilous medicinal plant, Capparis spinosa, to drought conditions. *Hortic. Environ. Biotechnol.* **2013**, *54*, 156–161. [CrossRef]
45. Grimalt, M.; Hernández, F.; Legua, P.; Almansa, M.; Amorós, A. Physicochemical composition and antioxidant activity of three Spanish caper (*Capparis spinosa* L.) fruit cultivars in three stages of development. *Sci. Hortic.* **2018**, *240*, 509–515. [CrossRef]
46. La Mantia, T.; Rühl, J.; Massa, B.; Pipitone, S.; Lo Verde, G.; Bueno, R.S. Vertebrate-mediated seed rain and artificial perches contribute to overcome seed dispersal limitation in a Mediterranean old field. *Restor. Ecol.* **2019**, *27*, 1393–1400. [CrossRef]

47. Barbera, G. Le câprier (*Capparis* spp.). In *Programme de Recherche Agrimed*; Guiseppe, B., Ed.; Commission des Communautés Européennes L-2920: Luxembourg, 1991; p. 62.
48. Fierotti, G. *I Suoli della Sicilia: Con Elementi di Genesi, Classificazione, Cartografia e Valutazione dei Suol*, 1st ed.; Flaccovio Dario: Palermo, Italy, 1997.
49. Tuttolomondo, T.; La Bella, S.; Cammalleri, I.; Gaglio, G. Indagine preliminare sul germoplasma di *Capparis spinosa* subsp. *rupestris* (Sm.) Nyman dell'isola di Linosa. *Ital. J. Agron.* **2009**, *4*, 513–517.
50. Lancashire, P.D.; Bleiholder, H.; Van Den Boom, T.; Langelüddeke, P.; Stauss, R.; Weber, E.; Witzenberger, A. A uniform decimal code for growth stages of crops and weeds. *Ann. Appl. Biol.* **1991**, *119*, 561–601. [CrossRef]
51. Tuttolomondo, T.; La Bella, S.; Hornsby, L. Il cappero: Analisi desk sui flussi di commercializzazione in Italia e nei principali Paesi competitors. *Ital. J. Agron.* **2009**, *4*, 789–796.
52. Tuttolomondo, T.; Iapichino, G.; Licata, M.; Virga, G.; Leto, C.; La Bella, S. Agronomic evaluation and chemical characterization of Sicilian *Salvia sclarea* L. accessions. *Agronomy* **2020**, *10*, 1114. [CrossRef]
53. Perrino, E.V.; Perrino, P. Crop wild relatives: Know how past and present to improve future research, conservation and utilization strategies, especially in Italy: A review. *Genet. Resour. Crop. Evol.* **2020**, *67*, 1067–1105. [CrossRef]
54. Landucci, F.; Panella, L.; Lucarini, D.; Gigante, D.; Donnini, D.; Kell, S.; Maxted, N.; Venanzoni, R.; Negri, V. A prioritized inventory of crop wild relatives and wild harvested plants of Italy. *Crop. Sci.* **2014**, *54*, 1628–1644. [CrossRef]
55. Boletín Oficial del Estado. Normas de calidad para el comercio exterior de alcaparras y alcaparrones. *BOE* **1984**, *100*, 11394–11397.
56. Inocencio, C.; Rivera, D.; Obon, C.; Alcaraz, F.; Barrena, J.A. A systematic revision of *Capparis* section *Capparis* (Capparaceae). *Ann. Mo. Bot. Gard.* **2006**, *93*, 122–149. [CrossRef]
57. Fici, S.; Gianguzzi, L. Diversity and conservation in wild and cultivated *Capparis* in Sicily. *Bocconea* **1997**, *7*, 437–443.
58. Aytaç, Z.; Kinaci, G.; Caylan, A. Yield and some morphological characteristics of Caper (*Capparis spinosa* L.) population cultivated at various slopes in Aegean ecological conditions. *Pak. J. Bot.* **2009**, *41*, 591–596.
59. Barbera, G.; Di Lorenzo, R. La coltura specializzata del cappero nell' isola di Pantelleria. *Inf. Agrar.* **1982**, *XXXVIII*, 22113–22117.
60. Shewfelt, R.L. What is quality? *Postharvest Biol. Technol.* **1999**, *15*, 197–200. [CrossRef]
61. Crisosto, C.H. Stone fruit maturity indices: A descriptive. *Postharvest News Inf.* **1994**, *5*, 65–68.
62. Melgarejo, P.; Legua, P.; Martinez, J.; Martinez-Font, R.; Hernandez, F. Preliminary characterization of sixty one caper clones (*Capparis spinosa* L.). *Acta Hortic.* **2009**, *818*, 155–160. [CrossRef]
63. Burghardt, L.T.; Edwards, B.R.; Donohue, K. Multiple paths to similar germination behavior in *Arabidopsis thaliana*. *New Phytol.* **2015**, *209*, 1301–1312. [CrossRef]
64. Awatef, R.; Hédia, H.; Sonia, H.; Mohamed, B. The use of morphological descriptors to study variability in wild populations of *Capparis spinosa* L. (Capparaceae) in Tunisia. *Afr. J. Ecol.* **2012**, *51*, 47–54. [CrossRef]

Article

Cultivating for the Industry: Cropping Experiences with *Hypericum perforatum* L. in a Mediterranean Environment

Silvia Lazzara [1], Alessandra Carrubba [2,*] and Edoardo Napoli [3]

[1] Consiglio per la Ricerca in Agricoltura e l'analisi dell'Economia Agraria—Difesa e Certificazione (CREA-DC), S.S. 113 km 245.500, 90011 Bagheria, Italy; silvia.lazzara@crea.gov.it
[2] Dip. di Scienze Agrarie, Alimentari e Forestali—Università di Palermo (SAAF), Viale delle Scienze Ed.4 Ingr, L, 90128 Palermo, Italy
[3] Consiglio Nazionale delle Ricerche—Istituto di Chimica Biomolecolare (CNR-ICB), Via Paolo Gaifami 18, 95126 Catania, Italy; edoardo.napoli@icb.cnr.it
* Correspondence: alessandra.carrubba@unipa.it; Tel.: +39-091-23862208

Abstract: *Hypericum perforatum* is an intensively studied medicinal plant, and much experimental activity has been addressed to evaluate its bio-agronomical and phytochemical features as far. In most cases, plant material used for experimental purposes is obtained from wild populations or, alternatively, from individuals grown in vases and/or pots. When *Hypericum* is addressed to industrial purposes, the most convenient option for achieving satisfactory amounts of plant biomass is field cultivation. Pot cultivation and open field condition, however, are likely to induce different responses on plant's metabolism, and the obtained yield and composition are not necessarily the same. To compare these management techniques, a 4-year cultivation trial (2013–2016) was performed, using three *Hypericum* biotypes obtained from different areas in Italy: PFR-TN, from Trento province, Trentino; PFR-SI, from Siena, Tuscany; PFR-AG, from Agrigento province, Sicily. Both managements gave scarce biomass and flower yields at the first year, whereas higher yields were measured at the second year (in open field), and at the third year (in pots). Plant ageing induced significant differences in phytochemical composition, and the total amount of phenolic substances was much higher in 2015 than in 2014. A different performance of genotypes was observed; the local genotype was generally more suitable for field cultivation, whereas the two non-native biotypes performed better in pots. Phytochemical profile of in-pots plants was not always reflecting the actual situation of open field. Consequently, when cultivation is intended for industrial purposes, accurate quality checks of the harvested material are advised.

Keywords: St. John's wort; *Hypericum perforatum*; secondary metabolites; cropping technique

1. Introduction

Hypericum perforatum L. (fam. *Hypericaceae*) is one of the most famous and widespread medicinal plants in the world. Due to its many pharmaceutical activities, ranging from antioxidant [1], anti-inflammatory [2], antiviral [3], antimicrobial [4], and antiproliferative [5], this species is traditionally used throughout the world for a number of internal and external applications. According to the European Pharmacopoeia, *Hypericum* drug (Hyperici herba) consists of the plant's dried aerial part and flowering tops, collected at flowering time [6,7]. Within continental Europe and the whole Mediterranean area, the *Hypericum* oleolite (Hyperici oleum) represents a very popular remedy against minor wounds and skin conditions, burns, and sunburst [8,9]. This extract is obtained by macerating the flowers in vegetable oil (mainly sunflower or olive oil), with a Drug Extract Ratio (DER) varying from 1:4 to 1:20 according to the given traditional recipe [6,7,10].

The interest towards the plant sharply rose in the early 1980s, when specific antidepressant and anxiolytic properties were discovered [11,12]. For a long time, hypericin was thought to be the main responsible for *Hypericum* antidepressant activity [13], and

Hypericum-based products started to be valued according to their content in hypericins. Later on, research enlightened the role played by hyperforin first [14], and then by other plant constituents, including many phenolic compounds. So far, a general consensus has been reached on the shared synergic pharmacological importance of the many constituents of *Hypericum* extract, that therefore should be properly considered a phytocomplex [15–17].

Despite its high commercial importance, however, the availability of *H. perforatum* raw material is presently rather limited. In Europe, the major production areas are located in Germany, Italy, and Romania [18], but collection from wild populations still forms a large part of the total *Hypericum* supply.

It appears that cultivating *Hypericum* for industrial purposes, i.e., aimed at achieving high and stable amounts of the desired active metabolites, could be a great resource for farmers. Nevertheless, the definition of a comprehensive set of information about the field management techniques still requires a great research effort, as many factors are known to deeply affect the yield and proportion of active compounds in *Hypericum* [19]. The available literature shows that scarce research indeed is addressed to the evaluation of the bio-agronomical and phytochemical response of *H. perforatum* to open field conditions. As a matter of fact, the majority of available papers are based on plant samples collected from the wild, or, when plants are cultivated, on individuals grown in constrained conditions, mostly in vases and pots. Although these kinds of experiments have many advantages—first of all reproducibility, some differences between pots cultivation and open field condition are likely to occur [20], especially when physiological, chemical, or yield response are evaluated on individually-grown plants. This issue can have striking consequences especially in Mediterranean environments, where the high variability of climatic and environmental conditions is expected to play an additional and important role on cultivated plants' metabolism. Hence, there is room for a straightforward research, aimed at comparing the phytochemical and biomass response of *H. perforatum* in pots and open field conditions. In this work we analyze the results of a cultivation trial, performed throughout the whole crop duration (2013–2016), of three *H. perforatum* biotypes, obtained from different Italian geographical areas, with contrasting cultivation methods.

2. Materials and Methods

2.1. Plants Management and Data Collection

Mature seeds of *H. perforatum* were retrieved in spring-summer 2012 from three different geographical areas of Italy: mount Bondone (TN-Trento province, Trentino), Massa Marittima (SI-near Siena, Tuscany), and San Biagio Platani (AG-Agrigento province, Sicily), representative of Northern, Central, and Southern Italy, respectively.

Since *Hypericum* seeds are usually considered "recalcitrant" to germinate [21], prior to sowing, seeds were submitted to a 1-week vernalization period at T = 4 °C [22]. In the second week of August 2012, seeds were sown in 104-holes expanded polystyrene trays and, after germination, plantlets were transferred to larger (about 5 cm diameter) plastic pots filled with a 1:1 sand-perlite mixed substrate for root establishment [23]. Three months after sowing (November 2012), one half of the obtained fully established plants (70 individuals per each biotype, i.e., a total of 210 plants) was transferred into larger pots (18 cm diameter), filled with a growth substrate composed of a mixture of peat, sand, and vermiculite (60%, 30%, and 10% in weight, respectively) and positioned into the facilities of CREA-DC in Bagheria (PA, Sicily, 38°5′ N,13°31′ E, 25 m a.s.l.). The remaining plants (70 individuals per each biotype) were transplanted in open field within the experimental farm "Sparacia" (Cammarata, AG, Sicily, 37°38′ N–13°46′ E; 415 m s.l.m.). The soil (Table S1) was a vertic-xerofluvent [24], characterized by a definite clayey texture, and scarcely endowed with nitrogen and organic matter. Climatic pattern recorded in Sparacia throughout the whole trial period is reported in Figure S1. Three experimental plots were set, one per each biotype, sized 10.80 m^2 (3.6 by 3.0 m); plants were arranged in rows 50 cm apart, at a distance of 40 cm one another (plant population: 5 plants m^{-2}). During the four years' timeframe and in both management systems, the crop was monitored until harvest time, keeping note of plants' phytosanitary state and general development conditions. No intervention against pests was needed, except for weed control. That was carried out manually once a year, in springtime, before the emission of flower buds. Fertilization consisted in a light N supply in organic form. A commercial pelletized organic fertilizer, containing 5% organic N, 37% total organic carbon, and 74% organic matter, was used; fertilizer supply, corresponding to 50 kg ha^{-1} N, was distributed only once in 2012, before transplant. In all cultivation years and in both experimental conditions, plants were watered throughout spring and summer, from the transplant (in the years after the first, from the restarting of vegetation) to full flowering time (i.e., harvest time). In doing this, the amount of administered water was managed in order to achieve and maintain field capacity and, therefore, to entirely satisfy crop requirements.

From 2013 to 2016 at flowering time (between late May and early June in field, and in mid-June in pots), all plants were cut at ground level, in order to allow a quick regrowth in the following year. Samples of five plants were randomly taken per treatment, and data on weight, height, and number of stems per plant were collected. Stems were considered flowering when containing at least one fully developed flower; hence, stems of all plants were sorted by flowering and vegetative stems (without flowers). Flowers were further picked up and weighed. All separate plant fractions were open-air dried in the dark for one week, and weighed again, in order to obtain the yields of herbal product.

2.2. Phytochemical Analyses

Analyses were conducted on samples of dried flowers obtained from all experimental sets in 2014 and 2015, except for 2016, when only samples from pot-managed plants were analyzed. In all treatments and years, the flowered tops (15–20 cm) of full-flowering individuals were collected, and after cutting, plant samples were stored in paper bags and dried at 20–25 °C in the dark for further analyses. The dried flowers collected from the different experimental conditions were finely crushed and aliquots (1 g) of powder were extracted with 20 mL of ethanol for 72 h under gently continuous stirring, avoiding light exposure due to the photo sensibility of some of the metabolites of interest. The resulting deep red colored suspensions were filtered on PTFE 0.45 filters (PALL Corporation), put into 2 mL amber vials, and sent to analytical determinations.

Determinations involved some of the most relevant phenolic compounds, belonging to the chemical families of naphthodianthrones (hypericins: hypericin, pseudohypericin, protohypericin, and protopseudohypericin); phloroglucinols (hyperforin and adhyperforin); cinnamic acids and derivatives (3-O-caffeoylquinic acid, 5-O-caffeoylquinic acid, p-coumaroilquinic acid, and p-coumaric acid); flavonols (quercetin, quercitrin, rutin, hyperoside, isoquercitrin, myricitrin, and myricetin derivative); dimers (biapigenin and amentoflavone); flavan-3-ols (catechin). Quantitative analyses were carried out following the procedure already described in previous works [25]. Briefly, polyphenol quantitative analysis was carried out on a Ultimate3000 instrument equipped with a binary high-pressure pump, a photodiode array detector (Thermo Scientific, Milan, Italy). The data were processed through the Chromeleon Chromatography Information Management System v. 6.80 (Thermo Fisher Scientific Inc., Waltham, MA, USA). All chromatographic runs were performed using a reverse-phase column (Gemini C_{18}, 250 × 4.6 mm, 5 µm, Phenomenex, Italy). Chromatographic runs were carried out with a gradient of 5%–90% Buffer B (2.5% formic acid in acetonitrile) in Buffer A (2.5% formic acid in water) over 50 min after which the system was maintained for 7 min at 100% Buffer B, with a constant solvent flow of 1 mL/min. Quantifications were carried out building calibration curves using the corresponding reference substance, if applicable, or a similar molecule with analogue chromophore.

The quantitative analysis of naphthodianthrones and acylphloroglucinols was carried out on a Hitachi Chromaster instrument, equipped with a binary high-pressure pump and a photodiode array detector. Data were processed through Agilent OpenLab CDS version A.04.05 (Agilent, Santa Clara, CA, USA). Chromatographic runs were performed using the same column as the polyphenols, and were carried out with the following gradient of Buffer B (acetonitrile) in Buffer A (ammonium acetate 20 mM in water): 0 min: 50% B; 25 min: 50% B; 35 min: 10% B; 45 min: 90% B; 50 min: 50% B. The solvent flow rate was 1 mL/min. Quantifications were carried out building calibration curves using the corresponding reference substance, if applicable, or a similar molecule with analogue chromophore. All analyses were carried out in triplicate.

2.3. Statistical Treatment of Data

Statistical analyses were performed by means of the statistical package Minitab® version 17.1.0 (Minitab Inc., State College, PA, USA, 2013). The GLM (General Linear Model) procedure was used, setting as dependent variables the data measured in all experiments, whereas the independent variables were "year," "management", and "*Hypericum* biotype," respectively. Data obtained in all trial years were preliminary submitted to the Levene's test for variance homogeneity, assessing a substantial equality of variances for some variables (number of stems per plant—including stems with flowers, stems without flowers, and total stems number—and fresh and dry weight of stems), and a significant non-homogeneity for plant height, fresh and dry weight of flowers per plant, and fresh and dry weight of total plant biomass. Hence, only in the former group of variables a complete ANOVA on pooled data was run, setting the "year" as a random factor, and "management" and "*Hypericum* biotype" as fixed factors. For the latter group of variables, separate ANOVA procedures were otherwise carried out throughout each experimental year. Chemical data were submitted to a Principal Component Analysis (PCA) by means of the PAST statistical package version 3.26b [26,27]. Values of hypericin, pseudohypericin, and hyperforin were submitted to a one-way ANOVA; because of the unbalanced structure of data, the analysis was run separately for each treatment year. The differences among means were appreciated through Tukey's post-hoc comparison test.

3. Results and Discussion
3.1. Plant Growth and Yield

In the first cultivation year, the height of plants was not different between the two managements (Table 1; Figure 1), whereas remarkable differences among biotypes emerged, with the highest mean value (67.8 cm) in the AG biotype and rather small values (about 30 cm) in the other two biotypes. In the second year, higher values were observed but the trend was similar: biotype from AG gained the maximum height value of the whole experiment (85.0 cm) and, in general, plants grown in pots reached higher values than those in open field. In the last two trial years, plant height seemed to stabilize on rather constant values; however, plants managed in pots, although not statistically significant, expressed an overall decrease in mean height.

The total number of stems per plant (vegetative + flowered) (Table 2) was significantly influenced by the different management systems in all years and in each different biotype (YxM and MxB interactions significant at $p \leq 0.05$). This result, combined with the observation of the mean YxMxB interactions (Figure 2) allows to drive some general consideration about the plants' response throughout the different years and conditions.

Table 1. Results of ANOVA (F values) for the height of plants at flowering time of 3 *H. perforatum* biotypes cultivated from 2014 to 2016 in open field and in pots.

Source of Variability	DF	2013	2014	2015	2016
Management (M)	1	<1 n.s.	33.67 ***	2.14 n.s.	3.52 n.s.
Biotype (B)	2	111.94 ***	162.66 ***	1.71 n.s.	<1 n.s.
M × B	2	<1 n.s.	<1 n.s.	<1 n.s.	1.03 n.s.
Error	24				
Total	29				

***: significant at $p \leq 0.001$; n.s.: not significant.

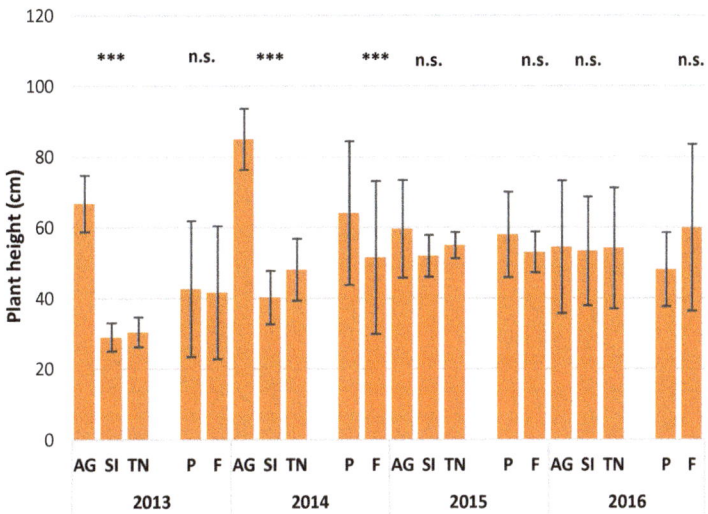

Figure 1. Height (cm) at flowering time of 3 *H. perforatum* biotypes cultivated from 2014 to 2016 with two management systems. Mean values by biotype (AG—Agrigento, SI—Siena, and TN—Trento) and plant management (P—pots and F—open field) across cultivation years. Error bars indicate standard deviation. Symbols above each group refer to the significance at ANOVA (***: $p \leq 0.001$; n.s.: not significant).

Table 2. Results of ANOVA (F values) for the number and the fresh (FW) and dry (DW) weight of stems per plant in 3 *H. perforatum* biotypes cultivated from 2014 to 2016 with two management systems.

Source of Variability	DF	Number of Stems (no.)			Weight of Stems (g)	
		with Flowers	Vegetative	Total	FW	DW
Year (Y)	3	<1 n.s.	2.65 n.s.	<1 n.s.	7.69 n.s.	4.15 n.s.
Management (M)	1	1.10	<1 n.s.	<1 n.s.	16.36 *	5.00 n.s.
Biotype (B)	2	<1 n.s.	15.73 **	<1 n.s.	2.37 n.s.	5.94 *
Y × M	3	3.27 n.s.	1.7 n.s.	4.87 *	<1 n.s.	<1 n.s.
Y × B	6	2.52 n.s.	1.53 n.s.	4.1 n.s.	1.31 n.s.	1.38 n.s.
M × B	2	10.16 *	1.4 n.s.	8.22 *	3.34 n.s.	4.81 n.s.
Y × M × B	6	3.28 ***	1.29 n.s.	1.67 n.s.	7.27 ***	5.40 ***
Error	96					
Total	119					

*: significant at $p \leq 0.05$; **: significant at $p \leq 0.01$; ***: significant at $p \leq 0.001$; n.s.: not significant.

Although ANOVA did not highlight significant differences on the Y×M×B interactions, it is worth noting that pots in 2015 allowed both the maximum (27.5 stems/plant, biotype TN) and the minimum (7.2 stems/plant, biotype SI) of the whole experiment. Nonetheless, on average, the highest number of stems per plant was reached under field conditions in 2014 and 2016, whereas the lowest could be observed in pots in 2014. In the ANOVA table of the weight of stems per plant (Table 2) a significant three-factor interaction (Y×M×B) shows up, underlining the outstanding differences in the behavior of the three biotypes as a consequence of the tested experimental factors. In all four trial years, total aerial plant biomass (stems + flowers), either fresh or air-dried, varied significantly according to all experimental factors, and in all cases but 2016, the M×B interaction was also significant (Table 3).

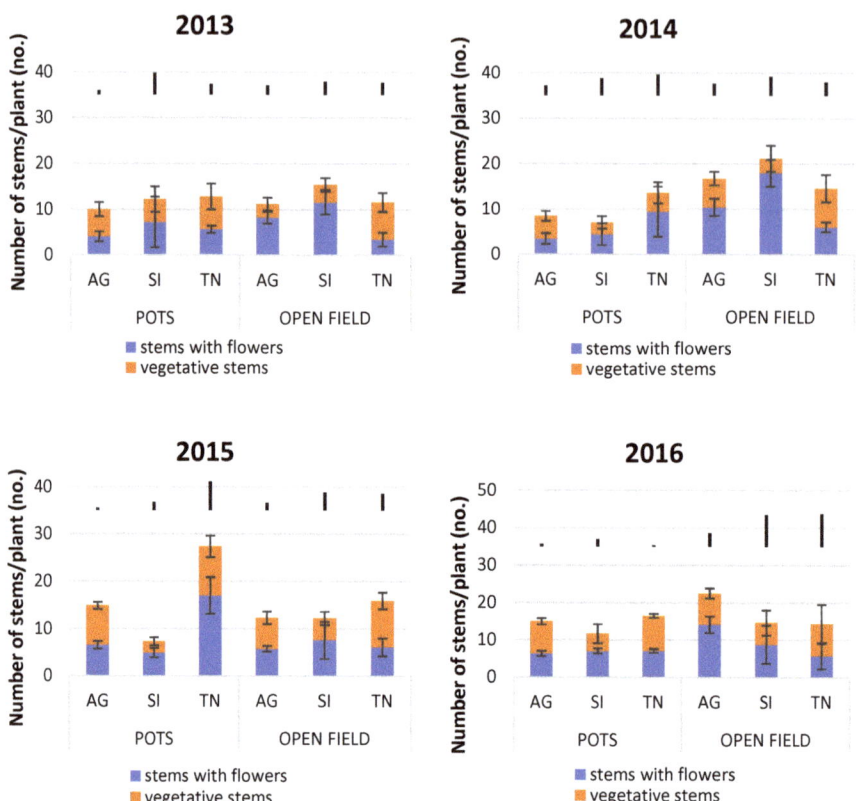

Figure 2. Average number per plant of stems with flowers (blue bars) and vegetative stems (orange) in 3 *H. perforatum* biotypes cultivated from 2014 to 2016 in pots and in open field. Error bars represent the standard deviations of each mean. Vertical bars indicate the standard deviations of the total number of stems (flowered + vegetative).

Table 3. Results of ANOVA (F values) for the fresh (FW) and dry (DW) weight of aerial plant biomass from 2014 to 2016 according to crop management (M) and biotype (B).

Source of Variability	DF	2013		2014		2015		2016	
		FW	DW	FW	DW	FW	DW	FW	DW
Management (M)	1	8.18 **	14.77 ***	1.96 n.s.	4.66 *	19.91 ***	5.94 *	11.12 **	6.58 *
Biotype (B)	2	52.00 ***	59.67 ***	4.84 *	10.65 ***	38.04 ***	28.79 ***	2.01 n.s.	2.31 n.s.
M × B	2	9.03 ***	12.97 ***	13.11 ***	8.05 **	13.25 ***	8.73 ***	<1 n.s.	<1 n.s.
Error	24								
Total	29								

*: significant at $p \leq 0.05$; **: significant at $p \leq 0.01$; ***: significant at $p \leq 0.001$; n.s.: not significant.

These results indicate that the effect of management methods varied in years and, within each year, among biotypes tested. Together with the lower height values (Figure 1), lower number of stems (Figure 2), and weight of plant biomass (Figure 3), accounted for an overall lower plant size in the first year. However, this general trend was differently pronounced according to the biotype and the cultivation management. In terms of plant biomass, the most productive year for plants grown in open field was the second (2014), whereas, an increased plant biomass was observed in the third trial year (2015) in pot cultivation. Higher herbage yields of *Hypericum* in the second year after sowing have been already reported by other authors [28]; however, various response patterns have been recognized between different *H. perforatum* genotypes [29], and this inherent variability may explain the different outcome obtained when the same genotype is cultivated with contrasting cultivation managements.

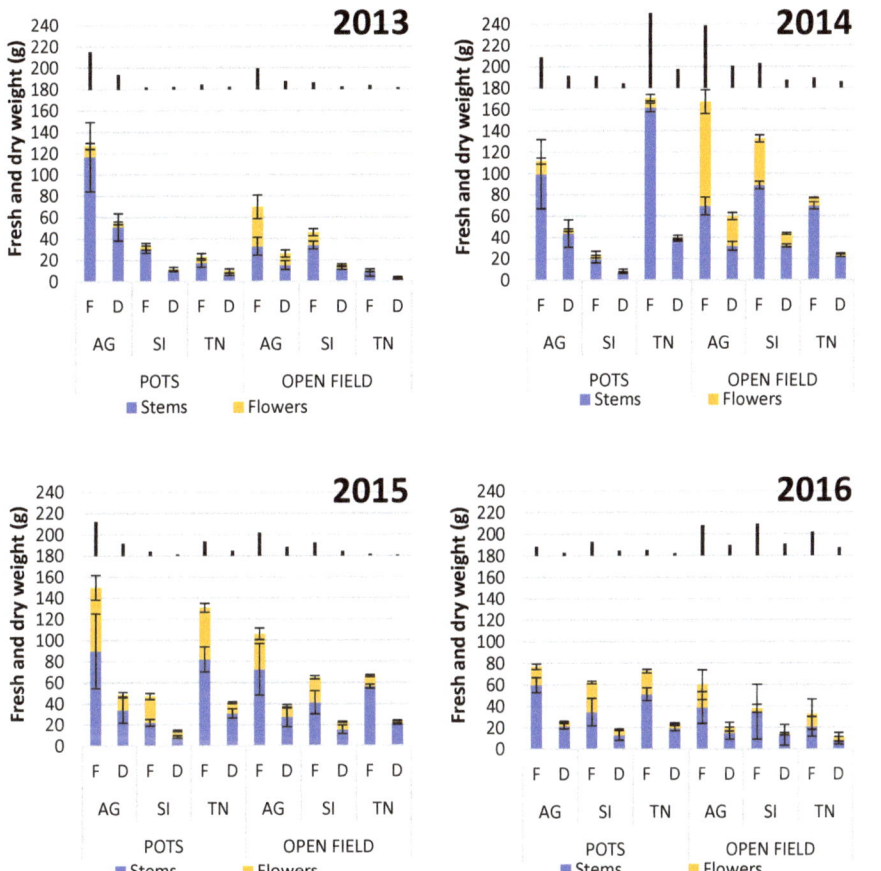

Figure 3. Fresh (F) and dry (D) weight of flowers (yellow) and stems (blue) in 3 *H. perforatum* biotypes cultivated from 2014 to 2016 in two management systems. Error bars represent the standard deviations of each mean. Vertical bars in the upper part of each graph indicate the standard deviations of the total aerial mass (flowers + stems) per plant.

All three biotypes reached the highest value of total aerial biomass in field cultivation, with the only exception of the TN biotype, which, in this respect, found the best cultivation conditions in pot. It must be observed that the TN biotype experienced the highest susceptibility to contrasting cropping conditions, showing both the highest (170.7 g/plant in pots in 2014) and the lowest (8.8 g/plant in open field in 2013) mean values of plant biomass throughout the whole experiment (Figure 3).

In the last trial year (2016), in the majority of experimental conditions—except for the SI biotype cultivated in a pot—plant biomass decreased, achieving values similar to those obtained in the first year. As expected, similarly to the whole aerial biomass, flower yields (Figure 3) were generally lower in the first year and higher in the second; in the remaining two years, flower yields decreased, until reaching values similar to those achieved in the first year. However, relevant exceptions may be found, as assessed by the highly significant MxB interactions in all trial years (Table 4). The highest yields (97.8 g fresh flowers/plant) were obtained in the second trial year (2014) by the AG biotype grown in open field; in most cases, open field conditions allowed highest flower yields in 2014, whereas, when grown in pot, the same biotypes achieved the highest flower yields in the following year.

Table 4. *Hypericum perforatum* L. results of ANOVA (F values) for the fresh and dry mass of flowers according to crop management (M) and biotype (B).

Source of Variability	DF	2013		2014		2015		2016	
		FW	DW	FW	DW	FW	DW	FW	DW
Management (M)	1	24.17 ***	6.42 *	31.26 ***	29.50 ***	104.10 ***	35.69 ***	8.56 **	1.32 n.s.
Biotype (B)	2	34.54 ***	20.89 ***	13.97 ***	13.60 ***	38.74 ***	48.31 ***	<1 n.s.	1.59 n.s.
M × B	2	18.34 ***	12.93 ***	11.56 ***	10.66 ***	25.34 ***	22.30 ***	5.81 **	2.96 n.s.
Error	24								
Total	29								

*: significant at $p \leq 0.05$; **: significant at $p \leq 0.01$; ***: significant at $p \leq 0.001$; n.s.: not significant.

3.2. Phytochemical Composition

The majority of the components in the extracts from *H. perforatum* flowers resulted phloroglucinols (hyperforins) and flavonols (myricetin derivates, rutin, myricitrin, hyperoside, isoquercitrin, quercitrin, and quercetin), adding up from 68% to 84% of total identified phenols. This feature is typical of *H. perforatum*, and allows a rather precise separation of this species from many others, even if taxonomically close [25,30].

The PCA on chemical data, grouped by families of compounds (Table 5; Figure 4), allowed in first instance a sharp partitioning of samples among years. All samples are closely distributed near the first Principal Component (PC), showing some dispersion towards negative values above all in 2014. The first PC allows very easily to discriminate samples collected in 2014 from those of 2015 and, to a lesser extent, 2016. Total phenolics content and phloroglucinols affect the composition of the first PC, whereas flavonols affect the second PC. Hence, total phenolics content and phloroglucinols appear to be a relevant factor in discriminating among years. The retrieved amount of both groups of compounds was indeed much lower in 2014 than in 2015 (Figure 5).

Table 5. Loadings, eigenvalues, and variance (%) accounted for by the 7 components found by PCA.

Loadings	PC 1	PC 2	PC 3	PC 4	PC 5	PC 6	PC 7
Cat	0.54750	0.59633	−0.01455	0.27531	−0.03179	0.51733	1.97×10^{-14}
Phlor	0.98032	−0.19132	−0.04742	−0.00513	−0.00964	−0.00125	3.02×10^{-16}
Napht	0.69665	−0.40432	0.58406	−0.09964	−0.01061	−0.00619	1.7×10^{-15}
Cinn	−0.48407	0.56480	−0.33371	−0.16055	0.55611	−0.01671	4.41×10^{-15}
Flav	−0.11142	0.98737	−0.07198	−0.06860	−0.05277	−0.00416	9.72×10^{-16}
Dim	0.66202	0.63131	0.11850	0.38590	−0.00336	−0.01434	2.23×10^{-15}
Phen	0.98775	0.15402	0.02353	0.00109	0.00800	0.00106	$−2.6 \times 10^{-16}$
Eigenvalues	1029.890	79.784	6.816	1.657	0.860	0.034	2.43×10^{-18}
% variance	92.03	7.13	0.61	0.15	0.08	0.00	2.17×10^{-19}

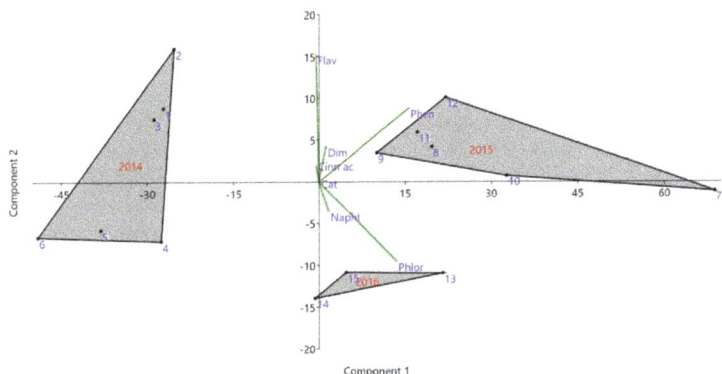

Figure 4. PCA biplot for the major groups of phenolic compounds detected in flowering tops of 3 *H. perforatum* biotypes, grown with two management systems and in three years. N = 15. Explained variance: PC1 = 92.1%; PC2 = 7.1%. Cinn—cinnamic acids; Phlor—phloroglucinols; Dim—dimers; Napht—naphthodianthrones; Cat—catechins; Flav—flavonols; Phen—phenols tot.

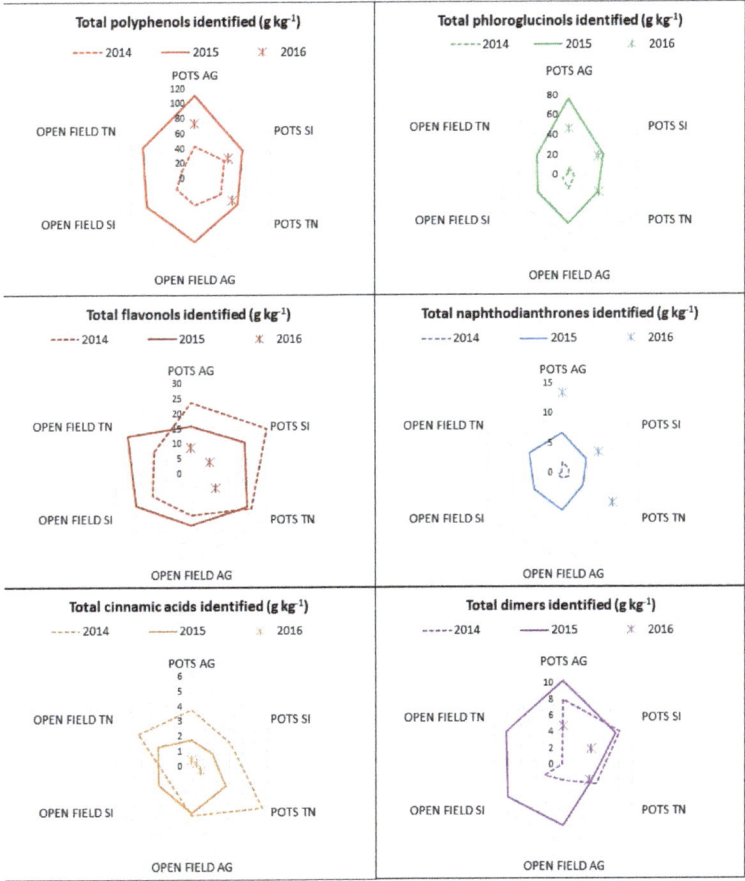

Figure 5. Radar diagrams of the detected content in metabolites, grouped by chemical family, in flowering tops of 3 biotypes of *H. perforatum*. Results from 2014, 2015, and 2016 cultivations in pots and in open field.

In 2015 and 2016, phloroglucinols (35 to 75 g kg^{-1} d.m.) shared more than 50% of total *Hypericum* phenolic content, whereas in 2014 they were less abundant (from 3% to 43%). Hyperforins in plant material mainly take two forms: hyperforin and its homologue adhyperforin. Both compounds are unstable in the presence of light, and are rapidly oxidized [31,32]. According to environmental and cropping conditions, hyperforin content in *H. perforatum* can range between 37–43 g kg^{-1} [9,25], and hyperforin high-yielding and low-yielding genotypes are often recognizable [9]. In this trial, hyperforins production (Figure 6) confirmed to be genotype-dependant, meaning the most high-yielding genotype (the AG biotype) ranked first under all experimental conditions; however, field cultivation seems to stabilize hyperforins yield of all genotypes.

Figure 6. Mean values of hyperforin content (g kg^{-1} d.m.) in flowering tops of 3 biotypes of *H. perforatum* (AG, SI, and TN), obtained in 2014, 2015, and 2016 from cultivations in pots (P) and in open field (F). Within each year, means that do not share a letter are significantly different at $p \leq 0.05$ (Tukey's test). Avg Y—yearly averages; Avg B—averages by biotype; Avg M—averages by cultivation management.

Hypericins (naphthodianthrones) (Figure 7) significantly increased from 2014 to 2016 (on average, from 1.0 to 10.1 g kg^{-1} d.m.), while maintaining rather stable values across biotypes and cultivation management. Within this chemical group, pseudohypericin was always more abundant (from 30% higher, to more than twice) than hypericin. As previously assessed [9], pseudohypericin and hypericin amounts were always linearly associated ($r = 0.82$), consistent with the hypothesis that they originate from the same precursors [33]. On average, the highest hypericins content (hypericin + pseudohypericin + the precursors protohypericin and protopseudohypericin) was measured within the local biotype (AG, 6.5 g kg^{-1} d.m.), which achieved the highest hypericins yield (13.4 g kg^{-1} d.m.) in 2016 and in pots cultivation. However, although a large variability in hypericins content was found, all analyzed samples showed values higher than the threshold value of 0.8‰, i.e., the minimum amount pointed out by the European Pharmacopoeia [7].

Flavonols (Figure 5) were predominant in 2014, whereas in the two following years they ranked second, after phloroglucinols. According to the European Pharmacopoeia [7], flavonols are mainly represented by glycosides of the flavonol quercetin (hyperoside, rutin, isoquercitrin, and quercitrin), and account for 2%–4% of phytochemical components in *Hypericum* herb. Research has showed that this chemical group has a great importance in determining *Hypericum* bioactivity, most importantly with regards to its significant antioxidant and radical-scavenging activity [25,34]. Although flavonols content was slightly higher in plants from open field than in pots (about 18.7 vs. 17.1 g kg^{-1} d.m.), the widest variations in this group of compounds were undoubtedly due to the year of cultivation. In flowers harvested in 2016, they averaged less than 9 g kg^{-1} d.m, compared to the 25.9 and

19.7 g kg^{-1} d.m. obtained, in the same management conditions (pots) in 2014 and 2015, respectively. Hyperoside, rutin, isoquercetin, and quercitrin, listed in decrescent order, were the most represented flavonols in the analyzed samples.

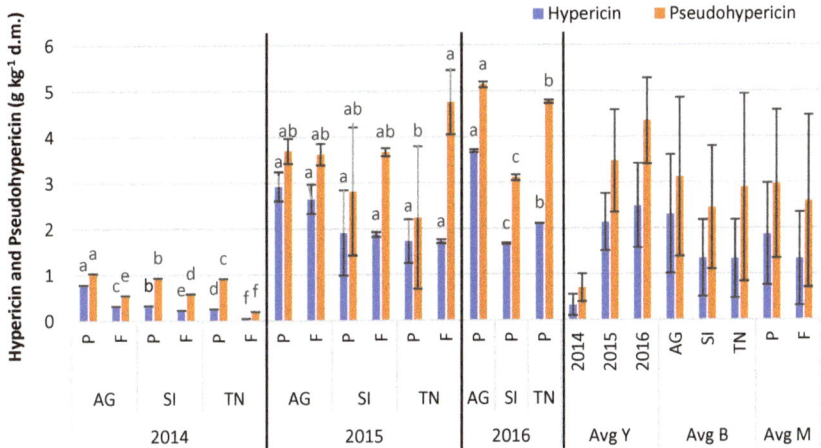

Figure 7. Mean values of hypericin and pseudohypericin content (g kg^{-1} d.m.) in flowering tops of 3 biotypes of *H. perforatum* (AG, SI, and TN), obtained in 2014, 2015, and 2016 from cultivations in pots (P) and in open field (F). Within each year and compound, means that do not share a letter are significantly different at $p \leq 0.05$ (Tukey's test). Avg Y—averages by year; Avg B—averages by biotype; Avg M—averages by cultivation management.

Cinnamic acids (3-O-caffeoylquinic acid, 5-O-caffeoylquinic acid, p-coumaroylquinic acid, and p-coumaric acid), involved in the antioxidant properties of the plant [25], showed a sharp decrease throughout experimental years (on average, 3.6, 2.4, and 0.5 g kg^{-1} d.m. in 2014, 2015, and 2016, respectively) (Figure 5). This decrease was more evident in plants cultivated in pots and less marked in plants derived from open field; in open field, two biotypes (AG and SI) exhibited amounts of cinnamic acids similar (biotype AG) and even higher (biotype SI) in 2015 with respect to 2014.

Although their mechanism of action is not perfectly elucidated yet, apigenin dimers (biapigenin and amentoflavone) are thought to be involved in *H. perforatum* pharmacological activity, being probably associated—together with phloroglucinols and naphthodianthrones—to their overall anxiolytic and antidepressant effect [35]. These compounds (Figure 5) were found in lower amounts in 2014 and 2016 (about 4.1 g kg^{-1} d.m.), whereas in 2015 their total content raised to 7.6 g kg^{-1} d.m. On average, their content was higher in the AG and SI biotypes (6.7 and 6.2 g kg^{-1} d.m., respectively), and lower in the biotype from Northern Italy (TN), averaging 4.4 g kg^{-1} d.m. Biapigenin (5.8 g kg^{-1} d.m in pots and 5.2 in open field) shared the most part of this chemical group, whereas amentoflavone never surpassed the amount of 0.28 g kg^{-1} d.m.; these values agree with previous results obtained on *H. perforatum* [25].

4. Conclusions

The trial evidenced a high variability in both biomass and phytochemical response of *H. perforatum* according to the growth conditions, also including the environmental effects, enclosed in the factor "year". Some general considerations—useful both for further research and to give preliminary indications to farmers in Mediterranean environments—can be driven.

Firstly, it appeared clear that, although classed as a perennial herb [36], cultivated *Hypericum* has a limited duration. In general, this was no longer than 2–3 years—although plants grown in pots seemed more suitable for longer stands—and herbal yields tended to thereafter stabilize on lower levels.

Secondly, noticeable differences showed up between the results obtained by the two management techniques. Under field conditions, *Hypericum* allowed satisfactory yield levels, in terms of total biomass and herbal product (flowers), in the second cultivation year. Contrastingly, cultivations in pots reached their best yield performances in the following growth year, both in the local biotype and in one of the non-native ones. Some differences could also be observed in the behavior of the biotypes. In general, the locally-obtained biotype performed best in field cultivation, whereas the cultivation in pots seemed more capable of meeting the requirements of non-native biotypes.

This last outcome seems to confirm the higher suitability to field cultivation of local populations, that probably have developed in time a better fitness to local environmental conditions. Of course, further research in different environments, adopting locally-selected biotypes, is necessary to confirm this hypothesis.

All biotypes tested showed hypericins levels satisfactory for marketing. However, from the phytochemical point of view, a remarkable variability was observed. A strong variability of chemical composition due to the effect of cultivation year was expected, as it was already assessed in *Hypericum* [37], as well as in many other medicinal and aromatic plants. Therefore, the cultivation of *Hypericum* requires a properly tuned cropping technique, along with a sound choice of the genotype to be cultivated. When cultivation is addressed to the industry, the choice of the most proper genotype is necessary, but this outstanding variability requires accurate post-harvest analyses to check the qualitative features of each production prior to commercialization, in order to verify if the harvested product meets the required industrial standards.

Finally, it must be observed that pots cultivation does not reflect the performance obtainable from field cultivations, often leading to a biased response. That means, a biotype that seems to achieve exceptionally high results in pots, does not necessarily keep this exceptional performance under field conditions, and vice-versa. Hence, studies performed on *Hypericum* in pots are not able to give a definite response on plants performance in open field, making accurate post-harvest analyses necessary.

Supplementary Materials: The following are available online at https://www.mdpi.com/article/10.3390/agriculture11050446/s1, Figure S1: Ten-day values of rainfall (mm) and minimum, maximum, and average temperatures (°C) recorded at Sparacia (Cammarata, AG, Sicily) from November 2012 to July 2016; Table S1: Main characteristics of the soil used for the trial at Sparacia (Cammarata, AG, Sicily).

Author Contributions: Conceptualization, S.L. and A.C.; methodology, S.L. and A.C.; validation, S.L., A.C. and E.N.; formal analysis, A.C.; investigation, S.L. and E.N.; data curation, S.L. and E.N.; writing—original draft preparation, A.C.; writing—review and editing, S.L. and E.N.; visualization, S.L., A.C. and E.N.; supervision, A.C. All authors have read and agreed to the published version of the manuscript.

Funding: This research received no external funding.

Institutional Review Board Statement: Not applicable.

Informed Consent Statement: Not applicable.

Data Availability Statement: The data presented in this study are available on request from the corresponding author.

Conflicts of Interest: The authors declare no conflict of interest.

References

1. Silva, B.A.; Ferreres, F.; Malva, J.O.; Dias, A.C.P. Phytochemical and antioxidant characterization of *Hypericum perforatum* alcoholic extracts. *Food Chem.* **2005**, *90*, 157–167. [CrossRef]
2. Šavikin, K.; Dobrić, S.; Tadić, V.; Zdunić, G. Antiinflammatory activity of ethanol extracts of *Hypericum perforatum* L., *H. barbatum* Jacq., *H. hirsutum* L., *H. richeri* Vill. and *H. androsaemum* L. in rats. *Phytother. Res.* **2007**, *21*, 176–180. [CrossRef]
3. Pu, X.; Liang, J.; Wang, X.; Xu, T.; Hua, L.; Shang, R.; Liu, Y.; Xing, Y. Anti-influenza A virus effect of *Hypericum perforatum* L. extract. *Virol. Sin.* **2009**, *24*, 19. [CrossRef]

4. Cecchini, C.; Cresci, A.; Coman, M.M.; Ricciutelli, M.; Sagratini, G.; Vittori, S.; Lucarini, D.; Maggi, F. Antimicrobial activity of seven *Hypericum* entities from Central Italy. *Planta Med.* **2007**, *73*, 564–566. [CrossRef]
5. Kacerovská, D.; Pizinger, K.; Majer, F.; Šmíd, F. Photodynamic therapy of nonmelanoma skin cancer with topical *Hypericum perforatum* extract—A pilot study. *Photochem. Photobiol.* **2008**, *84*, 779–785. [CrossRef] [PubMed]
6. EMA—European Medicines Agency; Committee on Herbal Medicinal Products (HMPC). Assessment Report on *Hypericum perforatum* L., Herba. 2009. Available online: https://www.ema.europa.eu/en/documents/herbal-report/assessment-report-hypericum-perforatum-l-herba_en.pdf (accessed on 13 February 2021).
7. EMA—European Medicines Agency; Committee on Herbal Medicinal Products (HMPC). Assessment Report on *Hypericum perforatum* L., Herba. 2018. Available online: https://www.ema.europa.eu/en/documents/herbal-report/draft-assessment-report-hypericum-perforatum-l-herba-revision-1_en.pdf (accessed on 13 February 2021).
8. Wölfle, U.; Seelinger, G.; Schempp, C.M. Topical Application of St. John's Wort (*Hypericum perforatum*). *Planta Med.* **2014**, *80*, 109–120. [CrossRef]
9. Lazzara, S.; Carrubba, A.; Napoli, E. Variability of Hypericins and Hyperforin in *Hypericum* Species from the Sicilian Flora. *Chem. Biodivers* **2020**, *17*, e1900596. [CrossRef]
10. ESCOP. *European Scientific Cooperative on Phytotherapy. Hyperici Herba–St. John's Wort. ESCOP Monographs*, 2nd ed. Online Series. 2018. Available online: https://escop.com/downloads/hypericum-2018/ (accessed on 13 February 2021).
11. Bombardelli, E.; Morazzoni, P. *Hypericum perforatum* . *Fitoterapia* **1995**, *66*, 43–68.
12. Vitiello, B. *Hypericum perforatum* extracts as potential antidepressants. *J. Pharm. Pharmacol.* **1999**, *51*, 513–517. [CrossRef] [PubMed]
13. Butterweck, V.; Petereit, F.; Winterhoff, H.; Nahrstedt, A. Solubilized hypericin and pseudohypericin from *Hypericum perforatum* exert antidepressant activity in the forced swimming test. *Planta Med.* **1998**, *64*, 291–294. [CrossRef]
14. Chatterjee, S.S.; Bhattacharya, S.K.; Wonnemann, M.; Singer, A.; Müller, W.E. Hyperforin as a possible antidepressant component of *Hypericum* extracts. *Life Sci.* **1998**, *63*, 499–510. [CrossRef]
15. Butterweck, V. Mechanism of Action of St John's Wort in Depression. What is known? *CNS Drugs* **2003**, *17*, 539–562. [CrossRef] [PubMed]
16. Nahrstedt, A.; Butterweck, V. Lessons learned from herbal medicinal products: The example of St. John's Wort. *J. Nat. Prod* **2010**, *73*, 1015–1021. [CrossRef] [PubMed]
17. Schmidt, M.; Butterweck, V. The mechanisms of action of St. John's wort: An update. *Wien Med. Wochenschr.* **2015**, *165*, 229–235. [CrossRef] [PubMed]
18. Crockett, S.L. Essential oil and volatile components of the genus *Hypericum* (Hypericaceae). *Nat. Prod. Commun.* **2010**, *5*, 1493–1506. [CrossRef]
19. Bruni, R.; Sacchetti, G. Factors affecting polyphenol biosynthesis in wild and field grown St. John's Wort (*Hypericum perforatum* L. Hypericaceae/Guttiferae). *Molecules* **2009**, *14*, 682–725. [CrossRef]
20. Poorter, H.; Bühler, J.; van Dusschoten, D.; Climent, J.; Postma, J. Pot size matters: A meta-analysis of the effects of rooting volume on plant growth. *Funct. Plant Biol.* **2012**, *39*, 839–850. [CrossRef]
21. Cyrak, C. Seed germination protocols for ex situ conservation of some *Hypericum* species from Turkey. *Am. J. Plant Physiol.* **2007**, *2*, 287–294. [CrossRef]
22. Lazzara, S.; Landini, E.; Saia, S.; Fascella, G.; Carrubba, A. Preliminary studies about sexual and vegetative propagation of *Hypericum perforatum* in a Mediterranean environment. In Proceedings of the X Convegno Nazionale sulla Biodiversità, Roma, Italy, 3–5 September 2014; pp. 206–211.
23. Fascella, G.; Airò, M.; Mammano, M.M.; Giardina, G.; Carrubba, A.; Lazzara, S. Rooting and acclimatization of micropropagated *Hypericum perforatum* L. native to Sicily. *Acta Hort.* **2017**, *1155*, 543–548. [CrossRef]
24. NRCS. *Keys to Soil Taxonomy*, 9th ed.; USDA NRCS, Soil Survey Staff: Washington, DC, USA, 2003; 332p.
25. Napoli, E.; Siracusa, L.; Ruberto, G.; Carrubba, A.; Lazzara, S.; Speciale, A.; Cimino, F.; Saija, A.; Cristani, M. Phytochemical profiles, phototoxic and antioxidant properties of eleven *Hypericum* species—A comparative study. *Phytochemistry* **2018**, *152*, 162–173. [CrossRef] [PubMed]
26. Hammer, Ø. PAST 4.03. 2020. Available online: https://www.nhm.uio.no/english/research/infrastructure/past/ (accessed on 13 February 2020).
27. Hammer, Ø.; Harper, D.A.T.; Ryan, P.D. PAST: Paleontological Statistics software package for education and data analysis. *Palaeontol. Electron.* **2001**, *4*, 1–9.
28. Kizil, S.; Inan, M.; Kirici, S. Determination of the best herbage yield and hypericin content of St. John's wort (*Hypericum perforatum* L.) under semi-arid climatic conditions. *Turk. J. Field Crops* **2013**, *18*, 95–100.
29. Pluhár, Z.S.; Bernáth, J.; Neumayer, É. Morphological, production biological and chemical diversity of St. John's Wort (*Hypericum perforatum* L.). *Acta Hort.* **2002**, *576*, 33–40. [CrossRef]
30. Giovino, A.; Carrubba, A.; Lazzara, S.; Napoli, E.; Domina, G. An integrated approach to the study of *Hypericum* occurring in Sicily. *Turk. J. Bot.* **2020**, *44*, 309–321. [CrossRef]
31. Maisenbacher, P.; Kovar, K.A. Analysis and stability of Hyperici oleum. *Planta Med.* **1992**, *58*, 351–354. [CrossRef] [PubMed]
32. Tatsis, E.C.; Boeren, S.; Exarchou, V.; Troganis, A.N.; Vervoort, J.; Gerothanassis, I.P. Identification of the major constituents of *Hypericum perforatum* by LC/SPE/NMR and/or LC/MS. *Phytochemistry* **2007**, *68*, 383–393. [CrossRef] [PubMed]
33. Karioti, A.; Bilia, A.R. Hypericins as potential leads for new therapeutics. *Int. J. Mol. Sci.* **2010**, *11*, 562–594. [CrossRef] [PubMed]

34. Rizzo, P.; Altschmied, L.; Ravindran, B.M.; Rutten, T.; D'Auria, J.C. The biochemical and genetic basis for the biosynthesis of bioactive compounds in *Hypericum perforatum* L., one of the largest medicinal crops in Europe. *Genes* **2020**, *11*, 1210. [CrossRef]
35. Michler, H.; Laakmann, G.; Wagner, H. Development of an LC-MS method for simultaneous quantitation of Amentoflavone and Biapigenin, the minor and major biflavones from *Hypericum perforatum* L., in human plasma and its application to real blood. *Phytochem. Anal.* **2011**, *22*, 42–50. [CrossRef]
36. Robson, N.K.B. Studies in the genus *Hypericum* L. (*Guttiferae*) 4 (2). Section 9. *Hypericum* sensu lato (part 2): Subsection 1. *Hypericum* series 1. *Hypericum. Bull. Br. Mus. Nat. Hist. Bot.* **2002**, *32*, 61–123. [CrossRef]
37. Büter, B.; Orlacchio, C.; Soldati, A.; Berger, K. Significance of genetic and environmental aspects in the field cultivation of *Hypericum perforatum*. *Planta Med.* **1998**, *64*, 431–437. [CrossRef] [PubMed]

Review

Hops (*Humulus lupulus* L.) as a Novel Multipurpose Crop for the Mediterranean Region of Europe: Challenges and Opportunities of Their Cultivation

Francesco Rossini [1], Giuseppe Virga [2], Paolo Loreti [1], Nicolò Iacuzzi [3,*], Roberto Ruggeri [1,*] and Maria Elena Provenzano [1]

[1] Department of Agriculture and Forest Science (DAFNE), University of Tuscia, Via San Camillo de Lellis, 01100 Viterbo, Italy; rossini@unitus.it (F.R.); info@drpagronomia.it (P.L.); provenzano.mariaelena@gmail.com (M.E.P.)
[2] Research Consortium for the Development of Innovative Agro-Environmental Systems (Corissia), Via della Libertà 203, 90143 Palermo, Italy; giuseppe.virga@corissia.it
[3] Department of Agricultural, Food and Forest Sciences, Università Degli Studi di Palermo, Viale delle Scienze 13, Building 4, 90128 Palermo, Italy
* Correspondence: nicolo.iacuzzi@unipa.it (N.I.); r.ruggeri@unitus.it (R.R.); Tel.: +39-0761-357561 (R.R.)

Abstract: The common hop (*Humulus lupulus* L.) is a dioecious perennial climbing plant, mainly known for the use of its female inflorescences (cones or, simply, "hops") in the brewing industry. However, the very first interest towards hops was due to its medicinal properties. Actually, the variety of compounds present in almost all plant parts were (and still are) used to treat or prevent several ailments and metabolic disorders, from insomnia to menopausal symptoms as well as obesity and even cancer. Although hops are predominantly grown for hopping beer, the increasing interest in natural medicine is widening new interesting perspectives for this crop. Moreover, the recent success of the craft beer sector all over the world, made the cultivated hop come out from its traditional growing areas. Particularly, in Europe this resulted in a movement towards southern countries such as Italy, which added itself to the already existing hop industry in Portugal and Spain. In these relatively new environments, a complete knowledge and expertise of hop growing practices is lacking. Overall, while many studies were conducted globally on phytochemistry, bioactivity, and the genetics of hops, results from public research activity on basic hop agronomy are very few and discontinuous as well. The objective of this article is to provide an overview of possible uses, phenology, and agronomic aspects of hops, with specific reference to the difficulties and opportunities this crop is experiencing in the new growing areas, under both conventional and organic farming. The present review aims to fill a void still existing for this topic in the literature and to give directions for farmers that want to face the cultivation of such a challenging crop.

Keywords: hops; *Humulus lupulus* L.; Mediterranean environment; trellising system; medicinal plant; industrial crop; hop shoots; powdery mildew; downy mildew; Japanese beetle

1. Introduction

The common hop (*Humulus lupulus* L.) is a dioecious perennial climbing plant, belonging to the *Cannabaceae* family and native to Northern temperate climates [1,2]. Only female plants are grown, as their inflorescences (strobiles or, more commonly, cones or "hops") are rich in constituents (mainly bitter principles and essential oils) used for both industrial and medicinal purposes. Although wild hops are classified as lianose phanerophytes [3], cultivated hops are forced to behave as hemicryptophytes. In fact, when cones are ripe for harvest, climbing bines are cut at the ground level, thus stimulating the resprouting from the rootstock in the following spring.

In the last ten years, annual hop production and area harvested worldwide have increased by 34% and 18%, respectively [4]. Both those indicators continuously raised from 2012 to 2019, achieving a total production of about 131,000 t and about 66,000 ha harvested, with an average cone yield near to 2 t·ha^{-1}. The top producing countries are the United States of America (a little less than 51,000 t), Germany (48,500 t), and Czech Republic (7150 t). However, the European Union (EU-27) is largely the main global hop producer, accounting for almost 50% of world hop production [4].

Hop cultivation area is limited mainly by the photoperiodic needs of the plant [2]. The lowest latitudes at which a commercial production is feasible are approximately 34–35° in both hemispheres and, even here, there are problems for the too short day length. Due to these considerations, including Ethiopia (which lies between the 3° N and 15° N Latitude) among the main hop producing countries [5], is an incorrect information. Moreover, for optimal growth, yield, and cone quality, hop plants need particular climatic conditions such as exposure to low temperatures during dormancy, mild temperature in spring, sufficient moisture from irrigation or rainfall throughout the season, and dry weather during harvest time [2,6,7]. For that reason, in Europe, hop cultivation historically developed in both continental and temperate regions, characterized by *Dfb* and *Cfb* climate (humid continental climate and oceanic climate, respectively), according to the Köppen–Geiger classification system [8]. Traditional European hop production areas are the Hallertau in Germany, the Žatec region in Czech Republic, Kent in the South East of England, the Lublin region in Eastern Poland, Savinja Valley, Ptuj, and the Koroška region in Slovenia, Alsace in France, the Poperinge area in Belgium [9,10].

The global growing popularity of craft beer in the last decades [11] fostered many brewpub owners and microbrewers to grow their own local hopyard, thus making hop overstep its typical cultivation borders. In Europe, this resulted in a movement towards southern countries such as Italy, which added itself to the already existing hop industry of Portugal and Spain. In these environments, a very limited knowledge about hop growing practices exists, since hop phenology, crop requirements, and hopyard management may be very different from those known and applied in the traditional hop lands. Overall, while many studies were conducted globally on phytochemistry, bioactivity, and genetics of hops ([12], and references therein), results from public research activity on basic hop agronomy are very few and discontinuous as well [13,14].

The objective of this article is to provide an overview of possible uses, phenology, and agronomic aspects of hops, with specific reference to the difficulties and opportunities this crop is experiencing in the new growing areas, under both conventional and organic farming. The review aims to fill a void still existing for this topic in the literature and to give directions for farmers that want to face the cultivation of such a challenging crop.

2. Uses

Even though hops are mainly known as one of the four ingredients of beer, their first use was as a medicinal plant [15]. Almost all plant parts are rich in bioactive compounds such as bitter acids, and flavonoids [16,17], having potent antimicrobial, antioxidant, and antifungal activity. These features coupled with the growing interest in natural health-promoting substances, open new interesting perspectives for hops beyond the beer industry [18,19]. During centuries and until now, from hop cultivation human beings obtained several products such as tea, cordage (using stems), food (cooking the young shoots), and even a preservative for bread [2,20,21]. In the following sections, all ways to use hops will be reviewed, with an emphasis on their health-promoting properties.

2.1. As Medicinal Crop

Since ancient times, hop has been recognized as a medicinal plant [22,23]. Over centuries, hops were used as remedy for leprosy, pulmonary tuberculosis, silicosis, indigestion, ear infection, and many other complaints ([24,25], and reference therein). Starting from the 19th century, hop preparations have always been recommended as a mild sedative useful for relief of mental stress and to aid sleep [26]. The potential sedative effect of hops was first supposed observing the bizarre characteristic known as "hop-picker fatigue". It was noticed that hop-pickers tired easily during their job and this was attributed to inhalation of the volatile oil of the hop plant or even to the transfer of hop resin from hands to mouth during harvesting and processing of cones [27,28]. Even now, pillows filled with hops (alone or together with valerian and lavender) are sold as an efficient remedy for insomnia. Despite this and different published researches suggesting hop sedating properties [29,30], there is still insufficient evidence from clinical studies to confirm the real effectiveness of hops in the treatment of sleeplessness [25]. However, a more recent study, conducted on 17 female nurses, concluded that, thanks to its hop components, a moderate consumption of non-alcoholic beer favors night-time rest [31].

Beside sedative effect, another medicinal property of hops was discovered thanks to hop-pickers: its estrogenic activity [32]. When hops were harvested by hand, menstrual disorders frequently happened among female pickers [33]. Initially, the estrogenic activity was attributed to xanthohumol [34], until a potent phytoestrogen, 8-prenylnaringenin (8-PN), was identified in hops and beer [33,35]. Nowadays, supplements containing hop phytoestrogens are often used for menopausal symptoms, breast enhancement, and bone health [36–38].

In the last 25 years, many studies have been conducted to demonstrate the anti-carcinogenic properties of hop secondary metabolites ([24,25,39], and reference therein). Xanthohumol was the compound mainly investigated for its properties of chemopreventive agent, but also bitter acids and essential oils proved to have an anti-cancer activity ([24,25], and reference therein).

In general, the strong antimicrobial, antioxidant, and antifungal effectiveness of hop extracts is well documented [12,15,40–42]. The bitter acids, particularly humulone and lupulone, are the main agents of the antibacterial and antifungal effects ([25], and reference therein).

Recently, treatment with iso-α-acid extracted from hops was found to soothe or prevent different ailments such as liver steatosis, inflammation, and fibrosis as well as metabolic disorders such as obesity and diabetes ([43,44], and reference therein).

2.2. As Industrial Crop

The use of hops in brewing has more than a thousand years of history, since the first records on hops as a cultivated plant were dated between the 8th and 9th century A.D. [2,45,46]. Hop cultivation started in Central Europe (in the regions of Bohemia, Slovenia, and Bavaria) and, from there, it gradually spread across Europe and, then, in the rest of the world [2]. Currently, hop is almost entirely grown for the brewing industry [47].

Even though hop cones contain a variety of compounds, such as resins, essential oils, proteins, polyphenols, lipids, waxes, cellulose, and amino acids, the brewing value of hops is mainly attributed to hop resins and essential oils. Both these components are synthesized in the glandular trichomes (lupulin glands, Figure 1) of hops [48]. Total hop resins can be divided into soft and hard resins. Soft resins are more important than hard ones in the brewing process and consist of the α-acids, β-acids, and the so-called uncharacterized soft resins. The α-acids are the source of the bittering agents of beer, the iso-α-acids, which also have a remarkable bacteriostatic activity during wort boiling. However, the dependence on pasteurization made the preservative function of the hops irrelevant in the modern and industrial brewing process.

On the basis of their chemical composition, hop varieties are simply classified as "bittering hops" or "aroma hops". Besides that, according to α-acid content, bittering hop cultivars are classified as "bittering hops", "high alpha", and "super high alpha". Similarly, the hop oil composition distinguishes "fine aroma hops" from "aroma hops" [47]. A further distinction was proposed to define the so-called "flavor hops" [47]. This kind of hops are characterized by adding a distinct taste and intense flavor to beer, and can come either from bittering or aroma cultivars [47].

Figure 1. Glandular trichomes from female hop bracts.

For many years, the main role of hops in brewing was to give a bitter taste to the beer, and, often, hops were purchased just for this reason. This is particularly still true for many big brewing companies that are more interested in the α-acid content than in other hops' characteristics. Conversely, current consumers' attitude shows a growing appreciation of flavor and perceived quality of beer [49,50], while less importance is given to bitterness. As a result, in the last decade, many farmers shifted their production to more aroma hops [10]. Unfortunately, many traditional aroma hop varieties are often characterized by low yields and high susceptibility to pathogens, thus making them unappealing for hop growers [51]. For this reason, much of the effort of the hop breeders is devoted to obtaining genotypes having quality traits similar to those of traditional varieties but with improved yield potential and disease resistance.

2.3. Other Uses

Picking and cooking the young and tender hop shoots emerging from rootstock in the spring, was probably the very first way to use hop plants. It seems that the first written document on this habit was the *Naturalis Historia* in which Pliny the Elder (23–79 A.D.) claims that *lupum salictarium* (probably, the young shoots of common hop), and other wild plants, are *eaque verius oblectamenta quam cibos*, namely, "more a pleasure than a food". Besides their culinary appreciation in many European regions (from Belgium to Spain, from England to Italy), young hop shoots, both wild and cultivated, have a very low fat content and were found to be a good source of dietary fiber, proteins, and vitamins C, E, and B9 [52–54]. Additionally, in a recent study carried out in Slovenia by Vidmar et al. (2019) [55], white hop shoots from different cultivars showed a better antioxidant activity than hop cones and leaves.

Since only the female inflorescences have interest in the industrial cultivation of hops, a large quantity of plant residue is left on the field after cone harvest. This leftover material consists almost exclusively of leaves and stems. Similarly to hemp (*Cannabis sativa* L., belonging to the same family of hops), hop stems are rich in bast fibers usable for paper- and rope-making [20]. Thus, applying the principles of circular economy to the hopyard management, dead stems could be efficiently re-used as training strings supporting living stems. However, it is worth mentioning that the retting process of the long, twisted stems of hops is difficult compared to straight stems of flax, hemp, and nettle [20].

Besides the pharmaceutical application, the well-known antimicrobial and antifungal activity of many hop compounds was exploited also for the biocontrol of pests and plant diseases. Specifically, bitter acids and essential oils were found to be effective control agents or repellants against the invasive mosquito *Aedes albopictus* (Skuse) (Diptera Culicidae) and fresh water snail *Physella acuta* (Draparnaud) (Mollusca Physidae) [56], insect pests of stored foods [57], the honeybee mite *Varroa destructor* (Anderson and Truman) [58], and the ascomycete fungus *Zymoseptoria tritici* (Desm.) Quaedvlieg & Crous [15].

Moreover, the efficiency of hop extracts as a natural food preservative was demonstrated for meat, cheese, and bread, thus meeting the increasing demand for products without additional chemicals [21,59,60].

3. Hops in the Mediterranean Basin

Even though hops belong to the spontaneous flora of many European countries surrounding the Mediterranean Basin, its cultivation never widespread significantly in that region. The reasons can be found in the following social, historical, and environmental aspects: (i) lower beer consumption as compared to central and northern Europe; (ii) strong competition with grape and political power of wine producers over the history; (iii) high irrigation requirement during crop cycle in a region well known for its water shortage issues.

3.1. Phenology

Hop phenology is divided into nine principal growth stages (Figure 2): (0) sprouting; (1) leaf development; (2) formation of side shoots; (3) elongation of bines; (5) inflorescence emergence; (6) flowering; (7) development of cones; (8) maturity of cones; (9) senescence and entry into dormancy [61]. A specific heat sum, expressed as growing degree days (GDD), is required for the completion of each phenophase in different crops and varieties. For this reason, studying hop phenology in new growing areas (e.g., southern European environments) can give useful directions for both selection of cultivars and temporal optimization of agronomic practices. To the best of our knowledge, just two studies, both in Italy, described the phenology of hops in the Mediterranean Basin [14,62]. Specifically, Rossini et al. (2016) [14], in a three-year open field study, calculated the GDD accumulation for the onset of each of the four key growing stages (sprouting, flowering, development of cones and maturity of cones) of twenty hop cultivars in Central Italy. They found that 'Cascade', 'Phoenix', and 'Yeoman' were always the earliest cultivars in each phenological stage, while 'Perle', 'EKG', and 'H. Spat' were the latest ones. Marceddu et al. (2020) [62], in their two-year pot experiment, showed the progress of development stages for three hop varieties in the semi-arid environment of Sicily. They calculated GDD accumulation starting from transplanting date in which plants were at the second pair of leaves unfolded (phase 1.2, according to BBCH scale, [61]). They found the hop growing cycle varying from 108 days for 'Chinook' in 2019 to 147 days for 'Nugget' in 2018. The duration of vegetative stages prevailed, in almost all cases, on that of reproductive ones. However, this study did not consider the first hop phenophase: the break of rootstock dormancy and the emergence of shoots from the soil. Conversely, this stage was accurately monitored by Ruggeri et al. (2018) [63] for nine commercial cultivars in a two-year open field study in Central Italy. The authors found that some cultivars such as 'Cascade', 'H. Magnum' and 'Yeoman' were consistently recorded as the earliest sprouting varieties, while 'Perle', 'H. Bitter', and 'Tettnager' as the latest ones. Conversely, 'Fuggle' showed the highest variability between the two years for the breaking of dormancy (79 GDD vs 301 GDD). The difference between the earliest and latest cultivar ('Cascade' and 'Perle', respectively) was about 111 GDD (20 days) in 2013 and 208 GDD (31 days) in 2014. The high variability in shoot emergence date between years led the authors to conclude that further studies should be conducted to find the model that best predicts the break of dormancy. Overall, recording the date of shoot emergence (growth stage 09) in Central Italy (42°26' N, 12°04' E, altitude 310 m a.s.l.) over five years, we observed the majority of cultivars emerging from the soil between the last decade of March and the first decade of April (Table 1).

Since hop is a short-day plant, the transition between vegetative and reproductive growth happens when day length drops below the critical threshold of 15–16 h [2,64].

The timing of flowering can significantly affect cone yield, because a complete vegetative growth (about 5–6 m height) prior to flower differentiation is pivotal to obtain high yields. Day length depends on latitude, and latitude varies approximately from 36° and 45° N in the Mediterranean regions of Europe. In that range, the longest day length of the year varies from 15 h to 16 h, thus allowing hop plants to reach an adequate vegetative growth before flowering. Observing the phenology of hops in Central Italy (42°26′ N, 12°04′ E, altitude 310 m a.s.l.) for many years and cultivars, we recorded flowering phase approximately from mid-June to mid-July; development of cones, from the end of June to the end of July; cones ripe for picking, from the last decade of August to the last decade of September.

Figure 2. Some phenological growth stages of hop: (**A**) shoot emergence; (**B**) leaf development; (**C**) flowering; (**D**) development of cones; (**E**) maturity of cones.

Table 1. Dates of shoot emergence recorded over five years for different hop commercial cultivars in Central Italy.

Cultivar	2013	2014	2015	2017	2018
Cascade	25 March	28 March	19 March	18 March	27 March
Challenger	22 March	2 April	31 March	16 March	26 March
Columbus	22 March	28 March	19 March	21 March	26 March
Fuggle	29 March	18 April	30 March	4 April	30 March
H. Aroma	8 April	14 April	27 March	22 March	26 March
H. Bitter	12 April	23 April	30 March	18 March	30 March
H. Magnum	29 March	31 March	23 March	18 March	26 March
H. Mittlefruh	6 April	10 April	23 March	16 March	29 March
HNB	15 April	23 April	31 March	28 March	28 March
H. Spat	6 April	23 April	31 March	24 March	30 March
H. Taurus	3 April	10 April	2 April	4 April	29 March
H. Tradition	3 April	24 April	31 March	4 April	26 March
Omega	8 April	14 April	27 March	22 March	27 March
Perle	15 April	29 April	31 March	29 March	6 April

3.2. Cultivation

Although many amateur hop growers consider hops as an easy-to-grow plant, managing a commercial-scale hopyard is markedly more challenging. Professional hop cultivation requires a well-designed structure, well-timed pruning, careful bine training, and the proper application of all the other agronomic practices to succeed in obtaining high yields and production quality.

3.2.1. Trellising System

Commercial hop production requires a trellising system to support plants. The conventional high trellising system consists of a permanent structure of poles and wire to which strings are attached each year to provide support for the hop bines [2]. Poles can be made of treated wood, iron, steel, or concrete. Pre-stressed concrete poles are much more expensive than wood ones, but they better withstand mechanical stress. Growers should be informed that the trellis is an engineered structure subject to heavy loads, including the plant (heavier when wet) and wind. The top height of such structures can vary, but usually ranges from 5 m to 8 m. The most common training system used for hop production worldwide is the V-trellis (Figure 3), and the same is also used in the Mediterranean region [14,65].

Figure 3. V-trellis system.

Since the standard high trellising system is very expensive and labor-intensive, alternative solutions have been searched over years. Low trellis is probably the most famous growing system proposed as a substitute of the typical high trellis. Overall, the low trellising system consists in growing hops on netting fastened to 3/4 m tall poles.

Generally, hops growing on low trellis yield less compared to those thriving on high trellis [66]. However, the benefits related to plant health (easier and more effective control of pests and diseases, [67]), harvest efficiency, worker safety, and cost for construction and maintenance, make low trellis very interesting for small-scale producers and worthy of deeper investigations in different environments and with suitable cultivars (e.g., dwarf or low-cone setting genotypes). Currently, no studies comparing low and high trellising systems, under Mediterranean climatic conditions, have been published. Additionally, our experience suggests that, probably, the most used 6 m high trellis is not the best system to train the hops under climatic conditions of Italian peninsula. In fact, we observed many hop bines grow taller than the top wire and develop horizontally more than required.

This behavior creates serious problems during harvest for bine intertwining and during cultivation for plant shading.

3.2.2. Cultivars and Propagation

According to the International Hop Growers' Convention (IHGC), commercial hop varieties produced in 2020 by 21 member countries were 291 (excluding experimental selections and no-named records): 69 labelled as bitter hops and 222 as aroma hops [68]. The most cultivated hop varieties in the top producing countries are 'Citra® HBC 394', 'Mosaic® HBC 369', 'Perle', 'Hall. Tradition', and 'Saaz' for aroma hops and 'Herkules', 'Hall. Magnum', 'Pahto® HBC 682', and 'CTZ' (stands for 'Columbus', 'Tomahawk', and 'Zeus') for bitter hops. As for the Mediterranean Basin, 'Nugget' is largely the predominant cultivar in Spain (90% of the total harvested area), and probably in Portugal too. In Italy, there are no published statistics but, presumably, 'Cascade' is the most planted cultivar for both its good adaptability to a wide range of pedoclimatic conditions and its common use in many brewing recipes. Other varieties that performed well under different experimental conditions were 'Hall. Magnum', 'Chinook', and 'Yeoman' [14,62,63]. Understandably, the agronomic data available from open field trials in such a new growing country are very few and collected on a limited number of plants (from three to five per replicate). Thus, there is an urgent need for more reliable data collected from commercial hopyards and in different environmental conditions. At the same time, screening of germplasm from wild hop populations is gaining interest for breeding programs having as main target the creation of different local varieties [69,70].

Vegetative propagation is the common way to multiply hops because it is both a quite simple method and a fast tool to obtain a clone of the mother plant. For these reasons, seeds are used just in breeding programs for the improvement of genetic variability. Rhizomes collected from mature rootstocks and one-node green cuttings from stems are the most commonly used materials for propagation (Figure 4). Rhizome cuttings can be planted directly into the soil or potted and then transplanted from 6 to 12 months later. While the limited availability of suitable rhizomes hampers the adoption of this method, the use of green cuttings allows a faster and more effective multiplication of hop plants.

Tissue culture is another tool to obtain large amounts of clonal plants in a short period [71]. Micropropagation was used in hops mainly for virus eradication [72], cryopreservation [73], and mass shoot production [74]. Very recently, trials conducted in Southern Italy evidenced that micropropagation protocol adopted for hops, although successful for many other crops, did not perfectly succeed in obtaining healthy and vigorous plants [62].

3.2.3. Soil Preparation and Planting Techniques

Although hops can thrive on a wide range of soils, those loose and well drained have to be preferred when an adequate water supply and manuring is provided [2]. Light-textured soils also facilitate cultivation and management operations in the hopyard during both winter and summertime. Hops can grow well also in clay soils but raised beds have to be realized and an accurate surface and subsurface drainage system has to be designed prior to planting hills [65]. In fact, while hops naturally thrive along riverside and moist riparian zones, they do not tolerate waterlogged soils. Even though the majority of nutrient-collecting roots develop in the top 0–30 cm of soils, the root system of hops extends far down over the years (from 1 m up to 2.25 m, [75]) and, thus, takes advantage from deep soils. Plowing and amending soil is a common practice before planting a new hopyard. Deep cultivation, or ripping, is suggested on compacted soils, as this will improve drainage and root penetration.

Figure 4. Vegetative propagation of hops: (**A**) rhizomes; (**B**) rooted cutting.

Potted plants are transplanted on the top of raised beds in spring or autumn, following different planting designs. Generally, plant spacing varies according to cultivar (plant vigor and cone setting), trellising system, location and available machinery [76,77]. The most used planting designs in hop producing countries consist of 2.5 m to 4 m row spacing and from 1 to 2 m plant spacing, accounting for 1700 to 4000 plants per hectare. Even though, plant spacing is one of the most important agro-technical issues in designing a hopyard [76], very few studies have been conducted on this aspect so far and, to the best of our knowledge, no one in the Mediterranean Basin.

3.2.4. Pruning, Training, and Stripping

Pruning is the spring removal of the upper part of rootstock, also known as "crown" (Figure 5). Generally, pruning is the first spring agrotechnical operation, aiming to remove shoots and crown buds. While multiple ways of mechanical and chemical pruning exist, the reasons to apply this agronomic practice are essentially two: (i) to reduce downy and powdery mildew inoculum [78] and (ii) to promote a more uniform regrowth from rootstock [63,77]. It was suggested that, depending on timing and severity of pruning, growers could eliminate one fungicide application per year [78]. Specifically, a four-week delayed pruning can significantly reduce powdery mildew incidence on leaves and cones [79]. Moreover, combining the mechanical and chemical intervention generally leads to a more thorough effect compared with the application of a single method [78,79]. In new growing areas, no herbicide is currently registered for this purpose, as such, mechanical pruning is the only option available.

Figure 5. Hop pruning: (**A**) not pruned rootstock; (**B**) pruned rootstock.

Once secondary shoots regrow from "crowned" rootstock and reach the height of 50–60 cm, it is time to train them. Training means wrapping hop bines in a clockwise direction around each string to promote vertical growth. As a rule, from two to four bines are trained up each of the supporting strings, as this results in the optimal growth for high yield and good quality [2]. Established rootstocks produce much more shoots than are needed [63] and, after training is completed, it is necessary to remove the exceeding. Surplus shoots can be collected and sold as vegetables, like asparagus spears (see Section 2.3). Indeed, leaving the excess shoots at the base of the plants would favor the development of mildew diseases and would weaken the vertical growth of trained bines.

Choosing the appropriate training date is pivotal for obtaining maximum hop yields. Moving away from the optimal date, production could be reduced by nearly 40% [80]. Understandably, the training date depends on the pruning date, with training occurring about 3–4 weeks later. While in traditional hop producing zones, the long experience with this crop led to the acquisition of a practical "know-when", pruning and training in new growing areas are mostly driven by growers' invention.

Leaf stripping (removing lower leaves and laterals) is another common management practice applied during the vegetative phase of hops. The aim is to increase air flow in the basal part of the plant, thus limiting the spread of downy mildew. Mechanical stripping can start when bines are at least 3 m high, while chemical defoliation is usually done when bines reach the top wire [2]. In the new growing areas, such as Italy, because of the absence of registered herbicides and thanks to the limited dimensions of the existing hopyards, stripping is carried out by manually removing the lower leaves up to 0.90–1 m from the ground. Very interesting is the way of stripping hopyards in New Zealand, the top producer of organic hops. In this country, sheep are often used to graze the hop gardens with the following benefits: direct control of weed flora, sucker pruning, lower bine defoliation and manure supply. Hopefully, this strategy could be exported in those areas of the Mediterranean Basin where growers are interested in organic hop production and a sheep farming system already exists.

3.2.5. Irrigation and Fertilization

Under Mediterranean climatic conditions, irrigation is unavoidable to have a marketable hop production. In Southern European environments, severe drought and heat often occur when hops are in phenological phases particularly sensitive to such stresses, i.e., flowering, cone formation and early ripening. Cone yield was found to significantly increase with crop transpiration (r^2 = 0.92) in Spain, with hop transpiration representing 92% of actual evapotranspiration during the mid-season [81]. In the same study, yield obtained in the rainfed plots was 37% lower than in the irrigated ones. The picture of hops as water intensive crop clashes with water scarcity that typically afflicts Mediterranean countries. Despite this, limited studies were conducted to investigate (i) the effect of deficit irrigation on hop yield and quality, (ii) the drought tolerance in hop cultivars, and (iii) the management of irrigation systems [82,83]. Deep knowledge on these aspects will be crucial for successful hop cultivation in a global climate change scenario [7,84].

The most used irrigation systems in the Mediterranean hopyards are (i) surface irrigation system by flooding the space between rows [65] and (ii) drip irrigation along rows [14,63,81]. The latter system is more expensive than the former one but it is more water efficient. Moreover, a drip irrigation system can allow targeted application of soluble or liquid fertilizers by fertigation and reduces the incidence of fungal diseases. Recently, Nakawuka et al., 2017 [82], reported that the use of a subsurface drip irrigation system can substantially reduce water use in hop production even before deficit irrigation is considered.

Hops do not thrive under acidic soil conditions and liming should be carried out when necessary to prevent the pH from falling below 6.5 [2]. Additionally, extreme pH values alter the absorption of many micronutrients. For example, when pH is acid, manganese (Mn) levels can become toxic in hop tissues [65], while zinc (Zn) deficiencies are often associated with alcaline soils (above 7.5, [6]).

Hop nutrient requirements vary depending on soil testing, cultivars, yield potential, and growing region. Nutrient application guidelines have yet to be developed for hop production in new growing zones. Thus, indications published in other countries currently are the best available reference. As for nitrogen (N), 160 kg·ha^{-1} (split in two applications) were recommended in Slovenia [85], 168 kg·ha^{-1} in the USA [86], and 135 kg·ha^{-1} in England [2]. In the Mediterranean basin, Afonso et al. (2020) [65] reported 150 kg·ha^{-1} of N (split in three applications) in Portugal, while 100 kg·ha^{-1} of N (split in two applications) was used in Italy [14].

The need for phosphorus (P) fertilization is often overestimated. Hops typically have low P requirements, with suggested P amount to be applied ranging from 67 to 110 kg·ha^{-1}, when soil P concentration into the soil is very low (0–30 ppm, [87]).

Hops take up 90–170 kg·ha^{-1} of potash (K), but 75% can return to the soil with vegetative plant parts [87]. With high soil contents or heavy applications of farmyard manure, no additional phosphate or potash is generally considered necessary [2].

With the increasing interest towards organic hops, alternative ways to improve soil fertility have been proposed such as the use of leguminous cover crops, compost, slurry, farmyard manure, and other soil amendments [2,6,88]. However, meeting crop nutrient demands (especially for N) remains a challenging task when using organic fertilization [89,90].

Fertilizer placement is also important when considering plant nutrient needs. Broadcasting fertilizer makes the nutrients more available for weeds and can encourage their growth and competition with crops [91]. Thus, the best way to feed hop plants is the localized applications of fertilizers or, even better, the use of fertigation systems. Brant et al. (2020) [75], studying the distribution of root system in hop plants, suggested to place the fertilizers at 50 cm far from the center of the rootstock in both sides.

Besides having an annual soil testing report, tissue analysis is a recommended method to determine plant nutrient needs and detect potential deficiencies [92]. Hop petioles from mature leaves should be collected during growing season and analyzed for macro- and micronutrient content. If results show deficiencies, foliar fertilization or fertigation must be applied to restore a correct plant nutrition.

Finally, it is worth mentioning that, despite the growing interest towards the application of plant biostimulants in other crops, studies conducted on this topic are very few for hops and no one in the Mediterranean Basin. In a recent study, Prochazka et al. (2018) [93] found that complex formulations of biologically active substances, such as fulvic and humic acids and phytohormones, appear to be highly effective for the optimization of hop production performance in Czech Republic.

3.2.6. Disease, Pest, and Weed Control

Our ten-year experience in hopyard management, under the Mediterranean climatic conditions of Italy, underscored that mildew pathogens, aphids, and spider mites were the constant and most problematic diseases and pests that can really compromise the hop production. It should be noted that in Northern Italy, where maize is largely cultivated, another pest that can damage hops is the corn borer (*Ostrinia nubilalis* Hübner). The restricted availability of plant protection products registered for hops severely limits the strategies for pest and disease control in both organic and conventional farming. Thus, hop protection management must focus on the proper agronomic practices, preventive intervention and biostimulants for induction of plant resistance [93]. These products determine a higher resistance of plant tissues (e.g., vaterite) and a higher synthesis of phytoalexins. Additionally, they are not dangerous for health and do not leave toxic residues into the soil. Hereafter, the main diseases and pests detected in our surveys will be shortly described. Lastly, some thoughts about weed control strategies in hopyard will be briefly discussed.

Downy mildew Caused by Pseudoperonospora humuli (Miyabe & Takah.) G. W. Wilson

Downy mildew can overwinter in infected hop rootstocks in the form of mycelium ([2], and references therein). Contaminated buds originate primary infected shoots (known as "basal spikes" due to their resemblance to a wheat spike) with a characteristic stunted form and pale down curled leaves [2,78]. These spikes are the primary source of infection each year. In our experience, this first infection can be limited by pruning. Infections of buds and leaves can develop at any time during the growing season, depending markedly upon weather conditions and the amount of inoculum being released [2]. This secondary infection can be controlled with copper-based products, when the hops flower, the developing burr, and cones may also become infected. Diseased cones turn brown and tend to break with mechanical harvesting.

Powdery mildew Caused by Podosphaera macularis (Wallr.) U. Braun & S. Takam.

Disease symptoms appear as powdery white colonies on leaves, buds, stems, and cones. Raised blisters can be visible when plant grows rapidly. In the same way of downy mildew, primary infection starts from diseased shoots (termed as "flag shoots") and then the pathogen spreads on young leaves, moving up the bine in sync with plant growth [78,79]. Disease development is favored by rapid plant growth, mild temperatures, high humidity, and cloudy weather [2]. Burrs and young cones are very susceptible. We have never recorded severe attacks of powdery mildew in Italy. However, where infections have been detected, sulfur-based products have well controlled the pathogen. It should be noted that recently, in Italy, orange oil-based products have been registered for the control of powdery mildew in hops, but we have not had any experience about their effectiveness.

Spider Mite (Tetranychus urticae C.L. Koch)

Together with aphids, spider mite (also known as two-spotted mite) is the key pest of hops in the Northern hemisphere [2]. In small numbers, the mites do not cause much damage but, under favorable conditions, they can multiply very rapidly and become a serious problem [2]. Since spider mites thrive in hot and dry environments, it is clear that it could represent a serious problem under Mediterranean climatic conditions. Personal observations on open-field trials suggested that *Beauveria bassiana*-based products and mineral white oil-based products were effective in controlling these pests. Ground cover mixes have been shown to host mite predators, thus regulating their presence [67].

Damson-hop aphid (Phorodon humuli Schrank)

This pest usually causes both direct damage, by feeding sap and debilitating the plants and an even worse indirect damage, by transmitting viruses. The main problem arises after cone formation, because sooty molds easily develop on the honeydew produced by the aphids, thus compromising the quality and commercial value of the hops [2]. Personal observations on open-field trials suggested that pyrethrum-based products and *Beauveria bassiana*-based products were effective in controlling these pests. Enhancing habitat to favor the presence of beneficial predators gave contrasting results in the reduction of aphid population [67].

The biotic stresses described until now are managed, both organically and conventionally, in all countries where hops are traditionally grown. In these countries, different options are available for hop growers to protect their crops. Understandably, in new growing countries such as Italy, products registered for use on hops are very few. In 2020, copper-based and sulfur-based products were temporarily allowed as an exception for the control of fungal diseases. A *Bacillus thuringensis*-based product was also registered for use on hops against the corn borer. Currently, there is nothing else as authorized product for hop protection.

Viroids

Even though we did not survey any stunted hop plant infected with viroids in our field trials, we touch this topic because viroids represent serious damaging pathogens in the neighboring hop producing countries such as Slovenia [94]. Hop latent viroid (HLVd) is the most diffuse viroid of hops and has been recently detected in many commercial cultivars sampled along the Italian peninsula [95]. Hop stunt viroid (HSVd), Apple fruit crinkle viroid (AFCVd), and Citrus bark cracking viroid (CBCVd) are the other viroids infecting hop plants around the world [96]. Among them, CBCVd seems to be the most aggressive type, causing yellowing, leaf malformation, bine cracking, and plant stunting after a post-inoculation period of only 14-months [94,97]. Since infected plants often show no visible symptoms (e.g., plants infected by HLVd), the vegetative propagation of hops easily diffuses the pathogens. The use of virus/viroid-free planting material is the most effective and obvious strategy to limit the spread of this plant disease.

Japanese beetle (Popillia japonica Newman)

Popillia japonica Newman (Coleoptera: Rutelidae) is a scarab beetle native to Japan and established in North America, the Azores and, more recently, in Northern Italy [98]. Japanese beetle is a highly polyphagous species that is listed in Annex IAII of the Directive 2000/29/EC. In our experience, we have been able to observe the damage caused by this pest in a hopyard located near Milan (Northern Italy). The presence of adults on hop plants was recorded in the period July–August 2019, and it was deleterious (Figure 6). Japanese beetle fed on the hop leaves, significantly reducing the photosynthetic activity of the plants. It seemed they were not interested in feeding on the flowers, but this was only an observation. Since the Japanese beetle is an alien species for our territory, the control of this pest is rather complicated. Solutions could involve agronomic practices applied to the meadows where the larvae overwinter, and the registration of pesticides (for organic and conventional farming) before this insect becomes a real emergency.

Figure 6. Japanese beetles (*Popillia japonica* Newman) feeding on hops.

Weed Control

Hop demand for nutrients and water is very high in the period of intensive growth, and weeds can greatly compete for these resources [99]. Obviously, competition for light is temporarily limited to the phenological phase of shoot emergence. Thus, weed management in dormant–early spring season is necessary to maintain a weed-free strip along the rootstocks [77]. Moreover, weed presence can interfere with some field operations such as spraying, training, and harvesting, thus reducing both the efficacy and efficiency of these agronomic practices. Mechanical cultivation, chemical control, and mulching are the most common ways to control weeds in hop production [6,13]. Generally, in conventionally managed hopyards, weed control involves mechanical tillage between rows and in-rows herbicide applications [99]. Conversely, organic hop growers rely mostly on a combination of tillage practices, the use of cover crops, mulching, and grazing sheeps [13,100]. We hypothesize that, in small-scale Mediterranean hopyards, scratching poultry also might be effective. Cover-cropping the inter-rows may be a useful way to control weeds in organic hopyards as well as to attract beneficial insects and enable machinery to drive in the field sooner after precipitations [13]. Ideal cover-crop species should be native, with low requirements and low competitive ability. Studies targeted to identify cover crops suitable for hopyard management in the Mediterranean environment are lacking [101].

Paradoxically, in new growing areas, such as Italy, where registered herbicides for use on hops does not exist yet, conventional and organic weed management coincide. Identifying the weed species and their life cycles are the crucial aspects for a successful weed management plan. In our experience, the most common and problematic hop weeds in Italy are the summer annuals like *Amaranthus retroflexus* L., *Solanum nigrum* L.,

Chenopodium album L., and the perennials such as *Cynodon dactylon* (L.) Pers., *Sorghum halepense* (L.) Pers., *Calystegia sepium* (L.) R. Br., and *Convolvulus arvensis* L.

Findings on weed management in the Mediterranean hopyards were not published so far.

3.2.7. Harvesting, Yield, and Post-Harvest Processing

Mature cones must be harvested, cleaned, and dried in the shortest time possible, to ensure optimum quality and storability. Depending on maturity timelines of cultivars, a typical hop harvest period can be approximately one month [77]. In Southern European environments, hop maturity generally occurs during the months of August and September [63,65].

The hop harvest takes place in two phases: cutting of the whole plant in the field (harvesting) and the separation of the cones from the bines (picking). For both operations, different types of machinery are used.

The global average yield for hops was 2.12 t·ha^{-1} in 2019, with some national yield reported in Table 2 [68]. Other yield reports for Southern Europe come from the very few open field studies conducted in Italy and Portugal [14,63,65]. In Italy, cone yield largely varied with cultivar and plant age, from about 30 g to 700 g per plant [63], while in Portugal, cultivar Nugget yielded from 75 g to 1193 g per plant [65]. However, it must be noticed that, in these first experimental studies, few plants were sampled per plot (from 3 to 6). Further studies with increased plot dimensions are needed to have a more realistic quantification of hop yield potential [85].

Moisture content of freshly harvested hops must be lowered from nearly 80% to 8–12%, to ensure optimum quality and storability [2]. There are different types of ovens used in the various traditional cultivation areas: tower, tunnel or trench. All systems are based on the circulation of hot and dry air (which can be conveyed to the hops mass from above or below) and on the extraction of cold and humid air. The air can be heated with different methods (gas, coal, petroleum derivatives, electricity, or solar). Generally, the process should not exceed 10–12 h with a maximum temperature of the hops mass of 45–50 °C.

Table 2. Average cone yield for some hop producing countries in 2019.

Country	2019 Hop Yield (t·ha^{-1})
Czech Republic	1.43
Germany	2.40
Poland	2.14
Slovenia	1.61
Spain	1.61
UK-England	1.77
USA	2.15

Finally, the dried hops need to be packed in food-grade plastic bags by means of presses dedicated to packaging hops. These bales can have a weight of 40–50 kg and it is good to store them in a cold room at 4–5 °C to avoid oxidation of the mass and the deterioration of the product.

All these machinery and processing systems have industrial dimensions, not economically sustainable for the small-scale production. For these reasons, most of the small Italian farms and companies have resorted to (i) the purchase of used machinery from Central European markets or (ii) the creation of smaller prototypes suitable for the small-scale production. Both solutions have pros and cons. Used machines are very cheap, they offer a certain reliability, even if, in most cases they are more than thirty years old. Conversely, they have the serious issue of not meeting with workplace safety standards. The second approach allows to obtain high-performance machines for the small-scale production (i.e., small, easily transportable, and multifunctional machines), fully compliant with safety regulations, but very expensive. This can only be justified for farms that do not only produce hops but also make beer and sold it directly on the market.

A separate discussion is warranted for the last post-harvest operation: the pelletization. To minimize deterioration, many brewers either keep their hops in cold store or have them processed into pellets [2]. A process that retains most of the character of the hop cones is to grind them to a powder, which is then compressed into pellets. Packing the pellets in vacuum or inert gas ensures they have a long shelf life with limited losses in brewing value [2]. Hop cones have a density varying between 130 kg·m^{-3} and 150 kg·m^{-3}. The greater density of pellets, around 500 kg·m^{-3}, reduces the transport volume and enables long storage stability through quality-preservation packing in a protective atmosphere. The pellets, known commercially as type 90 pellets, obtain their name from the ratio of approx. 90 kg of pellets obtained from 100 kg of hops. The lupulin-enriched hop pellets, also known as type 45 pellets, are obtained through mechanical enrichment of the lupulin gland when the hop is frozen ($-35\,°C$) [102].

The Italian market almost exclusively requires pelleted hops. In fact, modern micro- and industrial breweries are designed for the use of hop pellets. It is therefore evident that the Italian hop producers are equipping themselves to supply pelleted hops to national brewers. To obtain pellets from their hop cones, Italian producers send their hop cones to foreign pelletization centers or, with small machinery, pelletize directly in their farms.

While pellet has innumerable advantages in terms of storage space and shelf life, using hop cones can provide beer with a panorama of wider and more intense taste descriptors [14]. We recently verified that, with a relatively simple modification to the production systems, also the hop cones could be safely used in the homebrewing process.

The Italian hop supply chain need one or more modern and efficient pelletizing centers serving the market of industrial hops. Conversely, for small-scale hop growers, who produce their own craft beer, probably would be better to equip their breweries for the use of self-produced hop cones, in order to better transfer the "terroir" of their hop fields to their beers.

4. Quality Aspects and Sustainable Production

The brewing value of hops is mainly determined by the α-acid content and composition of essential oil. Results from research aiming to characterize quality in hops cultivated under Mediterranean climatic conditions are much more numerous than those from agronomic studies [70,103–105]. This is because the craft beer sector is mainly interested to assess the "terroir" that each growing environment can impress to hops and, consequently, to beer [105]. This approach is prone to leading to a half-truth, since the chemical composition of hops is not only affected by environmental conditions but also by variety, harvesting time, pest and disease control, and other agronomic practices [2,7,106–110]. Anyway, all authors consistently showed that quality attributes and sensory features of hops grown under Mediterranean environment are similar or even better than that coming from traditional growing zones [14,70,103].

Quality in crop production is often linked to the farming system [111]. In particular, many consumers believe that organic foods are healthier and taste better, in addition to being environmental-friendly [112]. A recent survey showed that, even if the average American consumer is not willing to pay a premium price for certified organic beer, buyers who regularly drink craft beers are willing to pay more for organic beers [113]. The authors suggested that microbreweries have an additional edge on the organic beer market, because craft beer consumers are more commonly looking for new flavors and styles of beers to try. In Italy, organic certification was found to be the fourth (on thirteen) most important beer attribute for craft beer drinkers, following taste, fermentation process, and color [114]. Analyzing beer consumer perceptions and preferences along ten years, in a just published systematic review, the authors suggested that consumer perceptions of new styles of beer and speciality beers (such as sustainable or organic beer) require additional attention from scholars and market analysts [115]. The authors also argued that these aspects might represent factors generating differentiation strategies in the beer market for the future.

Whereas growing hop organically is an option for the large majority of hop growers, it is practically the only way to manage hopyards in new growing areas such as Italy. Currently, since no chemicals are yet registered for the protection of hops in Italy, the only discriminating factor between organic and conventional hops is the use of chemical fertilizers. Our experience suggests that growing hop under organic farming conditions is feasible in Italy, but it requires a deep knowledge of all agronomic strategies that can either limit pathogen, pest and weed pressure, or increase soil fertility without resort to chemicals. Moreover, plant requirements in different phenological phases must be precisely known to timely apply the needed agronomic practices. Currently, finding this expertise is extremely rare in small-scale Italian hop production system, since hopyard managers are often brewers or pub owners and not professional farmers. This implies that an extensive and specialized consulting system must be created to help farmers that want to face this new challenging crop. However, to accomplish this goal, adequate public and private funding must be provided for the acquisition of reliable scientific data. Otherwise, the development of hop cultivation in the Mediterranean Basin will ever remain in its early stage.

5. Concluding Remarks and Future Perspectives

The growing popularity of craft beer and the increasing interest in natural medicine is widening appealing scenarios for hops, even in the new growing areas of Southern Europe. However, structural constraints and absence of registered products for plant protection heavily hinder the progress of cultivation in these emerging zones. One of the biggest challenges for new hop growers is finding harvesting and processing equipment that is affordable and scaled appropriately for their activity [77].

Currently, the major knowledge gaps are at the beginning of the hop supply chain. Although some information we have about the adaptation to the Mediterranean pedo-climatic conditions of different commercial hop cultivars, the effects that agronomic practices and trellising systems have on yield and quality of hops are mostly unknown. For this reason, reliable field data urgently need to optimize hop cultivation, thus avoiding the stagnation of such a promising crop. This means that public investments should increase and support research activity and hop growers in overcoming the abovementioned limitations. The aim should be to produce a high quality hops not only for the craft beer sector but also for the pharmaceutical and food industry.

Author Contributions: Conceptualization, F.R., P.L. and R.R.; methodology, F.R., R.R. and N.I.; validation of bibliography, G.V., N.I. and M.E.P.; writing—original draft preparation, R.R. and P.L.; writing—review and editing, F.R., M.E.P., G.V. and N.I.; visualization, P.L., M.E.P., G.V. and N.I.; supervision, F.R.; project administration, F.R. and R.R.; preparation of all figures, P.L. All authors have read and agreed to the published version of the manuscript.

Funding: This research received no external funding.

Institutional Review Board Statement: Not applicable.

Informed Consent Statement: Not applicable.

Acknowledgments: The authors wish to acknowledge Ljiljana Kuzmanović for providing the Figure 1.

Conflicts of Interest: The authors declare no conflict of interest.

References

1. Small, E. A numerical and nomenclatural analysis of morpho-geographic taxa of Humulus. *Syst. Bot.* **1978**, *3*, 37–76. [CrossRef]
2. Neve, R.A. *Hops*; Chapman & Hal: London, UK, 1991; ISBN 9789401053754.
3. Pignatti, S. *Flora d'Italia*; Edagricole-New Business Media: Devon, UK, 2017.
4. FAOSTAT. Available online: http://www.fao.org/faostat/en/#data/QC/visualize (accessed on 2 April 2021).

5. Carbone, K.; Macchioni, V.; Petrella, G.; Cicero, D.O. Exploring the potential of microwaves and ultrasounds in the green extraction of bioactive compounds from *Humulus lupulus* for the food and pharmaceutical industry. *Ind. Crops Prod.* **2020**, *156*, 112888. [CrossRef]
6. Sirrine, R.J.; Rothwell, N.; Lizotte, E.; Goldy, R.; Marquie, S.; Brown-Rytlewski, D.E. Sustainable hop production in the Great Lakes region. In *Michigan State University Extension Bulletin E-3083; Michigan State University Extension: January 2010*. Available online: https://www.uvm.edu/sites/default/files/media/Sirrine-Sustainable-Hop-Production-in-the-Great-Lakes-Region.pdf (accessed on 23 May 2021).
7. Mozny, M.; Tolasz, R.; Nekovar, J.; Sparks, T.; Trnka, M.; Zalud, Z. The impact of climate change on the yield and quality of Saaz hops in the Czech Republic. *Agric. For. Meteorol.* **2009**, *149*, 913–919. [CrossRef]
8. Beck, H.E.; Zimmermann, N.E.; McVicar, T.R.; Vergopolan, N.; Berg, A.; Wood, E.F. Present and future köppen-geiger climate classification maps at 1-km resolution. *Sci. Data* **2018**, *5*, 180214. [CrossRef] [PubMed]
9. Pavlovic, M. Production character of the eu hop industry. *Bulg. J. Agric. Sci.* **2012**, *18*, 233–239.
10. Šrédl, K.; Prášilová, M.; Svoboda, R.; Severová, L. Hop production in the Czech Republic and its international aspects. *Heliyon* **2020**, *6*, e04371. [CrossRef]
11. Pokrivčák, J.; Supeková, S.C.; Lančarič, D.; Savov, R.; Tóth, M.; Vašina, R. Development of beer industry and craft beer expansion. *J. Food Nutr. Res.* **2019**, *58*, 63–74.
12. Bocquet, L.; Sahpaz, S.; Hilbert, J.L.; Rambaud, C.; Rivière, C. *Humulus lupulus* L., a very popular beer ingredient and medicinal plant: Overview of its phytochemistry, its bioactivity, and its biotechnology. *Phytochem. Rev.* **2018**, *17*, 1047–1090. [CrossRef]
13. Turner, S.F.; Benedict, C.A.; Darby, H.; Hoagland, L.A.; Simonson, P.; Robert Sirrine, J.; Murphy, K.M. Challenges and opportunities for organic hop production in the United States. *Agron. J.* **2011**, *103*, 1645–1654. [CrossRef]
14. Rossini, F.; Loreti, P.; Provenzano, M.E.; De Santis, D.; Ruggeri, R. Agronomic performance and beer quality assessment of twenty hop cultivars grown in central Italy. *Ital. J. Agron.* **2016**, *11*, 180–187. [CrossRef]
15. Bocquet, L.; Rivière, C.; Dermont, C.; Samaillie, J.; Hilbert, J.L.; Halama, P.; Siah, A.; Sahpaz, S. Antifungal activity of hop extracts and compounds against the wheat pathogen Zymoseptoria tritici. *Ind. Crops Prod.* **2018**, *122*, 290–297. [CrossRef]
16. Abram, V.; Čeh, B.; Vidmar, M.; Hercezi, M.; Lazić, N.; Bucik, V.; Možina, S.S.; Košir, I.J.; Kač, M.; Demšar, L.; et al. A comparison of antioxidant and antimicrobial activity between hop leaves and hop cones. *Ind. Crops Prod.* **2015**, *64*, 124–134. [CrossRef]
17. Alonso-Esteban, J.I.; Pinela, J.; Barros, L.; Ćirić, A.; Soković, M.; Calhelha, R.C.; Torija-Isasa, E.; de Cortes Sánchez-Mata, M.; Ferreira, I.C.F.R. Phenolic composition and antioxidant, antimicrobial and cytotoxic properties of hop (*Humulus lupulus* L.) seeds. *Ind. Crops Prod.* **2019**, *134*, 154–159. [CrossRef]
18. Astray, G.; Gullón, P.; Gullón, B.; Munekata, P.E.S.; Lorenzo, J.M. *Humulus lupulus* L. as a natural source of functional biomolecules. *Appl. Sci.* **2020**, *10*, 5074. [CrossRef]
19. Srečec, S.; Zechner-Krpan, V.; Petravič-Tominac, V.; Košir, I.; Čerenak, A. Importance of medical effects of xanthohumol, hop (*Humulus lupulus* L.) bioflavonoid in restructuring of world hop industry. *Agric. Conspec. Sci.* **2012**, *77*, 61–67.
20. Lukešová, H.; Andersen, H.L.; Kolínová, M.; Holst, B. Is It Hop? Identifying hop fibres in a european historical context. *Archaeometry* **2019**, *61*, 494–505. [CrossRef]
21. Nionelli, L.; Pontonio, E.; Gobbetti, M.; Rizzello, C.G. Use of hop extract as antifungal ingredient for bread making and selection of autochthonous resistant starters for sourdough fermentation. *Int. J. Food Microbiol.* **2018**, *266*, 173–182. [CrossRef]
22. Koetter, U.; Biendl, M. Hops (*Humulus lupulus*): A review of its historic and medicinal uses. *HerbalGram* **2010**, *87*, 46–59.
23. Dostálek, P.; Karabín, M.; Jelínek, L. Hop phytochemicals and their potential role in metabolic syndrome prevention and therapy. *Molecules* **2017**, *22*, 1761. [CrossRef]
24. Karabín, M.; Hudcová, T.; Jelínek, L.; Dostálek, P. Biologically active compounds from hops and prospects for their use. *Compr. Rev. Food Sci. Food Saf.* **2016**, *15*, 542–567. [CrossRef]
25. Zanoli, P.; Zavatti, M. Pharmacognostic and pharmacological profile of *Humulus lupulus* L. *J. Ethnopharmacol.* **2008**, *116*, 383–396. [CrossRef] [PubMed]
26. EMA. Hop Strobile. Available online: http://www.ema.europa.eu/docs/en_GB/document_library/Herbal_-_Summary_of_assessment_report_for_the_public/2016/10/WC500213787.pdf (accessed on 18 April 2017).
27. Tyler, V.E. *The New Honest Herbal. A Sensible Guide to Herbs and Related Remedies*; G.F. Stickley Co.: Philadelphia, PA, USA, 1987.
28. Schulz, V.; Hänsel, R.; Blumenthal, M.; Tyler, V.E. *Rational Phytotherapy A Reference Guide for Physicians and Pharmacists*; Springer: Berlin/Heidelberg, Germany, 2004.
29. Franco, L.; Sánchez, C.; Bravo, R.; Rodriguez, A.; Barriga, C.; Juánez, J.C. The sedative effects of hops (*Humulus lupulus*), a component of beer, on the activity/rest rhythm. *Acta Physiol. Hung.* **2012**, *99*, 133–139. [CrossRef]
30. Schiller, H.; Forster, A.; Vonhoff, C.; Hegger, M.; Biller, A.; Winterhoff, H. Sedating effects of *Humulus lupulus* L. extracts. *Phytomedicine* **2006**, *13*, 535–541. [CrossRef] [PubMed]
31. Franco, L.; Sánchez, C.; Bravo, R.; Rodríguez, A.B.; Barriga, C.; Romero, E.; Cubero, J. The sedative effect of non-alcoholic beer in healthy female nurses. *PLoS ONE* **2012**, *7*, e37490. [CrossRef]
32. Chadwick, L.R.; Pauli, G.F.; Farnsworth, N.R. The pharmacognosy of *Humulus lupulus* L. (hops) with an emphasis on estrogenic properties. *Phytomedicine* **2006**, *13*, 119–131. [CrossRef]
33. Milligan, S.; Kalita, J.C.; Heyerick, A.; Rong, H.; De Cooman, L.; De Keukeleire, D. Identification of a potent phytoestrogen in hops (*Humulus lupulus* L.) and beer. *J. Clin. Endocrinol. Metab.* **1999**, *83*, 2249–2252. [CrossRef] [PubMed]

34. Verzele, M. Centenary review—100 years of hop chemistry and its relevance to brewing. *J. Inst. Brew.* **1986**, *92*, 32–48. [CrossRef]
35. Milligan, S.; Kalita, J.C.; Pocock, V.; Van de Kauter, V.; Stevens, J.F.; Deinzer, M.L.; Rong, H.; De Keukeleire, D. The endocrine activities of 8-prenylnaringenin and related hop (*Humulus lupulus* L.) flavonoids. *J. Clin. Endocrinol. Metab.* **2000**, *85*, 4912–4915. [CrossRef]
36. Karabin, M.; Hudcova, T.; Jelinek, L.; Dostalek, P. Biotransformations and biological activities of hop flavonoids. *Biotechnol. Adv.* **2015**, *33*, 1063–1090. [CrossRef] [PubMed]
37. Aichinger, G.; Beisl, J.; Marko, D. The hop polyphenols xanthohumol and 8-prenyl-naringenin antagonize the estrogenic effects of fusarium mycotoxins in human endometrial cancer cells. *Front. Nutr.* **2018**, *5*, 85. [CrossRef]
38. Tronina, T.; Popłonski, J.; Bartmanska, A. Flavonoids as phytoestrogenic components of hops and beer. *Molecules* **2020**, *25*, 4201. [CrossRef]
39. Wang, S.; Dunlap, T.L.; Howell, C.E.; Mbachu, O.C.; Rue, E.A.; Phansalkar, R.; Chen, S.N.; Pauli, G.F.; Dietz, B.M.; Bolton, J.L. Hop (*Humulus lupulus* L.) extract and 6-prenylnaringenin induce P450 1A1 catalyzed estrogen 2-hydroxylation. *Chem. Res. Toxicol.* **2016**, *29*, 1142–1150. [CrossRef]
40. Bocquet, L.; Sahpaz, S.; Bonneau, N.; Beaufay, C.; Mahieux, S.; Samaillie, J.; Roumy, V.; Jacquin, J.; Bordage, S.; Hennebelle, T.; et al. Phenolic compounds from *Humulus lupulus* as natural antimicrobial products: New weapons in the fight against methicillin resistant *Staphylococcus aureus*, *Leishmania mexicana* and *Trypanosoma brucei* strains. *Molecules* **2019**, *24*, 1024. [CrossRef]
41. Arruda, T.R.; Pinheiro, P.F.; Silva, P.I.; Bernardes, P.C. A new perspective of a well-recognized raw material: Phenolic content, antioxidant and antimicrobial activities and α- and β-acids profile of Brazilian hop (*Humulus lupulus* L.) extracts. *LWT Food Sci. Technol.* **2021**, *141*, 110905. [CrossRef]
42. Kontek, B.; Jedrejek, D.; Oleszek, W.; Olas, B. Antiradical and antioxidant activity in vitro of hops-derived extracts rich in bitter acids and xanthohumol. *Ind. Crops Prod.* **2021**, *161*, 113208. [CrossRef]
43. Hege, M.; Jung, F.; Sellmann, C.; Jin, C.; Ziegenhardt, D.; Hellerbrand, C.; Bergheim, I. An iso-α-acid-rich extract from hops (*Humulus lupulus*) attenuates acute alcohol-induced liver steatosis in mice. *Nutrition* **2018**, *45*, 68–75. [CrossRef]
44. Ponticelli, M.; Russo, D.; Faraone, I.; Sinisgalli, C.; Labanca, F.; Lela, L.; Milella, L. The promising ability of *Humulus lupulus* L. Iso-α-acids vs. diabetes, inflammation, and metabolic syndrome: A systematic review. *Molecules* **2021**, *26*, 954. [CrossRef]
45. Edwardson, J.R. Hops: Their botany, history, production and utilization. *Econ. Bot.* **1952**, *6*, 160–175. [CrossRef]
46. Kopp, P.A. The global hop: An agricultural overview of the brewer's gold. In *The Geography of Beer*; Patterson, M.W., Pullen, N.H., Eds.; Springer: Dordrecht, The Netherlands, 2014; pp. 77–88. ISBN 978-94-007-7786-6.
47. Almaguer, C.; Schönberger, C.; Gastl, M.; Arendt, E.K.; Becker, T. *Humulus lupulus*—A story that begs to be told. A review. *J. Inst. Brew.* **2014**, *120*, 289–314. [CrossRef]
48. Patzak, J.; Krofta, K.; Henychová, A.; Nesvadba, V. Number and size of lupulin glands, glandular trichomes of hop (*Humulus lupulus* L.), play a key role in contents of bitter acids and polyphenols in hop cone. *Int. J. Food Sci. Technol.* **2015**, *50*, 1864–1872. [CrossRef]
49. Aquilani, B.; Laureti, T.; Poponi, S.; Secondi, L. Beer choice and consumption determinants when craft beers are tasted: An exploratory study of consumer preferences. *Food Qual. Prefer.* **2015**, *41*, 214–224. [CrossRef]
50. Calvo-Porral, C.; Rivaroli, S.; Orosa-Gonzalez, J. How consumer involvement influences beer flavour preferences. *Int. J. Wine Bus. Res.* **2020**, *32*, 537–554. [CrossRef]
51. Haunold, A.; Nickerson, G. Factors Affecting Hop Production, Hop Quality, and Brewer Preference. Available online: https://www.morebeer.com/articles/factors_affecting_hop_production (accessed on 22 February 2021).
52. Sanchez-Mata, M.C.; Cabrera Loera, R.D.; Morales, P.; Fernandez-Ruiz, V.; Camara, M.; Diez Marqués, C.; Pardo-de-Santayana, M.; Tardio, J. Wild vegetables of the Mediterranean area as valuable sources of bioactive compounds. *Genet. Resour. Crop Evol.* **2012**, *59*, 431–443. [CrossRef]
53. Tardío, J.; de Cortes Sánchez-Mata, M.; Morales, R.; Molina, M.; García-Herrera, P.; Morales, P.; Díez-Marqués, C.; Fernández-Ruiz, V.; Cámara, M.; Pardo-de-Santayana, M.; et al. Ethnobotanical and food composition monographs of selected mediterranean wild edible plants. In *Mediterranean Wild Edible Plants—Ethnobotany and Food Composition Tables*; de Cortes Sánchez-Mata, M., Tardío, J., Eds.; Springer: New York, NY, USA, 2016; p. 478.
54. Rossini, F.; Virga, G.; Loreti, P.; Provenzano, M.E.; Danieli, P.P.; Ruggeri, R. Beyond beer: Hop shoot production and nutritional composition under mediterranean climatic conditions. *Agronomy* **2020**, *10*, 1547. [CrossRef]
55. Vidmar, M.; Čeh, B.; Demšar, L.; Ulrih, N.P. White hop shoot production in Slovenia: Total phenolic, microelement and pesticide residue content in five commercial cultivars. *Food Technol. Biotechnol.* **2019**, *57*, 525–534. [CrossRef]
56. Bedini, S.; Flamini, G.; Cosci, F.; Ascrizzi, R.; Benelli, G.; Conti, B. *Cannabis* sativa and *Humulus lupulus* essential oils as novel control tools against the invasive mosquito *Aedes albopictus* and fresh water snail *Physella acuta*. *Ind. Crops Prod.* **2016**, *85*, 318–323. [CrossRef]
57. Bedini, S.; Flamini, G.; Girardi, J.; Cosci, F.; Conti, B. Not just for beer: Evaluation of spent hops (*Humulus lupulus* L.) as a source of eco-friendly repellents for insect pests of stored foods. *J. Pest. Sci.* **2015**, *88*, 583–592. [CrossRef]
58. Iglesias, A.; Mitton, G.; Szawarski, N.; Cooley, H.; Ramos, F.; Meroi Arcerito, F.; Brasesco, C.; Ramirez, C.; Gende, L.; Eguaras, M.; et al. Essential oils from *Humulus lupulus* as novel control agents against *Varroa destructor*. *Ind. Crops Prod.* **2020**, *158*, 113043. [CrossRef]

59. Kramer, B.; Thielmann, J.; Hickisch, A.; Muranyi, P.; Wunderlich, J.; Hauser, C. Antimicrobial activity of hop extracts against foodborne pathogens for meat applications. *J. Appl. Microbiol.* **2015**, *118*, 648–657. [CrossRef]
60. Larson, A.E.; Yu, R.R.Y.; Lee, O.A.; Price, S.; Haas, G.J.; Johnson, E.A. Antimicrobial activity of hop extracts against Listeria monocytogenes in media and in food. *Int. J. Food Microbiol.* **1996**, *33*, 195–207. [CrossRef]
61. Rossbauer, G.; Buhr, L.; Hack, H.; Hauptmann, S.; Klose, R.; Meier, U.; Stauss, R.; Weber, E. Phänologische Entwicklungsstadien von Kultur-Hopfen (*Humulus lupulus* L.). *Nachrichtenbl. Deut. Pflanzenschutzd.* **1995**, *47*, 249–253.
62. Marceddu, R.; Carrubba, A.; Sarno, M. Cultivation trials of hop (*Humulus lupulus* L.) in semi-arid environments. *Heliyon* **2020**, *6*, e05114. [CrossRef]
63. Ruggeri, R.; Loreti, P.; Rossini, F. Exploring the potential of hop as a dual purpose crop in the Mediterranean environment: Shoot and cone yield from nine commercial cultivars. *Eur. J. Agron.* **2018**, *93*, 11–17. [CrossRef]
64. Krebs, C. Hops: A viable alternative crop for the central/southern plains? *Crop. Soils* **2019**, *52*, 4–6. [CrossRef]
65. Afonso, S.; Arrobas, M.; Rodrigues, M.Â. Soil and plant analyses to diagnose hop fields irregular growth. *J. Soil Sci. Plant Nutr.* **2020**, *20*, 1999–2013. [CrossRef]
66. Jezek, J.; Krivánek, J.; Pokorný, J. Trials with growing hops on low trellis in the czech republic in 2009–2011. *Acta Hortic.* **2013**, *1010*, 199–203. [CrossRef]
67. Campbell, C.A.M. Influence of companion planting on damson hop aphid *Phorodon humuli*, two spotted spider mite *Tetranychus urticae*, and their antagonists in low trellis hops. *Crop Prot.* **2018**, *114*, 23–31. [CrossRef]
68. IHGC (International Hop Growers' Convention). Available online: http://www.hmelj-giz.si/ihgc/obj.htm (accessed on 2 April 2021).
69. Mongelli, A.; Rodolfi, M.; Ganino, T.; Marieschi, M.; Dall'Asta, C.; Bruni, R. Italian hop germplasm: Characterization of wild *Humulus lupulus* L. genotypes from Northern Italy by means of phytochemical, morphological traits and multivariate data analysis. *Ind. Crops Prod.* **2015**, *70*, 16–27. [CrossRef]
70. Mongelli, A.; Rodolfi, M.; Ganino, T.; Marieschi, M.; Caligiani, A.; Dall'Asta, C.; Bruni, R. Are *Humulus lupulus* L. ecotypes and cultivars suitable for the cultivation of aromatic hop in Italy? A phytochemical approach. *Ind. Crops Prod.* **2016**, *83*, 693–700. [CrossRef]
71. Peredo, E.L.; Arroyo-García, R.; Revilla, M.Á. Epigenetic changes detected in micropropagated hop plants. *J. Plant Physiol.* **2009**, *166*, 1101–1111. [CrossRef]
72. Postman, J.D.; DeNoma, J.S.; Reed, B.M. Detection and elimination of viruses in USDA hop (*Humulus lupulus*) germoplasm collection. *Acta Hortic.* **2005**, *668*, 143–148. [CrossRef]
73. Reed, B.M.; Okut, N.; D'Achino, J.; Narver, L.; DeNoma, J. Cold storage and cryopreservation of hops (*Humulus* L.) shoot cultures through application of standard protocols. *Cryo Lett.* **2003**, *24*, 389–396.
74. Gurriarán, M.J.; Revilla, M.A.; Tamés, R.S. Adventitious shoot regeneration in cultures of *Humulus lupulus* L. (hop) cvs. Brewers gold and nugget. *Plant Cell Rep.* **1999**, *18*, 1007–1011. [CrossRef]
75. Brant, V.; Krofta, K.; Kroulík, M.; Zábranský, P.; Procházka, P.; Pokorný, J. Distribution of root system of hop plants in hop gardens with regular rows cultivation. *Plant Soil Environ.* **2020**, *66*, 317–326. [CrossRef]
76. Kořen, J. Influence of plantation row spacing on quality and yield of hops. *Plant Soil Environ.* **2007**, *53*, 276–282. [CrossRef]
77. Dodds, K. *Hops—A Guide for New Growers*; NSW Department of Primary Industries: Orange, NSW, Australia, 2017; ISBN 978-1-76058-007-0.
78. Gent, D.H.; Nelson, M.E.; Grove, G.G.; Mahaffee, W.F.; Turechek, W.W.; Woods, J.L. Association of spring pruning practices with severity of powdery mildew and downy mildew on hop. *Plant Dis.* **2012**, *96*, 1343–1351. [CrossRef]
79. Probst, C.; Nelson, M.E.; Grove, G.G.; Twomey, M.C.; Gent, D.H. Hop powdery mildew control through alteration of spring pruning practices. *Plant Dis.* **2016**, *100*, 1599–1605. [CrossRef]
80. Rybacek, V. *Hop Production*, 1st ed.; Elsevier Science: Amsterdam, The Netherlands, 1991; ISBN 9780444987709.
81. Fandino, M.; Olmedo, J.L.; Martinez, E.M.; Valladares, J.; Paredes, P.; Rey, B.J.; Mota, M.; Cancela, J.J.; Pereira, L.S. Assessing and modelling water use and the partition of evapotranspiration of irrigated hop (*Humulus lupulus*), and relations of transpiration with hops yield and alpha-acids. *Ind. Crops Prod.* **2015**, *77*, 204–217. [CrossRef]
82. Nakawuka, P.; Peters, T.R.; Kenny, S.; Walsh, D. Effect of deficit irrigation on yield quantity and quality, water productivity and economic returns of four cultivars of hops in the Yakima Valley, Washington State. *Ind. Crops Prod.* **2017**, *98*, 82–92. [CrossRef]
83. Gloser, V.; Baláz, M.; Jupa, R.; Korovetska, H.; Svoboda, P. The response of *Humulus lupulus* to drought: The contribution of structural and functional plant traits. *Acta Hortic.* **2013**, *1010*, 149–154. [CrossRef]
84. Potopová, V.; Lhotka, O.; Možný, M.; Musiolková, M. Vulnerability of hop-yields due to compound drought and heat events over European key-hop regions. *Int. J. Climatol.* **2021**, *41*, E2136–E2158. [CrossRef]
85. Bavec, F.; Čeh Brežnik, B.; Brežnik, M. Hop yield evaluation depending on experimental plot area under different nitrogen management. *Plant Soil Environ.* **2003**, *49*, 163–167. [CrossRef]
86. Senske, A.M. *Optimization of N Fertilization for Hops (Humulus lupulus) in Iowa Soils*; Iowa State University: Ames, IA, USA, 2020.
87. Gingrich, C.; Hart, J.; Christensen, N. Hops. In *Fertilizer Guide 79*; OSU-Extension Service: Corvallis, OR, USA, 2000.
88. Čeh, B. Impact of slurry on the hop (*Humulus lupulus* L.) yield, its quality and n-min content of the soil. *Plant Soil Environ.* **2014**, *60*, 267–273. [CrossRef]

89. Pang, X.P.; Letey, J. Organic farming challenge of timing nitrogen availability to crop nitrogen requirements. *Soil Sci. Soc. Am. J.* **2000**, *64*, 247–253. [CrossRef]
90. Rossini, F.; Provenzano, M.E.; Sestili, F.; Ruggeri, R. Synergistic effect of sulfur and nitrogen in the organic and mineral fertilization of durum wheat: Grain yield and quality traits in the Mediterranean environment. *Agronomy* **2018**, *8*, 189. [CrossRef]
91. Kirkland, K.J.; Beckie, H.J. Contribution of nitrogen fertilizer placement to weed management in spring wheat (*Triticum aestivum*). *Weed Technol.* **1998**, *12*, 507–514. [CrossRef]
92. Sirrine, R. The Importance of Testing Hop Fertility. Available online: https://www.canr.msu.edu/news/the_importance_of_testing_hop_fertility (accessed on 2 April 2021).
93. Procházka, P.; Štranc, P.; Pazderů, K.; Vostřel, J.; Řehoř, J. Use of biologically active substances in hops. *Plant Soil Environ.* **2018**, *64*, 626–632. [CrossRef]
94. Jakse, J.; Radisek, S.; Pokorn, T.; Matousek, J.; Javornik, B. Deep-sequencing revealed Citrus bark cracking viroid (CBCVd) as a highly aggressive pathogen on hop. *Plant Pathol.* **2015**, *64*, 831–842. [CrossRef]
95. Gargani, E.; Ferretti, L.; Faggioli, F.; Haegi, A.; Luigi, M.; Landi, S.; Simoni, S.; Benvenuti, C.; Guidi, S.; Simoncini, S.; et al. A survey on pests and diseases of Italian hop crops. *Italus Hortus* **2018**, *24*, 1–17. [CrossRef]
96. Matoušek, J.; Siglová, K.; Jakše, J.; Radišek, S.; Brass, J.R.J.; Tsushima, T.; Guček, T.; Duraisamy, G.S.; Sano, T.; Steger, G. Propagation and some physiological effects of Citrus bark cracking viroid and Apple fruit crinkle viroid in multiple infected hop (*Humulus lupulus* L.). *J. Plant Physiol.* **2017**, *213*, 166–177. [CrossRef] [PubMed]
97. Štajner, N.; Radišek, S.; Mishra, A.K.; Nath, V.S.; Matoušek, J.; Jakše, J. Evaluation of disease severity and global transcriptome response induced by Citrus bark cracking viroid, Hop latent viroid, and their co-infection in hop (*Humulus lupulus* L.). *Int. J. Mol. Sci.* **2019**, *20*, 3154. [CrossRef]
98. European and Mediterranean Plant Protection Organization. PM 9/21(1) *Popillia japonica*: Procedures for official control. *EPPO Bull.* **2016**, *46*, 543–555. [CrossRef]
99. Lipecki, J.; Berbeć, S. Soil management in perennial crops: Orchards and hop gardens. *Soil Tillage Res.* **1997**, *43*, 169–184. [CrossRef]
100. Delahunty, K.M.; Johnston, J.C. Strategies for weed management in organic hops, a perennial crop. *Agron. J.* **2015**, *107*, 634–640. [CrossRef]
101. Scarici, E.; Ruggeri, R.; Provenzano, M.E.; Rossini, F. Germination and performance of seven native wildflowers in the Mediterranean landscape plantings. *Ital. J. Agron.* **2018**, *13*, 163–171. [CrossRef]
102. HVG. Pellets. Available online: https://hvg-germany.de/en/hop-processing/hop-pellets (accessed on 2 April 2021).
103. Forteschi, M.; Porcu, M.C.; Fanari, M.; Zinellu, M.; Secchi, N.; Buiatti, S.; Passaghe, P.; Bertoli, S.; Pretti, L. Quality assessment of Cascade Hop (*Humulus lupulus* L.) grown in Sardinia. *Eur. Food Res. Technol.* **2019**, *245*, 863–871. [CrossRef]
104. Mozzon, M.; Foligni, R.; Mannozzi, C. Brewing quality of hop varieties cultivated in central Italy based on multivolatile fingerprinting and bitter acid content. *Foods* **2020**, *9*, 541. [CrossRef]
105. Rodolfi, M.; Chiancone, B.; Liberatore, C.M.; Fabbri, A.; Cirlini, M.; Ganino, T. Changes in chemical profile of Cascade hop cones according to the growing area. *J. Sci. Food Agric.* **2019**, *99*, 6011–6019. [CrossRef]
106. Matsui, H.; Inui, T.; Oka, K.; Fukui, N. The influence of pruning and harvest timing on hop aroma, cone appearance, and yield. *Food Chem.* **2016**, *202*, 15–22. [CrossRef]
107. Inui, T.; Okumura, K.; Matsui, H.; Hosoya, T.; Kumazawa, S. Effect of harvest time on some in vitro functional properties of hop polyphenols. *Food Chem.* **2017**, *225*, 69–76. [CrossRef]
108. Mackinnon, D.; Pavlovič, V.; Čeh, B.; Naglič, B.; Pavlovič, M. The impact of weather conditions on alpha-acid content in hop (*Humulus lupulus* L.) cv. Aurora. *Plant Soil Environ.* **2020**, *66*, 519–525. [CrossRef]
109. Lafontaine, S.; Caffrey, A.; Dailey, J.; Varnum, S.; Hale, A.; Eichler, B.; Dennenlöhr, J.; Schubert, C.; Knoke, L.; Lerno, L.; et al. Evaluation of variety, maturity, and farm on the concentrations of monoterpene diglycosides and hop volatile/nonvolatile composition in five *Humulus lupulus* Cultivars. *J. Agric. Food Chem.* **2021**, *69*, 4356–4370. [CrossRef] [PubMed]
110. De Keukeleire, J.; Janssens, I.; Heyerick, A.; Ghekiere, G.; Cambie, J.; Roldán-Ruiz, I.; Van Bockstaele, E.; De Keukeleire, D. Relevance of organic farming and effect of climatological conditions on the formation of α-acids, β-acids, desmethylxanthohumol, and xanthohumol in hop (*Humulus lupulus* L.). *J. Agric. Food Chem.* **2007**, *55*, 61–66. [CrossRef] [PubMed]
111. Reeve, J.; Hoagland, L.; Villalba, J.J.; Carr, P.M.; Atucha, A.; Cambardella, C.; Davis, D.R.; Delate, K. Organic farming, soil health, and food quality: Considering possible links. In *Advances in Agronomy*; Sparks, D.L., Ed.; Academic Press: Cambridge, MA, USA, 2016; pp. 319–367.
112. Monier-Dilhan, S.; Bergès, F. Consumers' motivations driving organic demand: Between selfinterest and sustainability. *Agric. Resour. Econ. Rev.* **2016**, *45*, 522–538. [CrossRef]
113. Waldrop, M.E.; McCluskey, J.J. Does information about organic status affect consumer sensory liking and willingness to pay for beer? *Agribusiness* **2019**, *35*, 149–167. [CrossRef]
114. Lerro, M.; Marotta, G.; Nazzaro, C. Measuring consumers' preferences for craft beer attributes through Best-Worst Scaling. *Agric. Food Econ.* **2020**, *8*, 1. [CrossRef]
115. Capitello, R.; Todirica, I.C. Understanding the behavior of beer consumers. In *Case Studies in the Beer Sector*; Capitello, R., Maehle, N., Eds.; Elsevier Inc.: Amsterdam, The Netherlands, 2021; pp. 15–36.

Article

Pre-Germination Treatments, Temperature, and Light Conditions Improved Seed Germination of *Passiflora incarnata* L.

Luciana G. Angelini [1,2], Clarissa Clemente [1] and Silvia Tavarini [1,2,*]

[1] Department of Agriculture, Food and Environment, University of Pisa, Via del Borghetto 80, 56124 Pisa, Italy; luciana.angelini@unipi.it (L.G.A.); clarissa.clemente@phd.unipi.it (C.C.)
[2] Interdepartmental Research Center "Nutraceuticals and Food for Health", University of Pisa, Via del Borghetto 80, 56124 Pisa, Italy
* Correspondence: silvia.tavarini@unipi.it

Citation: Angelini, L.G.; Clemente, C.; Tavarini, S. Pre-Germination Treatments, Temperature, and Light Conditions Improved Seed Germination of *Passiflora incarnata* L. *Agriculture* 2021, *11*, 937. https://doi.org/10.3390/agriculture11100937

Academic Editors: Mario Licata, Antonella Maria Maggio, Salvatore La Bella and Teresa Tuttolomondo

Received: 30 August 2021
Accepted: 24 September 2021
Published: 28 September 2021

Publisher's Note: MDPI stays neutral with regard to jurisdictional claims in published maps and institutional affiliations.

Copyright: © 2021 by the authors. Licensee MDPI, Basel, Switzerland. This article is an open access article distributed under the terms and conditions of the Creative Commons Attribution (CC BY) license (https://creativecommons.org/licenses/by/4.0/).

Abstract: Perennial medicinal and aromatic plants (MAPs) may represent interesting, environmentally friendly crops for the Mediterranean environments. Among MAPs, *Passiflora incarnata* L. (maypop) represents a very promising crop for its wide adaptability to diverse climatic conditions, low input requirements, and high added-value due to its unique medicinal properties. The main problem in *P. incarnata* large-scale cultivation is the poor seed quality with erratic and low seed germination, due to its apparent pronounced seed dormancy. Therefore, the aim of this work was to investigate different chemical and physical treatments for overpassing seed dormancy and enhancing seed germination rates of *P. incarnata*. The effects of (i) different pre-germination treatments (pre-chilling, gibberellic acid—GA$_3$, leaching, scarification, non-treated control), (ii) light or darkness exposure, and (iii) temperature conditions (25, 30, and 35 °C constant and 20–30 °C alternating temperatures) have been examined in seed germination percentage and mean germination time of three *P. incarnata* accessions (F2016, FF2016, and A2016) grown in field conditions in Central Italy. Data showed that the pre-germination treatments generally stimulated faster germination compared to the control, with the best results obtained in the dark and with high temperatures. These findings are useful for the choice of the most suitable seed pre-germination treatment that can facilitate stable, high and agronomically acceptable germination rates in *P. incarnata*.

Keywords: maypop; medicinal and aromatic plants; crop diversification; sustainability

1. Introduction

Among conservation agriculture practices, the introduction of perennials in crop rotations has been proposed as a viable opportunity to improve the long-term sustainability and productivity of systems thanks to the reduction in tillage, the protection of the soil surface, and the decrease in erosion and runoff. As a consequence, a considerable improvement in soil organic matter and nutrient cycling, as well as the overall physical and biological health of the soil, can be achieved. In this context, perennial medicinal and aromatic plants (MAPs) may represent interesting environmentally friendly crops for Mediterranean countries. In recent years, the attraction of MAPs as worthy farm crops has grown due to the demand created by consumer interest for these plants for culinary, medicinal, and other anthropogenic applications. Among MAPs, *Passiflora incarnata* could represent an interesting crop for Mediterranean systems, due to its perennial cycle and its potential agronomic benefits. *Passiflora* is a genus belonging to Passifloraceae's family, consisting of more than 500 species, which mostly live in tropical and subtropical regions, except for *P. incarnata*, which is native to temperate North America (southeast of the USA) and it has been introduced into Australasia, Bermuda, Europe, and Hawaii [1]. *P. incarnata* (maypop) is mainly cultivated for its pharmaceutical and cosmetic properties. It was historically used as a sedative and anxiolytic plant and for the treatment of insomnia in North America; as an analgesic, anti-spasmodic, anti-asthmatic, wormicide, and sedative in Brazil; as a sedative and narcotic in Iraq; in diseased conditions like dysmenorrhea, epilepsy, insomnia, neurosis, and neuralgia in Turkey; to cure hysteria and neurasthenia in Poland; and for morphine deaddiction in the traditional system of medicine in India [2–5]. *P. incarnata* contains, in its leaves, flavonoids (mainly C-glycosides of apigenin and luteolin), which probably are responsible of the pharmacological effects, and alkaloids—based on the β-carboline ring system, namely harmane, harmol, harmine, harmalol, and harmaline. The main active ingredients include chrysin, vitexin and isovitexin, schaftoside, isoschaftoside, coumerin, and umbelliferone, with considerable variation in qualitative and quantitative composition according to the source [6–9]. The presence of gynocardin (a cyanogenic glycoside) and essential oil in traces, comprising more than 150 components, have been also revealed [6–8]. Various other constituents have been identified, including γ-benzo-pyrone derivative maltol, carbohydrates such as raffinose, sucrose, D-glucose, and D-fructose [10].

P. incarnata is successfully cultivated in Florida, Guatemala, and Italy. In Italy, *P. incarnata* is grown mostly in the central regions of the country, where it behaves as perennial spring–summer crop with a stand duration of 5–7 years [11]. In winter, the aerial part of the plant dies and, at the beginning of springtime, there is the vegetative upturn from the bottom of the plant.

P. incarnata is one of the most important medicinal plants in Italy, where it is cultivated on a total area of approximately 150–180 hectares, of which 50 ha operate under organic farming conditions, with a production of 800–1000 tons/year. The main problem in its large-scale cultivation is the poor seed quality, with erratic and low seed germination, due to its apparent pronounced seed dormancy. This makes it difficult to grow *P. incarnata* crops from seeds, so the nursery reproduction is generally carried out by cuttings, with a substantial increase in the cultivation costs. Little is known about the seed germination behavior of *Passiflora* species, and no information is reported in the "International Rules for Seed Testing" [12] regarding minimum germination requirements or optimal conditions for germination. Dormancy and germination rates are factors that can control the number of progenies a plant can make, and they are variable both in space and time, with large environmental effects [13]. Several studies seem to have confirmed that *Passiflora* spp. have exogenous dormancy due to a combination of both mechanical and chemical factors. De Oliveira et al. [14] and Torres [15] highlighted that the semi-domesticated passionfruit have strong dormancy effects. However, the presence of other kinds of dormancy cannot be excluded, depending on the species, such as a physical dormancy due to the impermeable seed coat or even a physiological dormancy. Although for some *Passiflora* species, a combination of physical and physiological dormancy has been highlighted, studies regarding *P. incarnata* are very limited and not conclusive. In order to remove seed dormancy in *P. incarnata*, mechanical scarification was the widely used method; nevertheless, it never gave satisfactory results with potential risk of embryo damaging [16]. On the contrary, chemical scarification appeared to be more effective in overpassing dormancy in this species, but an extended period of soaking could reduce seed viability and germination rate [17]. Furthermore, among pre-germination treatments tested on *P. incarnata* seeds, different light and temperature conditions have been tested [18,19], highlighting that the seed negative photoblasty of this species increased at suboptimal temperature. To the best of our knowledge, pre-chilling treatment on *Passiflora* spp. has been never tested before, even though it is a well-known procedure for some medicinal and aromatic plants [20].

Therefore, with the objective to improve knowledge about seed germination and dormancy mechanisms of *P. incarnata* for enhancing nursery production, this work aimed to discover better and appropriate pre-germination treatments for overpassing seed dormancy and enhancing seed germination rates. For this purpose, the responses of seed lots of three *P. incarnata* accessions grown in 2016 in Central Italy with different treatments (pre-chilling, GA_3, leaching, scarification, non-treated control), different light or darkness exposure, and different temperature conditions (25, 30, and 35 °C constant temperatures and 20–30 °C alternating temperatures) have been examined.

2. Materials and Methods

2.1. Plant Materials

The experiments were carried out at the Seed Research and Testing Laboratory of the Department of Agriculture, Food, and Environment (DAFE) of the University of Pisa.

The seeds of three *P. incarnata* accessions, namely F2016, FF2016, and A2016, were kindly supplied by F.I.P.P.O. (Federazione Italiana Produttori Piante Officinali) and by Aboca s.r.l. company (Sansepolcro, Arezzo, Italy). Mature fruits were collected during 2016 from plants grown in an open field in Central Italy (Tuscan-Umbrian, Val Tiberina, Italy), under the same pedo-climatic conditions and with an organic management system. The planting had been carried out in 2014 by transplanting the seedlings on a clay loamy soil. The crop was carried out without irrigation since rainfall in the area was able to satisfy crop water requirements.

During the growing period, between the flowering and ripening stages, until fruit harvesting (from August to October 2016), rainfall and temperatures (maximum and minimum air temperature) were recorded using a weather station located nearby the cultivation area (Figure 1).

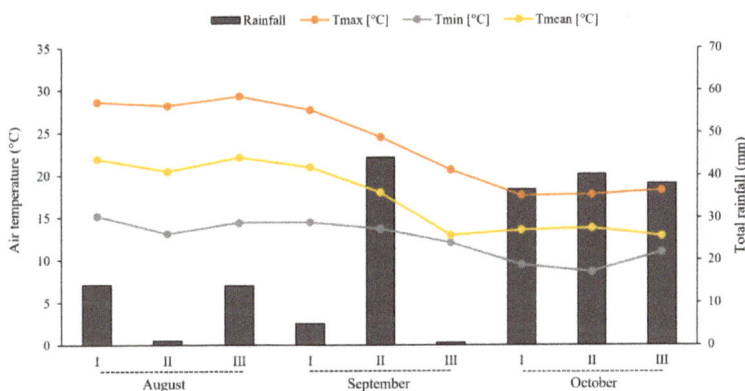

Figure 1. Cumulative (mm) precipitation and air temperatures (Tmax, Tmin, and Tmean) throughout the period between flowering and the harvesting of fruits. Data are reported for each decade (I, II, and III), from August to October 2016.

2.2. Pre-Germination Treatments

Fruits were harvested and soaked in tap water for a couple of days until maceration occurred. At the end of maceration process, the seeds were separated from the pulp, washed with tap water at room temperature, air dried, and cleaned with sieves and flows of air (Figure 2). Subsequently, seeds were stored in darkness at 4–5 °C and 60% relative humidity for 6 months.

Different pre-germination treatments in a completely random block design have been examined. In detail, the treatments were: (i) pre-chilling, (ii) Gibberellic acid (GA$_3$), (iii) leaching, (iv) scarification, and (v) non-treated control. For each of these treatments, different light (or darkness) and temperature conditions (25, 30, and 35 °C constant temperatures and 20–30 °C alternating temperatures) were also examined. For each *P. incarnata* accession, four replications of 50 seeds each for every pre-germination treatment and for the control, were used.

Figure 2. Dehydrated fruit of *Passiflora incarnata*, with seeds enclosed in the red and mucilaginous aril (on the left). Seeds used in experiments (on the right).

The seeds were placed in 12 cm Petri dishes and incubated in climatic cabinets. Preliminary tetrazolium tests, according to the International Seed Testing Association (ISTA) [12], were conducted to estimate the seed viability of each accession.

For pre-chilling treatment, seeds in groups of 25 were placed in Petri dishes between two sheets of filter paper and moistened with 5 mL of tap water, then they were put in a refrigerator at 4–5 °C for 4 days. The hormone treatment was conducted using GA_3 (200 ppm or 0.2 g/L) prepared starting from gibberellic acid, 90% gibberellin A3 basis (TLC) (Fluka Biochemika, Buchs, Switzerland). Seeds were placed between two sheets of filter paper, moistened with 5 mL GA_3 solution. The third treatment consisted of seed leaching in tap water for 8 h, which were then put in Petri dishes between two sheets of filter paper moistened with only 3 mL of tap water, because they were already partially imbibed. Finally, mechanical scarification was carried out by rubbing seeds manually on sandpaper to damage the hard outer layers, after which they were put in Petri dishes between two sheets of filter paper moistened with 5 mL of tap water. The dishes were put in climatized cabinets (Officine Meccaniche K.W. Mod. 1040, Siena, Italy) at different temperature and light conditions: 25, 30, and 35 °C, both in complete light or dark conditions or under 20–30 °C alternating temperature with a photoperiod of 16–8 h and 8–16 h. The light was provided by cool white-light fluorescent lamps Osram L18 W/20 (10 µmol photons s^{-1} m^{-2} photosynthetically active radiation) (Osram GmbH, München, Germany).

2.3. Germination Test and Measurements

Prior to the germination test, thousand seed weight, for each accession, was assessed according to ISTA [12]. Germination was monitored every two or three days up to 30 days as a function of temperature. Germination ended with the appearance of cotyledons. Germinated seeds were counted, and germination counts were stopped when final germination percentages were reached.

Germination percentage (G %) and mean of germination time (MGT) were calculated according to following equations:

- G (%) = $S_{NG}/S_{NO} \times 100$; where S_{NG} is the number of germinated seeds and S_{NO} is the number of experimental seeds with viability, respectively.
- MGT = $\Sigma (n \times d)/N$, where n = number of germinated seeds per day; d = number of days needed for germination, and N = total number of germinated seeds.

2.4. Statistical Analyses

Data of both seed germination and MGT tests were subjected to an analysis of variance (ANOVA) using the statistical software Costat Cohort V6.201 (CoHort Software, Monterey, CA, USA). For each accession, the effect of pre-treatments (A, i.e., prechilling, GA_3, leaching, scarification, and control) and the temperature/light treatments (B, i.e., different constant and alternating temperatures and light/dark conditions) and their reciprocal interactions

(A × B) were analyzed by two-way ANOVA. Means were separated on the basis of a least significant difference (LSD) test only when the ANOVA F-test per treatment was significant at the 0.05 probability level [21]. For germination percentage, data arcsin transformation $[(x + 0.5)/n]^{1/2}$ was performed before variance analysis.

3. Results

The 1000 seeds weights (TSW), evaluated before starting the experiment, are reported in Table 1. TSW significantly varied depending on accession, with the highest value reached by F2016, followed by A2016 and, finally, FF2016.

Table 1. Determination of the thousand seed weight (TSW) for each investigated accession of *P. incarnata* (F2016, FF2016, and A2016).

Accession	TSW (g)
F2016	32.36 ± 0.14 a
FF2016	31.07 ± 0.26 c
A2016	31.79 ± 0.11 b

Values are mean (±standard deviation) of four replications. Different letters mean statistically significant differences at $p \leq 0.05$, according to LSD post hoc test.

The main results of the two-way ANOVA, performed to assess the effects of pre-germination treatments (A), temperature/light (T&L) exposure (B), and A × B interaction on germination percentage and MGT, are reported in Table 2. The results showed that germination percentage was significantly affected by pre-germination treatments and T&L exposure, as well as by their reciprocal interactions, in all *P. incarnata* accessions. On the other hand, pre-germination treatments did not show any significant effect on MGT values of any accession. T&L exposure and the interaction between the two variability factors, instead, played a key role in affecting mean germination time in all three accessions.

Table 2. Effect of pre-germination treatment (A), temperature/light exposure (B), and their interactions (A × B) on germination percentage and mean germination time (days) of *P. incarnata* seeds, separately for each accession.

Accession	Source of Variation	Germination (%)	MGT [†] (Days)
F2016	pre-germination treatments (A)	**	NS
	T&L exposure (B)	**	***
	A × B	**	***
FF2016	pre-germination treatments (A)	**	NS
	T&L exposure (B)	**	***
	A × B	**	*
A2016	pre-germination treatments (A)	**	NS
	T&L exposure (B)	**	***
	A × B	**	***

[†] MGT data at 25 °C were not included. The significance of variability factors (A and B) and their interactions (A × B) according to the F-test is reported as follows: NS, not significant; *, significant at $p \leq 0.05$; **, significant at $p \leq 0.01$; ***, significant at $p \leq 0.001$.

In Tables 3–5, the differences in germination percentage, separately for each *P. incarnata* accession, are reported. Regarding the effect of temperature and light/dark (L/D) conditions, no germination was achieved for the control at 25 °C both in light and dark conditions, confirming that this temperature value represents, for maypop, the below threshold for germination. In F2016 accession (Table 3), the highest germination rate was achieved at 35 °C/D, while the lowest one was observed at 25 °C/L. Considering the effect of pre-germination treatments, significantly higher germination percentages were registered with prechilling, GA₃, and leaching in comparison with the control. Scarification worsened germination rate, with values equal to control. Taking into account the effect of

AxB interaction, interestingly, the best conditions were found at 35 °C/D in control and after prechilling, GA$_3$, and leaching. It is important to note that, in all other conditions (25 °C/D; 30 °C/D&L; 35 °C/L), the pre-germination treatments significantly increased the germination rates in comparison with the control, except for scarification.

Table 3. Effect of the pre-germination treatments (A), temperature/light conditions (B), and their reciprocal interactions (A × B) on germination percentage of *P. incarnata* F2016 accession.

F2016 Accession	Prechilling	GA$_3$	Leaching	Scarification	Control	Mean T/L Conditions
25 °C/L	2.67 [l-o]	0.00 [o]	0.00 [o]	1.33 [no]	0.00 [o]	0.8 F
25 °C/D	8.00 [h-n]	16.00 [f-i]	9.33 [g-m]	2.67 [l-o]	0.00 [o]	7.2 E
30 °C/L	34.67 [d-f]	20.00 [e-i]	24.00 [e-h]	17.33 [f-i]	12.00 [g-l]	21.6 D
30 °C/D	49.33 [cd]	73.33 [ab]	76.00 [ab]	25.33 [e-g]	40.00 [de]	52.8 B
35 °C/L	70.67 [bc]	69.33 [bc]	81.33 [ab]	46.67 [d]	40.00 [de]	61.6 B
35 °C/D	80.00 [ab]	80.00 [ab]	84.00 [ab]	69.33 [bc]	90.00 [a]	80.7 A
20/30 °C–16/8 h	78.67 [ab]	25.33 [e-g]	17.33 [f-i]	12.00 [g-l]	21.33 [e-h]	30.9 CD
20/30 °C–8/16 h	46.67 [d]	29.33 [d-f]	46.67 [d]	33.33 [d-f]	8.00 [h-n]	32.8 C
Mean pre-treatments	46.3 A	39.2 A	42.3 A	26.0 B	26.4 B	

A two-way ANOVA test was used to evaluate the effect of the interaction between pre-germination treatments (A) and temperature/light conditions (B) (A × B). Means followed by the same letter are not significantly different at $p \leq 0.05$ based on LSD test. Lower-case letters indicate A × B interaction, upper-case letters indicate the effect of pre-germination treatments (A) and temperature/light conditions (B).

Table 4. Effect of the pre-germination treatments (A), temperature/light conditions (B), and their reciprocal interactions (A × B) on germination percentage of *P. incarnata* FF2016 accession.

FF2016 Accession	Prechilling	GA$_3$	Leaching	Scarification	Control	Mean T/L Conditions
25 °C/L	2.67 [st]	6.67 [rs]	1.33 [st]	4.00 [st]	0.00 [t]	2.9 F
25 °C/D	22.67 [n-q]	28.00 [l-q]	13.33 [qr]	26.67 [m-q]	0.00 [t]	18.1 E
30 °C/L	20.00 [o-q]	34.67 [g-o]	24.00 [m-q]	21.33 [n-q]	20.00 [o-q]	24.0 D
30 °C/D	57.33 [a-f]	70.67 [ab]	53.33 [b-g]	64.00 [a-d]	60.00 [a-e]	61.1 A
35 °C/L	69.33 [abc]	42.67 [e-m]	46.67 [d-l]	49.33 [d-i]	48.00 [d-i]	51.2 B
35 °C/D	50.67 [c-h]	69.33 [a-c]	74.67 [a]	58.67 [a-f]	60.00 [a-e]	62.7 A
20/30 °C–16/8 h	74.67 [a]	30.67 [i-p]	17.33 [pq]	18.67 [opq]	20.00 [o-q]	32.3 CD
20/30 °C–8/16 h	38.67 [f-n]	42.67 [e-m]	33.33 [h-p]	33.33 [h-p]	20.00 [o-q]	33.6 C
Mean pre-treatments	42.0 A	40.7 AB	33.0 C	34.5 BC	28.5 D	

A two-way ANOVA test was used to evaluate the effect of the interaction between pre-germination treatments (A) and temperature/light conditions (B) (A × B). Means followed by the same letter are not significantly different at $p \leq 0.05$ based on LSD test. Lower-case letters indicate A × B interaction, upper-case letters indicate the effect of pre-germination treatments (A) and temperature/light conditions (B).

Table 5. Effect of the pre-germination treatments (A), temperature/light conditions (B), and their reciprocal interactions (A × B) on germination percentage of *P. incarnata* A2016 accession.

A2016 Accession	Prechilling	GA$_3$	Leaching	Scarification	Control	Mean T/L Conditions
25 °C/L	0.00 [s]	4.00 [p-s]	1.33 [rs]	4.00 [p-s]	0.00 [s]	1.9 E
25 °C/D	0.00 [s]	6.67 [o-r]	6.67 [n-r]	4.00 [p-s]	0.00 [s]	3.5 E
30 °C/L	13.33 [i-p]	22.67 [f-l]	20.00 [h-n]	18.67 [i-n]	8.00 [m-q]	16.5 D
30 °C/D	65.33 [a]	50.67 [a-c]	37.33 [c-g]	28.00 [e-l]	48.00 [a-d]	45.9 B
35 °C/L	36.00 [c-h]	55.33 [a-c]	48.00 [a-d]	49.33 [a-c]	40.00 [c-f]	45.7 B
35 °C/D	53.33 [a-c]	60.00 [ab]	65.33 [a]	65.33 [a]	60.00 [ab]	60.8 A
20/30 °C–16/8 h	64.00 [ab]	16.00 [i-n]	18.67 [h-n]	13.33 [i-p]	21.33 [g-m]	26.7 C
20/30 °C–8/16 h	16.00 [i-n]	45.33 [b-e]	29.33 [d-i]	24.00 [f-l]	8.00 [m-q]	24.5 C
Mean pre-treatments	31.0 A	32.6 A	28.3 A	25.8 A	23.2 B	

A two-way ANOVA test was used to evaluate the effect of the interaction between pre-germination treatments (A) and temperature/light conditions (B). Means followed by the same letter are not significantly different at $p \leq 0.05$ based on LSD test. Lower-case letters indicate A × B interaction, upper-case letters indicate the effect of pre-germination treatments (A) and temperature/light conditions (B).

For FF2016 accession (Table 4), the highest germination was obtained under both 35 °C/D and 30 °C/D, while, as observed for F2016, the lowest germination occurred by adopting the 25 °C/L condition. Differently to what was observed for F2016 accession, all the pre-germination treatments significantly enhanced the germination percentage in FF2016 seeds. Considering AxB interaction, a significant improvement of germination, going from the control to pre-chilling under alternating temperatures of 20/30 °C (both photoperiods), as well as under 35 °C/L, was observed (Table 4). Furthermore, the combination between leaching and 35 °C/D conditions provided the highest germination percentage, followed by the seeds subjected to GA_3 and scarification treatments under the same T&L conditions. For A2016 accession (Table 5), a similar trend as described for FF2016 was detected. In fact, under 35 °C/D, the best germination conditions occurred, followed by 30 °C both in light and dark. On the contrary, as observed for the other accessions, under 25 °C (light and dark), the worst germination values were recorded. Once again, considering the AxB interaction, the best conditions able to enhance germination percentage were due to the combination of 20/30 °C 16/8 h and prechilling, and the combination of 35 °C/D and leaching.

All these findings revealed that, among accessions, the untreated/control seeds of F2016 had the highest germination rate (germination percentage up to 90%) when exposed to 35 °C under dark conditions. This behavior confirmed that, for *P. incarnata*, the optimal germination can be achieved at 35 °C in the dark. In such conditions, the untreated seeds of FF2016 and A2016 achieved lower germination percentages (around to 60%) than F2016. The higher values observed for control seeds of F2016 were expected on the basis of 1000-seed weight (Table 1). In FF2016 and A2016 accessions, pre-germination treatments were absolutely necessary in order to improve the germination process.

MGT values showed that germination peaks usually occurred within two weeks. Beyond this time, seeds sporadically sprouted. As a general trend, the time required for germination (Figure 3) decreased progressively from light to dark conditions, depending on pre-germination treatments, accession, and temperature. In F2016, this behavior was particularly evident under 35 °C, while no differences were observed between light and dark at 30 °C. Conversely, in FF2016 and A2016, a strong decrease in MGT was observed from light to dark, both at 30 °C and 35 °C. MGT lasted roughly ten days in light conditions (as mean value among accessions and pre-germination treatments) but fell to about one week in dark conditions.

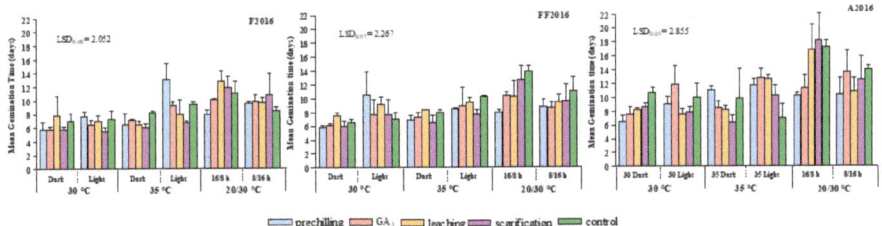

Figure 3. Effect of the interaction (A × B) between different pre-germination treatments (A) and temperature/light conditions (B) on mean germination time (MGT) of *P. incarnata* F2016, FF2016, and A2016 accessions.

On the contrary, alternating temperature (20/30 °C) did not improve germination energy, except when combined with prechilling. In fact, the interaction "pre-chilling × 20/30 °C (16/8 h)" conditions resulted in MGT reduction in comparison with the control (−200%). Alternating 20/30 °C temperatures, not associated with pre-chilling, gave the longer germination times, as well as 35 °C under light conditions. Furthermore, in all accessions, the adoption of an 8/16 h photoperiod significantly decreased MGT only in the control and in seeds subjected to scarification and leaching in comparison with a 16/8 h photoperiod.

Finally, considering the average values over the three accessions, all the pre-germination treatments generally stimulated a faster germination compared to control. Among the tested treatments, scarification seemed to lead to a quick germination process, even if germination rate was not elevated.

4. Discussion

In the effort to improve and promote the cultivation of *P. incarnata*, the effects of temperature, dark/light conditions, and pre-germination treatments on the germination percentage and MGT of its seeds were investigated. The 1000 seeds weight (TSW) was also evaluated as it represents one of the most important components in determining seed quality. In fact, it is generally reported that seed germinability is positively related to seed mass. Larger seeds germinated to a higher percentage thanks to their greater reserves, which enable a greater tolerance to a range of hazards, including shade, drought, and physical damage [22–24].

Definitively, the obtained results confirmed that *P. incarnata* seeds are photoblastically negative and have pronounced heat requirements for germination. Optimal germination percentages, in fact, were achieved with 35 °C in darkness, for each accession. In such conditions, a significant and strong decline in MGT was also detected, confirming the tropical origins of this species [25]. On the contrary, very low values were observed at 25 °C, more pronounced under light conditions, for each pre-germination treatment and seed accession. Data showed a significant interaction between complete light/dark exposition and temperatures, underlining the fact that the light exposition has an inhibitory effect on the germination of *P. incarnata* seeds. Among pre-germination treatments, pre-chilling, GA_3, and leaching appeared to be the most effective in enhancing normal seedling germination. Only for A2016, scarification gave similar results to pre-chilling, GA_3, and leaching treatments. On the contrary, in the other two accessions, under scarification, the dead seeds percentage considerably increased, probably due to embryo damaging. A significant interaction between pre-chilling and temperature was observed with significantly higher germination values than control (+330%) at 20/30 °C (16/8 h).

Previous studies investigated the effect of different combinations of light and different temperature regimes with the aim to improve seed germination in *P. incarnata*. In this regard, Benvenuti et al. [18] tested combinations of white light (or darkness) and temperature (20, 25, 30, 35, and 40 °C), or subjected *P. incarnata* seeds to different sequences of light treatments (succession of red and far-red light, with 5 min each one) after 12 h of dark incubation at 30 °C. These authors found that the germination threshold was surprisingly high, both in darkness and light conditions (25.4 °C and 23.9 °C, respectively), while no germination was observed at 20 °C. In addition, these authors observed that a suboptimal temperature (lower than 35 °C) and far/far-red light both produced extremely low levels of germination (around 5%). Zucarelli et al. [19] investigated the effect of alternate temperatures of 20–30 °C and 30–20 °C for periods of 16 and 8 h, simulating the photoperiod and highlighting that alternating temperatures of 30–20 °C promoted the highest germination rates. The results of these studies also demonstrated that high temperatures progressively decreased the time required for germination, independently from light conditions. Similarly, in our study, MGT decreased as result of the combination of high temperatures (35 °C) and dark conditions, regardless of the pre-germination treatment.

In the literature, several studies have underlined the presence of dormancy in *Passiflora* spp. [16,26] and, specifically in *P. incarnata*, a combination of physical and physiological dormancy has been detected [27]. In general, all seed pre-germination treatments led to enhanced water and oxygen exchange across seed coat layers. However, scarification did not have the best result, as was supposed to be the case with physical dormancy. Otherwise, results obtained with GA_3 and leaching seem to confirm a possible physiological dormancy in *P. incarnata*, as hormones and inhibitors were stimulated/removed, respectively.

Interestingly, our study, for the first time, pointed out that pre-chilling is an efficient treatment to improve germination in *P. incarnata* seeds. Pre-chilling enhanced germination as well, probably because this pre-treatment stimulated a variation in abscisic acid/gibberellic acid rate (ABA/GA$_3$) and free gibberellins biosynthesis [28]. Macchia et al. [29] found that prechilling for 7–15 days in light or in darkness hardly affected percentage germination of *Echinacea angustifolia* seeds, but significantly increased the rate of germination. On the contrary, GA$_3$ treatment was not useful for this species.

The time conclusively required for seed germination decreased progressively with increasing temperatures, but only under dark conditions, while in complete light conditions, no variation was observed and MGT values remained almost constant with increasing temperatures. Optimal germination times were achieved at 35 °C in dark conditions. Similarly to germination percentages, even in MGT, alternating temperature (20/30 °C) did not improve germination energy, except when combined with pre-chilling treatment.

5. Conclusions

The use of high-quality seeds is an important key factor in modern agriculture to obtain a successful nursery and crop production, and particularly to enhance food security. In fact, for a rapid and uniform crop establishment, the selection of good quality seeds with improved vigor is of primary importance in order to enhance this critical and yield-defining stage. Currently, in Italy, there are difficulties in the large-scale cultivation of *Passiflora incarnata*, starting from the seed, due to dormancy. An enhancement of seed performance during germination and seedling emergence is required for an efficient and competitive nursery production. Little is known about the conditions under which the germination process takes place and the treatments necessary to remove dormancy, since no guidelines have been reported by the International Seed Testing Association (ISTA) for this species. Thus, this study identified the main and most significant pre-germination treatments and environmental parameters for improving the seed germination percentage and germination energy of *P. incarnata* under controlled conditions. This makes possible to achieve stable and agronomically satisfactory germination rates, thereby reducing seed propagation costs for this species.

Author Contributions: Conceptualization, L.G.A.; methodology, S.T. and C.C.; software, S.T.; formal analysis, C.C. and S.T.; investigation, L.G.A., C.C., and S.T.; data curation, C.C. and S.T.; writing—original draft preparation, L.G.A. and S.T.; writing—review and editing, L.G.A. and S.T.; supervision, L.G.A. All authors have read and agreed to the published version of the manuscript.

Funding: This research received no external funding.

Institutional Review Board Statement: Not applicable.

Informed Consent Statement: Not applicable.

Data Availability Statement: Data sharing not applicable.

Acknowledgments: The authors wish to express their gratitude to Lucia Ceccarini for her technical assistance and to Giulia Lauria whose bachelor thesis was the subject of this work.

Conflicts of Interest: The authors declare no conflict of interest.

References

1. McGuire Christopher, M. *Passiflora incarnata* (Passifloraceae): A New Fruit Crop. *Econ. Bot.* **1999**, *53*, 161–176. [CrossRef]
2. Dhawan, K.; Dhawan, S.; Sharma, A. *Passiflora*: A review update. *J. Ethnopharmacol.* **2004**, *94*, 1–23. [CrossRef] [PubMed]
3. Bourin, M.; Bougerol, T.; Guitton, B.; Broutin, E. A combination of plant extracts in the treatment of outpatients with adjustment disorder with anxious mood: Controlled study versus placebo. *Fundam. Clin. Pharmacol.* **1997**, *11*, 127–132. [CrossRef]
4. Akhondzadeh, S.; Naghavi, H.R.; Vazirian, M.; Shayeganpour, A.; Rashidi, H.; Khani, M. Passionflower in the treatment of generalized anxiety: A pilot double-blind randomized controlled trial with oxazepam. *J. Clin. Pharm. Ther.* **2001**, *26*, 363–367. [CrossRef] [PubMed]
5. Miyasaka, L.S.; Atallah, A.N.; Soares, B.G. Passiflora for anxiety disorder. *Cochrane Database Syst. Rev.* **2007**, *4*, 24. [CrossRef] [PubMed]
6. Bradley, P.R. British Herbal Compendium. *Bournem. Br. Herb. Med. Assoc.* **1992**, *1*, 35.
7. Wichtl, M. *Herbal Drugs and Phytopharmaceuticals*, 3rd ed.; Medpharm Scientific Publishers: Stuttgart, Germany, 2004; pp. 430–433.

8. Barnes, J.; Anderson, L.; Phillipson, D. *Herbal Medicines*; Pharmaceutical Press Publication: London, UK, 2007; pp. 461–469.
9. Spinella, M. Herbal medicines and epilepsy: The potential for benefit and adverse effects. *Epilepsy Behav.* **2001**, *2*, 524–532. [CrossRef]
10. Patel, S.S.; Soni, H.; Mishra, K.; Singhai, A.K. Recent updates on the genus Passiflora: A review. *Int. J. Res. Phytochem. Pharmacol.* **2011**, *1*, 1–16.
11. Fuoco, V.; Primavera, A. *Passiflora incarnata* L.: La produzione in Italia. In *Category Archives: Settore Erboristico–Approfondimenti*; CEC Publisher: Milan, Italy, 2015.
12. ISTA. *International Rules for Seed Testing*; The International Seed Testing Association: Bassersdorf, Switzerland, 2015.
13. Smýkal, P.; Vernoud, V.; Blair, M.W.; Thompson, R.D. The role of the testa during development and in establishment of dormancy of the legume seed. *Front. Plant Sci.* **2014**, *5*, 351.
14. De Oliveira, M.; São José, A.; Hojo, N.; Magalhães, O.; Novaes, F. Superação de dormência de maracujá-do-mato (*Passiflora cincinnata* Mast.). *Rev. Brasil. Fruticul.* **2010**, *32*, 584–590. [CrossRef]
15. Torres, M. Seed dormancy and germination of two cultivated species of *Passifloraceae*. *Bol. Cient. Mus. His. Nat.* **2018**, *22*, 15–27.
16. Delanoy, M.; Van Damme, P.; Scheldeman, X.; Beltran, J. Germination of *Passiflora mollissima* (Kunth) L. H. Bailey, *Passiflora tricuspis* Mast. and *Passiflora nov* sp. Seeds. *Sci. Hort.* **2006**, *110*, 198–203. [CrossRef]
17. Rego, M.M.; Rego, E.R.; Nattrodt, L.P.U.; Barroso, P.A.; Finger, F.L.; Otoni, W.C. Evaluation of different methods to overcome in vitro deed dormancy from yellow passion fruit. *Afr. J. Biotechnol.* **2014**, *13*, 3657–3665. [CrossRef]
18. Benvenuti, S.; Simonelli, G.; Macchia, M. Elevated temperature and darkness improve germination in *Passiflora incarnata* L. seed. *Seed Sci. Technol.* **2001**, *29*, 533–541.
19. Zucarelli, V.; Henrique, L.A.V.; Ono, E.O. Influence of light and temperature on the germination of *Passiflora incarnata* L. seeds. *J. Seed Sci.* **2015**, *37*, 162–167. [CrossRef]
20. Aghilian, S.; Hosseini, M.K.; Anvarkhah, S. Evaluation of seed dormancy in forty medicinal plant species. *Int. J. Agric. Crop Sci.* **2014**, *7*, 760–768.
21. Gomez, A.A.; Gomez, K.A. Statistical procedures for agricultural research. In *Statistical Procedures for Agricultural Research*; John Wiley & Sons: Hoboken, NJ, USA, 1984; p. 680.
22. Leishman, M.R.; Westoby, M. The role of seed mass in seedling establishment in dry soil conditions- experimental evidence from semi-arid species. *J. Ecol.* **1994**, *82*, 249–258. [CrossRef]
23. Leishman, M.R.; Wright, I.J.; Moles, A.T.; Westoby, M. The evolutionary ecology of seed size. In *Seeds: The Ecology of Regeneration in Plant Communities*, 2nd ed.; CABI Publishing: Wallingford, UK, 2000; pp. 31–57.
24. Elliott, R.H.; Mann, L.W.; Olfert, O.O. Effects of seed size and seed weight on seedling establishment, seedling vigour and tolerance of summer turnip rape (*Brassica rapa*) to flea beetles, *Phyllotreta* spp. *Can. J. Plant Sci.* **2007**, *87*, 385–393. [CrossRef]
25. Killip, E.P. *Supplemental Notes on the American Species of Passifloraceae with Descriptions of New Species*; Bulletin of the United States National Museum; Smithsonian Institution: Washington, DC, USA, 1960; pp. 19–20.
26. Salazar, A.; Ramírez, C. Fruit maturity stage and provenance affect seed germination of *Passiflora mollissima* (banana passion fruit) and *P. ligularis* (sweet granadilla), two commercially valuable tropical fruit species. *Seed Sci. Technol.* **2017**, *45*, 383–397. [CrossRef]
27. Veiga-Barbosa, L.; Mira, S.; González-Benito, M.E.; Souza, M.M.; Meletti, L.M.M.; Pérez-García, F. Seed germination, desiccation tollerance and cryopreservation of Passiflora species. *Seed Sci. Technol.* **2013**, *41*, 89–97. [CrossRef]
28. Rascio, N.; Carfagna, S.; Esposito, S.; La Rocca, N.; Lo Gullo, M.A.; Trost, P.; Vona, V. *Elementi di Fisiologia Vegetale*; EdiSES: Naples, Italy, 2014; pp. 329–354.
29. Macchia, M.; Angelini, L.G.; Ceccarini, L. Methods to overcome seed dormancy in *Echinacea angustifolia* DC. *Sci. Hortic.* **2001**, *89*, 317–324. [CrossRef]

Article

Enhancement of Interplanting of *Ficus carica* L. with *Taxus cuspidata* Sieb. et Zucc. on Growth of Two Plants

Xue Yang [1,2,3,4,5], Yuzheng Li [1,3], Chunying Li [1,2,3,4,5,*], Qianqian Li [1,2,3,4,5], Bin Qiao [1,2,3,4,5], Sen Shi [1,2,3,4,5] and Chunjian Zhao [1,2,3,4,5]

[1] Key Laboratory of Forest Plant Ecology, Ministry of Education, Northeast Forestry University, Harbin 150040, China; klp20yx@nefu.edu.cn (X.Y.); liyuzheng@3sbio.com (Y.L.); lqq21@nefu.edu.cn (Q.L.); klp20qb@nefu.edu.cn (B.Q.); klp20ss@nefu.edu.cn (S.S.); zcj@nefu.edu.cn (C.Z.)
[2] College of Chemistry, Chemical Engineering and Resource Utilization, Northeast Forestry University, Harbin 150040, China
[3] Engineering Research Center of Forest Bio-Preparation, Ministry of Education, Northeast Forestry University, Harbin 150040, China
[4] Collaborative Innovation Center for Development and Utilization of Forest Resources, Harbin 150040, China
[5] Heilongjiang Provincial Key Laboratory of Ecological Utilization of Forestry-Based Active Substances, Northeast Forestry University, Harbin 150040, China
* Correspondence: lcy@nefu.edu

Abstract: Medicinal-agroforestry systems are one of the multi-functional medicinal plant production systems, gaining attention as a sustainable alternative to traditional monoculture systems. In this study, three planting patterns were established which included: (1) monoculture *F. carica* (MF); (2) monoculture *T. cuspidata* (MT); and (3) interplanting *F. carica* with *T. cuspidata* (IFT). The differences of growth biomass, photosynthesis, soil nutrients, soil enzyme activities, soil microorganisms, and main secondary metabolites of *F. carica* and *T. cuspidata* under the above three models were investigated. Compared with the MF and MT patterns, IFT pattern for 5 months significantly increased the plant growth biomass, photosynthesis, soil organic carbon, total nitrogen, and secondary metabolites content. The activities of acid phosphatase, sucrase, protease, polyphenol oxidase, urease, dehydrogenase, and catalase in soil of IFT were significantly higher than MF and MT patterns. Results showed that IFT pattern is preferred compared to the MF and MT patterns. Our result will help to provide a feasible theoretical basis for the large-scale establishment of *F. carica* and *T. cuspidata* mixed forests and obtain high-quality medicine sources for extracting important active ingredients, psoralen and paclitaxel, which are crucial to the long-term sustainable development and production of medicinal plants.

Keywords: *Ficus carica* L.; *Taxus cuspidata* Sieb. et Zucc.; medicinal-agroforestry system; soil enzyme activity; secondary metabolites; photosynthesis; sustainability

1. Introduction

Ficus carica L. is a species of *Ficus* plant in the Moraceae family. It is a perennial deciduous shrub with rapid growth and broad leaves [1,2]. *F. carica* is a robust and highly productive traditional medicinal plant that can adapt well to weather changes. The most important secondary metabolite in *F. carica* is psoralen [3]. Because of its strong physiological activity, *F. carica* is considered to have anti-bacterial, anti-viral, anti-tumor, and blood sugar lowering effects [4,5]. *F. carica* is cultivated widely in China, primarily distributed in Xinjiang, Fujian, Shandong, and other places.

Taxus cuspidata Sieb. et Zucc. is a species of *Taxus* in the Taxaceae family, which is a national first-level key protected wild plant that grows slowly and prefers a moist and cool environment [6,7]. *T. cuspidata* is in an endangered state due to its reproductive characteristics, habitat conditions, man-made destruction, and other factors, as well as the need for moderate shade treatment in the early growth period (1 to 3 years). Paclitaxel is contained in the roots, stems, and leaves of *T. cuspidata*, which has a broad-spectrum anti-cancer effect and is considered to be the most promising new anti-cancer drug, so its demand is increasing with each passing year [8]. *T. cuspidata* resources in China account for most of the worldwide *T. cuspidata* resources, but from the perspective of the worldwide demand for raw materials for paclitaxel extraction, its resource reserves are still far from sufficient. Therefore, it is urgent to adopt appropriate methods for large-scale cultivation of *T. cuspidata* plantation.

A medicinal-agroforestry system is a traditional way of land use which combines different plants according to the different suitable environments and the characteristics of spatial location, such as deep and shallow roots, growth rate, etc., thus forming a stable and efficient artificial compound ecosystem that promotes itself [9]. In addition, the medicinal-agroforestry system is a multi-functional medicinal plant production system and is gaining attention as a sustainable alternative to traditional monoculture systems. Compared to monoculture patterns, the medicinal-agroforestry system is considered to be a more beneficial land use practice, contributing to improved soil quality and soil biodiversity [10,11]. Medicinal-agroforestry systems, which promote ecological diversity and sustainability while also providing social, economic, and environmental benefits, deserve to be vigorously pursued.

There are no reports on interplanting *F. carica* with *T. cuspidata*. In this study, according to the characteristics of slow growth, deep root system, and shade-loving growth characteristics of *T. cuspidata*, the heliophilous plant *F. carica* with medium shallow root system and fast growth was selected for interplanting. In the cultivation practice of *T. cuspidata*, due to the slow growth rate of *T. cuspidata* seedlings and intolerance to strong light, it is often necessary to build shade sheds to block part of the sunlight. The *F. carica* seedlings grow rapidly and can be quickly uplifted in the year of the transplanting and grow larger leaves [1]. *F. carica* are interplanted with the *T. cuspidata*, so that the crowns of the *F. carica* and *T. cuspidata* are located on the upper and lower layers, respectively. *F. carica* not only decreases the damage on *T. cuspidata* caused by strong light but also reduces the cost of building shade sheds.

This study investigated the effects of different planting patterns on the growth of two plants. Indicators for proving the effects on plant growth include the growth biomass, photosynthesis, soil nutrients, soil microbial communities, and secondary metabolites. It is expected to provide a feasible theoretical basis for formulating a high-efficiency interplanting patterns of *F. carica* with *T. cuspidata* and realizing sustainable land use.

2. Materials and Methods

2.1. Location of the Experiment and Plant Materials

This study was performed from April to September 2020 at the Xiazhuang experimental field located in Rongcheng City, Shandong Province, China (37°23′ N; 122°52′ E) (Figure 1). The study area has a warm temperate monsoon humid climate with the annual average temperature of 12.4 °C, the annual average precipitation of 800 mm, and the annual average humidity of 89%. It has an annual average sunshine of 2600 h, the annual evaporation of 1930.7 mm, and the frost-free period was 208 days. Compared with the inland areas at the same latitude, it is characterized by abundant rainfall, moderate annual temperature, and mild climate. The soil was classified as typical brown soil and has a paddy soil type with topsoil (0–20 cm) that the soil nutrient content was as follows: pH 5.5; contained organic matter 55.2 mg·kg^{-1}; total nitrogen 4.6 g·kg^{-1}; alkali nitrogen 61.8 mg·kg^{-1}; available phosphorus 38.2 mg·kg^{-1}; and available potassium 319.7 mg·kg^{-1}.

A *F. carica* annual cutting seedling was selected and seedling height was 40–50 cm. The selected *T. cuspidata* were four-year-old seedlings and the seedling height was 25–35 cm. They were all grown at the Xiazhuang experimental field located in Rongcheng City, Shandong Province, China (37°23′ N; 122°52′ E).

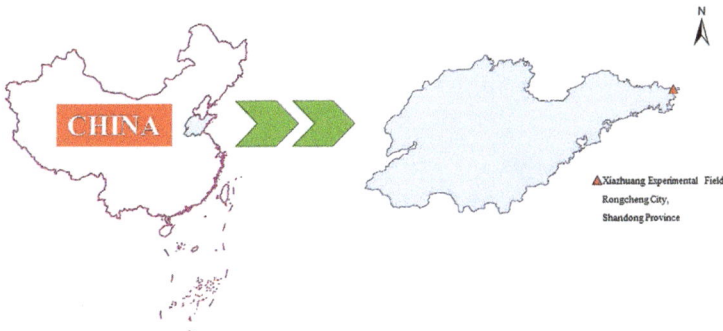

Figure 1. Location of medicinal-agroforestry system in Xiazhuang County, Rongcheng City, Shandong Province, China.

2.2. Experimental Design and Treatments

In this study, three planting patterns were established, which included: (1) monoculture *F. carica* (MF); (2) monoculture *T. cuspidata* (MT); and (3) interplanting *F. carica* with *T. cuspidata* (IFT) (Figure 2). All systems adopt the same agronomic management practices. The experiment was arranged in plots in a randomized design. Four plots were randomly set up for each pattern in this experiment. The area of each plot was 20 × 50 m^2, and the distance between adjacent plots was at least 10 m. In MF, *F. carica* was planted with a row and plant spacing of 0.8 m × 1.0 m. In MT, *T. cuspidata* was planted with a row and plant spacing of 0.8 m × 1.0 m. In IFT, a *T. cuspidata* was planted between every two adjacent *F. carica*, and the spacing between the adjacent *T. cuspidata* and *F. carica* was 0.8 m; the row spacing was 1.0 m.

F. carica and *T. cuspidata* were transplanted after a period of slow seedling and the plants began to grow normally. The following experimental indicators were measured on the 15th day of each month from May to September 2020.

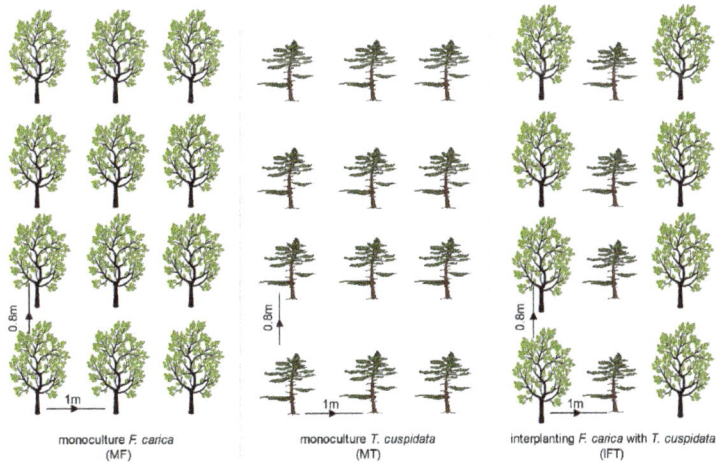

Figure 2. Experimental design of three planting patterns.

2.3. Determination of Plant Biomass

F. carica growth measurement: We used a tape measure to measure the length of all the branches of the current year. We used a traveling caliper to measure the diameter of the base, base to end, and at the end position of the current year branches. The branch is approximately regarded as composed of several circular truncated cones whose height does not exceed 5 cm. By calculating the volume of each circular truncated cone, the current year branch volume of *F. carica* can be calculated. The total volume (cm^3) of *F. carica* branch in the current year was used as an index to evaluate the growth of *F. carica*.

T. cuspidata growth measurement: The plant height and base diameter were measured by measuring tape and vernier caliper, respectively, and these indicators were used as the evaluation indexes of *T. cuspidata* growth.

2.4. Determination of Plant Photosynthesis

When the temperature and humidity were suitable on a sunny day, the apical mature leaves of *F. carica* and *T. cuspidata* with good growth and consistent leaf orientation were selected to ensure that the tested leaves received good light. The photosynthetic parameters such as the net photosynthetic rate (Pn, $\mu mol \cdot m^{-2} \cdot s^{-1}$), stomatal conductance (Gs, $mmol \cdot m^{-2} \cdot s^{-1}$), and intercellular carbon dioxide concentration (Ci, $\mu mol \cdot mol^{-1}$) were measured using a Yaxin-1102 portable photosynthesis apparatus at the same time every month.

2.5. Soil Sampling and Soil Chemical Analysis

2.5.1. Soil Sampling

Soil samples from the depth of 0–20 cm were randomly collected from the three patterns (MF, MT, and IFT). Each pattern consisted of four sub-samples, repeated four times. After collection, the soil samples were sealed in labeled vacuum plastic bags and sent to the laboratory as early as possible. Each soil sample was divided into two parts: one was air drying, grinding, and 0.20 mm sieving to determine soil organic carbon (SOC), total nitrogen (TN), and enzyme activity analysis, and the other was stored in a refrigerator at 4 °C for microbial determination.

2.5.2. Determination of Soil Organic Carbon and Total Nitrogen

In short, SOC was measured using the hydrated potassium thermo-dichromate oxidation method [12], while the Kjeldahl digestion method was used to determine TN [13].

2.5.3. Determination of Soil Microbial Community

The abundance of soil microbial community was calculated by conventional plate colony counting. Bacteria were cultured on beef extract-peptone medium (beef extract, 3.0 g; peptone,10.0 g; NaCl, 5.0 g; agar, 15–25 g; metalaxyl and propamocarb 1.25 g; water, 1000 mL; pH 7.4–7.6 for 1 L) [14], while fungi were cultured on Martin medium ($KH_2PO_4 \cdot 3H_2O$, 0.1 g; $MgSO_4 \cdot 7H_2O$, 0.05 g; 0.1% Bengal red solution, 0.33 mL; distilled water, 100 mL; natural pH 2%; sodium deoxycholate solution, 2 mL; streptomycin solution (10,000 units·mL^{-1}) 0.33 mL for 1 L) [15]. For the cultivation of bacteria and fungi, we took 50 μL 10^{-6} and 10^{-4} soil suspension, respectively, and added them to the corresponding medium. Each treatment was repeated 4 times. After solidification, they were placed in a 28 °C constant temperature incubator. Bacteria and fungi were cultured for 3–4 days and 3–5 days, respectively.

2.5.4. Dynamic Changes of Soil Enzyme Activities

Seven kinds of soil enzyme activities (acid phosphatase, sucrase, protease, polyphenol oxidase, urease, dehydrogenase, and catalase) were detected. Acid phosphatase activity was determined by the disodium phosphate benzene colorimetric method [16], and the 3,5-dinitrosalicylic acid method was applied to assay the sucrase activity [17]. Protease activity was assayed according to ninhydrin colorimetric method [18]. The activity of polyphenol oxidase in the soil was determined by spectrophotometry [19]. Urease activity, dehydrogenase activity, and catalase were measured by colorimetric analysis of sodium phenate-sodium hypochlorite [20], the triphenyl tetrazolium chloride reduction [21], and the potassium permanganate titration method [22], respectively.

2.6. Determination of Secondary Metabolites Content

2.6.1. Determination of Psoralen in F. carica

The fresh leaves were collected from the sample plot and dried at 60 °C for 24 h, then ground into powder. Dried powder samples were weighed (1 g) and dissolved in 25 mL of methanol. Each sample was extracted for 40 min by an ultrasonic cleaner. After filtration, the filtrate residue was ultrasonic with 25 mL methanol for 40 min, repeated twice. The filtrate was combined and evaporated to dryness, redissolved with methanol to 10 mL, and centrifuged to obtain the supernatant at 1200 rpm for 20 min, which was injected into a high-performance liquid chromatography (HPLC) system. Chromatographic conditions: The mobile phase was A: methanol and B: ACN-water (15:85, v/v), injection volume was 20 µL, the flow rate was 1.0 mL min^{-1}, detection wavelength was 310 nm, and an analysis cycle was 30 min [23].

2.6.2. Determination of Paclitaxel in T. cuspidata

The preparation method of the T. cuspidata extract is the same as 2.6.1. Chromatographic conditions: The mobile phase was methanol/acetonitrile/water (25:35:40, $v/v/v$) run over a period of 30 min at flow rate was 1.0 mL min^{-1}. Injection volume was 20 µL, detection wavelength was 232 nm [24].

2.7. Statistical Analysis

All experiments were conducted as completely randomized design in four replications. All of the data were analyzed using SPSS 22.0; one-way ANOVA and Duncan's multiple range test ($p < 0.01$) were used to further deal with the experimental data differences between treatments. Figures were created with Origin Pro 9.0.

3. Results

3.1. Variation of Plant Biomass in Different Planting Patterns

3.1.1. Variation of F. carica Growth in Different Planting Patterns

Variation of total volume of F. carica new branch under different planting patterns are shown in Table 1. The new branch total volume of IFT for 1–5 months was significantly ($p < 0.01$) higher than MF. The biomass of IFT for 5 months in the same year was 96.8 cm^3, which was 23.5% higher than that of MF. In terms of the total volume of F. carica new branch, the IFT pattern was better than MF.

Table 1. Variation of total volume of F. carica new branch under different planting patterns.

Month	Planting Patterns	
	Interplanting (cm^3)	Monoculture (cm^3)
5	45.8 ± 1.1 Ea	40.5 ± 2.0 Eb
6	63.4 ± 1.3 Da	56.1 ± 1.9 Db
7	75.9 ± 1.7 Ca	67.2 ± 2.1 Cb
8	85.7 ± 1.8 Ba	73.8 ± 1.4 Bb
9	96.8 ± 2.1 Aa	78.4 ± 1.8 Ab

Different uppercase letters indicate significant differences between different treatments time at $p < 0.01$, while different lowercase letters indicate significant differences in different planting patterns $p < 0.01$.

3.1.2. Variation of T. cuspidata Growth in Different Planting Patterns

Variation of T. cuspidata plant height under different planting patterns is shown in Table 2. Regardless of the planting pattern, the plant height of T. cuspidata increased significantly monthly ($p < 0.01$), and there was no significant difference in the growth rate of plant height between MT and IFT planted for 1 month ($p > 0.01$). The growth rate of plant height of IFT planted for 2–5 months was significantly ($p < 0.01$) higher than MT. The growth rate of plant height of IFT for 5 months was 121.7%, which was 1.2 times the growth rate of MT.

Table 2. Variation of *T. cuspidata* plant height under different planting patterns.

Month	Interplanting		Monoculture	
	Plant Height (cm)	Growth Rate of Plant Height (%)	Plant Height (cm)	Growth Rate of Plant Height (%)
4	34.1 ± 1.9 Fa	—	30.2 ± 1.5 Fb	—
5	40.9 ± 1.7 Ea	19.9 ± 1.9 Ea	34.7 ± 2.4 Eb	14.9 ± 2.4 Ea
6	47.4 ± 2.1 Da	39.0 ± 1.6 Da	40.2 ± 1.3 Db	33.1 ± 0.4 Db
7	57.9 ± 1.8 Ca	69.8 ± 1.8 Ca	48.7 ± 1.7 Cb	61.3 ± 0.7 Cb
8	68.5 ± 2.4 Ba	100.8 ± 1.3 Ba	57.5 ± 1.6 Bb	90.4 ± 1.8 Bb
9	75.6 ± 1.6 Aa	121.7 ± 1.2 Aa	61.8 ± 1.3 Ab	104.6 ± 0.9 Ab

Different uppercase letters indicate significant differences between different treatments time at $p < 0.01$, while different lowercase letters indicate significant differences in different planting patterns $p < 0.01$.

Variation of *T. cuspidata* basal diameter under different planting patterns is shown in Table 3. The growth rate of basal diameter of IFT planted for 2–5 months was significantly ($p < 0.01$) higher than that of MT. The growth rate of basal diameter of IFT for 5 months was 45.3%, which was 2.1 times the growth rate of MT. Considering the growth of plant height and basal diameter of *T. cuspidata*, IFT pattern is better than MT.

Table 3. Variation of *T. cuspidata* basal diameter under different planting patterns.

Month	Interplanting		Monoculture	
	Basal Diameter (mm)	Growth Rate of Basal Diameter (%)	Basal Diameter (mm)	Growth Rate of Basal Diameter (%)
4	14.54 ± 0.11 Fa	—	14.46 ± 0.17 Fa	—
5	15.38 ± 0.32 Ea	5.8 ± 2.1 Ea	14.98 ± 0.14 Eb	3.6 ± 7.8 Eb
6	16.89 ± 0.28 Da	16.2 ± 5.4 Da	15.71 ± 0.24 Db	8.6 ± 5.7 Db
7	18.53 ± 0.23 Ca	27.4 ± 2.2 Ca	16.63 ± 0.33 Cb	15.0 ± 8.0 Cb
8	20.23 ± 0.37 Ba	39.1 ± 3.2 Ba	17.35 ± 0.26 Bb	20.0 ± 4.1 Bb
9	21.13 ± 0.25 Aa	45.3 ± 1.2 Aa	17.63 ± 0.22 Ab	21.9 ± 7.3 Ab

Different uppercase letters indicate significant differences between different treatments time at $p < 0.01$, while different lowercase letters indicate significant differences in different planting patterns $p < 0.01$.

3.2. Variation of Plant Photosynthesis in Different Planting Patterns
3.2.1. Variation of F. carica Photosynthesis in Different Planting Patterns

Variation of *F. carica* Pn, Gs, and Ci in different planting patterns was investigated and results were shown in Figure 3. The Pn in the IFT pattern was better than MF, which increased significantly ($p < 0.01$) by 22.0%, 14.4%, 14.3%, 12.1%, and 8.8% from May to September. Compared with MF, the Gs of IFT was significantly increased ($p < 0.01$) by 10.2% and 11.1% from August to September, but there was no significant difference from May to July. The Ci of IFT pattern was 3.8%, 4.1%, 4.4%, 3.8%, and 3.7% (significantly, $p < 0.01$) higher than those of the MF from May to September, respectively. These results indicated that the IFT pattern had the advantage of enhancing the photosynthesis of *F. carica*.

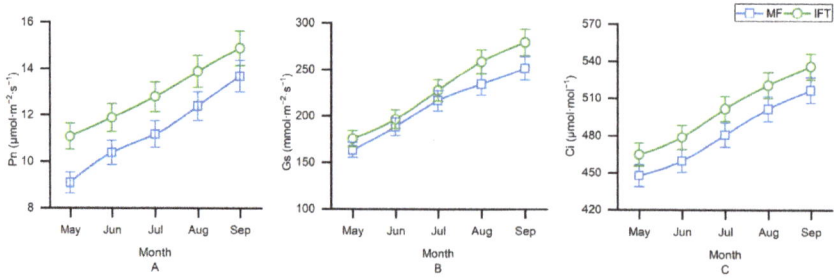

Figure 3. Variation of net photosynthetic rate (**A**), stomatal conductance (**B**), and intercellular carbon dioxide concentration (**C**) of *F. carica* under different planting patterns. Values are reported as Mean ± SD, $n = 4$.

3.2.2. Variation of T. cuspidata Photosynthesis in Different Planting Patterns

Figure 4 shows that the variation of Pn, Gs, and Ci of *T. cuspidata* under different planting patterns. Regardless of the planting pattern, the Pn, Gs, and Ci of *T. cuspidata* increased significantly monthly and reached the highest point in September. Compared with MT, the Pn, Gs, and Ci of IFT were significantly ($p < 0.01$) increased by 10.0%, 6.7%, and 9.0% in September, respectively. In terms of photosynthesis, the IFT pattern was better than MT.

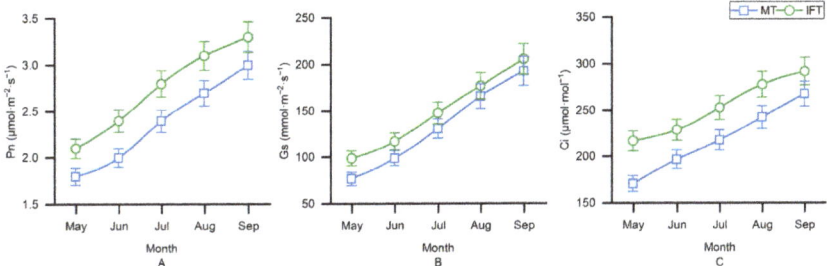

Figure 4. Variation of net photosynthetic rate (**A**), stomatal conductance (**B**), and intercellular carbon dioxide concentration (**C**) of *T. cuspidata* under different planting patterns. Values are reported as mean ± SD, $n = 4$.

3.3. Variation of Soil Chemical Analysis in Different Planting Patterns

3.3.1. Variation of Soil Organic Carbon and Total Nitrogen in Different Planting Patterns

Both monoculture and interplanting patterns had significant effects on the contents of SOC and TN (Figure 5). The SOC content ranged from 2.02 mg g^{-1} to 3.45 mg g^{-1} among all planting patterns and SOC content increased as time increase, where it was observed as the highest in September. For SOC content, IFT had the highest content among all planting patterns. Overall, the SOC contents among all planting patterns were in the order of IFT > MT > MF (Figure 5A).

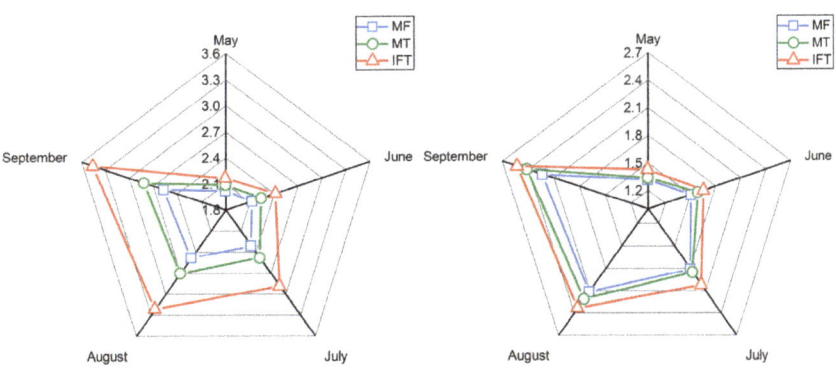

Figure 5. Variability in (**A**) soil organic carbon (SOC) and (**B**) total nitrogen (TN) in different planting patterns. Values are reported as mean ± SD, $n = 4$.

Similar results compared with SOC, the TN showed the same trend between the monoculture and interplanting patterns. The TN content ranged from 1.13 mg g^{-1} to 2.52 mg g^{-1} among all planting patterns. For TN content, IFT had the highest content among all planting patterns. Overall, the TN contents among all planting patterns were in the order of IFT > MT > MF (Figure 5B). In terms of SOC and TN content, the IFT pattern was better than MF and MT.

3.3.2. Variation of Soil Microbial Community in Different Planting Patterns

Variability in bacteria, fungus, and bacteria/fungus in soil under different planting patterns is shown in Figure 6. Regardless of the planting pattern, the number of bacteria increased significantly monthly and reached the highest point in September. Overall, the number of bacteria were in the order of MT > MF > IFT (Figure 6A). The amount of fungus in the IFT pattern was 13.8% significantly higher than MF in September. Compared with MT, the amount of fungus in IFT was significantly increased ($p < 0.01$) by 22.2% in September. Overall, the amount of fungus was in the order of IFT > MF > MT (Figure 6B). The ratio of bacteria/fungus reached the lowest value under IFT pattern in August, which was 0.89. Overall, the ratio of bacteria/fungus was in the order of IFT > MT > MF (Figure 6C).

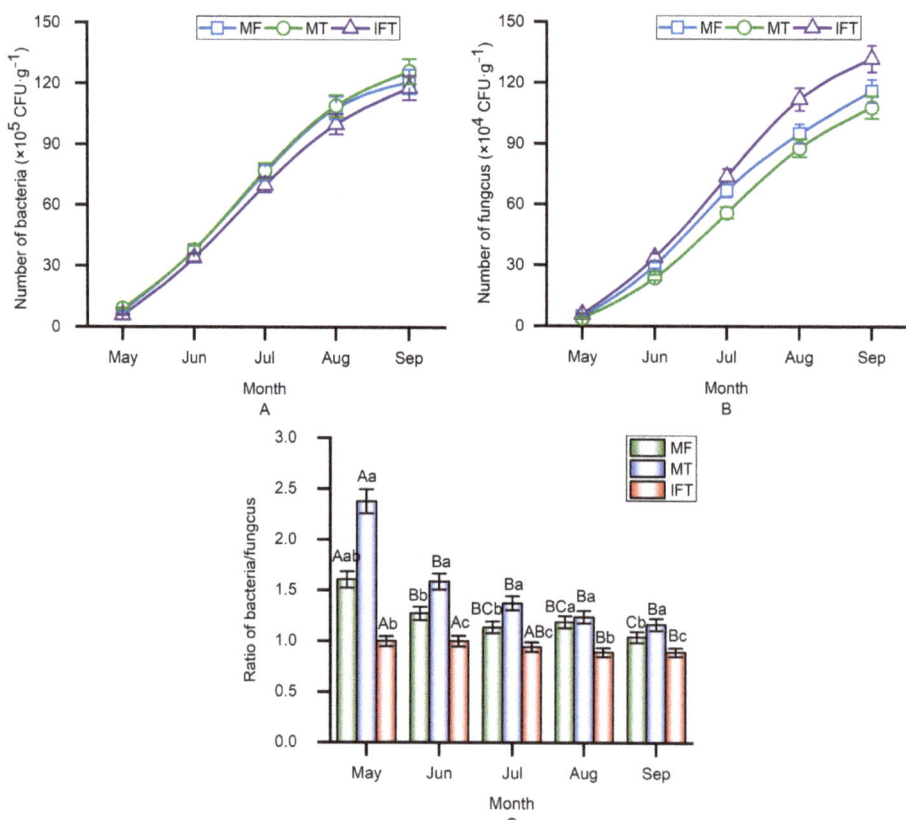

Figure 6. Variability in (**A**) bacteria, (**B**) fungus, and (**C**) bacteria /fungus in soil under different planting pat terns. Values are reported as mean ± SD, $n = 4$. Different uppercase letters indicate significant differences between different treatments time at $p < 0.01$, while different lowercase letters indicate significant differences in different planting patterns $p < 0.01$.

3.3.3. Variation of Soil Enzyme Activities in Different Planting Patterns

Figure 7 shows the variability in soil enzyme activities of acid phosphatase, sucrase, protease, polyphenol oxidase, urease, dehydrogenase, and catalase with time under three planting patterns. Regardless of the planting pattern, seven soil enzyme activities increased significantly monthly and were observed to be the highest in September. The seven enzyme activities of IFT pattern were 14.4%, 60.0%, 5.1%, 47.3%, 15.4%, 9.2%, and 30.0% higher than that of MF in September, respectively. Compared with MT, the soil corresponding enzyme activities of IFT were significantly increased ($p < 0.01$) by 34.8%, 52.4%, 13.8%, 131.0%, 48.4%, 15.3%, and 47.7% in September, respectively. In terms of soil enzyme activities, the IFT pattern was better than MF and MT.

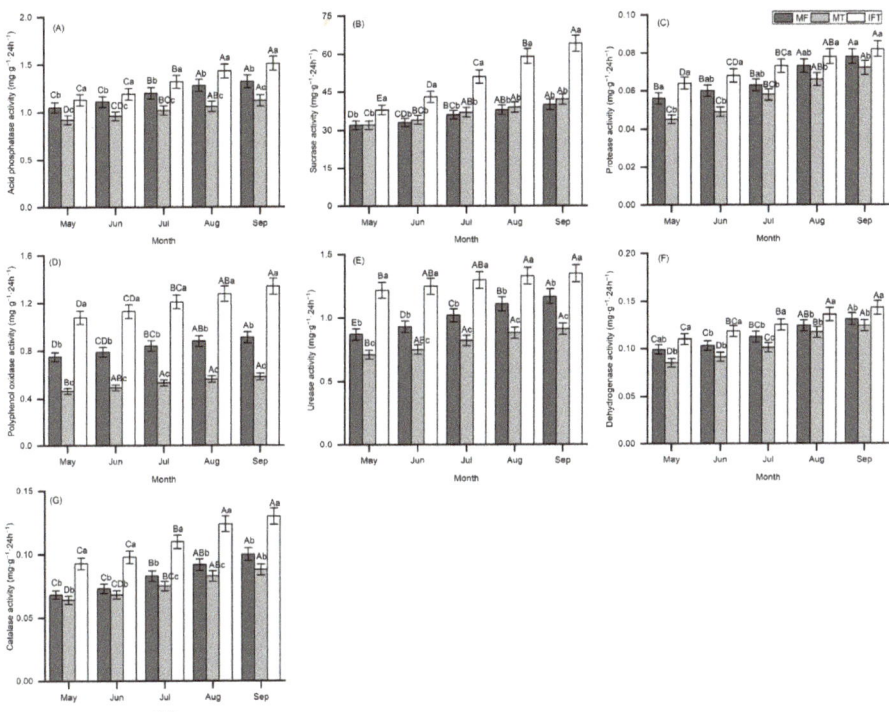

Figure 7. Variability in (**A**) acid phosphatase, (**B**) sucrase, (**C**) protease, (**D**) polyphenol oxidase, (**E**) urease, (**F**) dehydrogenase, and (**G**) catalase bacteria soil enzyme activities under different planting patterns. Values are reported as mean ± SD, n = 4. Different uppercase letters indicate significant differences between different treatments time at $p < 0.01$, while different lowercase letters indicate significant differences in different planting patterns $p < 0.01$.

3.4. Variation of Secondary Metabolites Content in Different Planting Patterns

The psoralen content in *F. carica* and paclitaxel content in *T. cuspidata* under different planting patterns are shown in Figure 8. The psoralen content in *F. carica* planted for 1–5 months in IFT was significantly ($p < 0.01$) higher than that of MF. The psoralen content in *F. carica* of IFT pattern for 5 months was 4.12 mg g^{-1}, which was 1.1 times of MF (Figure 8A).

The paclitaxel content in *T. cuspidata* planted for 1–5 months was significantly ($p < 0.01$) higher than that of MT. The paclitaxel content significantly increased as time increased, where it was observed as the highest in September. In addition, the paclitaxel content of IFT was 34.3% higher than that of MT in September (Figure 8B). These results indicated that IFT pattern had the advantage of improving the secondary metabolites content.

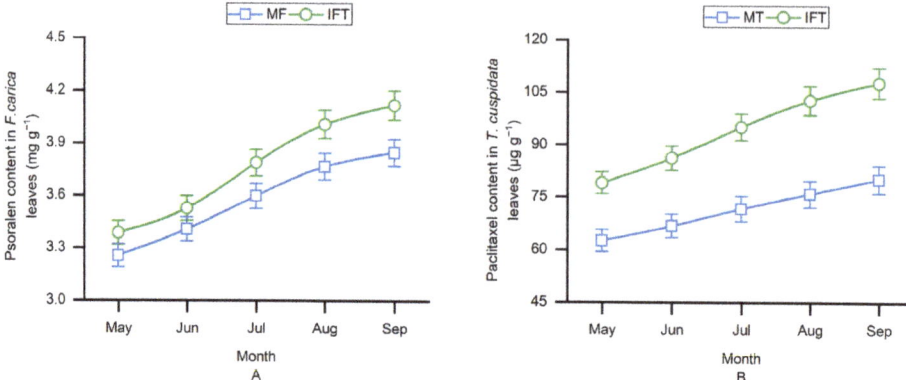

Figure 8. Variability in (**A**) psoralen content in *F. carica* and (**B**) paclitaxel content in *T. cuspidata* under different planting patterns. Values are reported as mean ± SD, $n = 4$.

4. Discussion

Compared with the monoculture pattern, the growth biomass of *F. carica* and *T. cuspidata* under interplanting pattern were significantly improved. It was reported that interplanting can effectively improve the growth of some plants [25,26]. *F. carica* could weaken the damage of strong light to *T. cuspidata* under IFT pattern, which could partly explain the promoting effect of *F. carica* on *T. cuspidata* growth under the interplanting pattern. The photosynthetic index under IFT pattern was significantly higher than that of MF, which indicated that there was more dry matter accumulation in the interplanting pattern [27]. In theory, interplanting did not significantly change the light, temperature, water, air, or other ecological factors that might affect the growth of *F. carica*, so it was speculated that the change of soil factors caused by interplanting pattern was the important factor promoting the growth of *F. carica* and *T. cuspidata*.

In our study, we found that the most significant effect of IFT was the increase of SOC and TN contents. Great SOC accumulation in the presence of the interplanting pattern was ascribed to enhance SOC input from higher plant biomass production [28]. The higher SOC and TN contents in the interplanting pattern might be a sign of improved soil nutrient cycling and accumulation [29–31]. The presence of high diversity interplanting pattern promoted soil fertility and storage of SOC and TN contents compared to monoculture pattern. Another possible reason was that interplanting can provide more comprehensive ground coverage and better water use efficiency [32,33]. The combination of different plant species affects the composition of the soil nutrient; in particular, plant diversity has a significantly higher impact on soil nutrients than any other factor, which is likely due to the more developed roots of the IFT pattern and the surrounding soil and results in more favorable nutrients for root development [34–36]. Thus, it was speculated that the IFT pattern increases the soil nutrient conversion cycle and might help the suppression of plant disease.

Our research suggests that the amount of bacteria in soil tends to be lower than that of the monoculture pattern following the IFT while the amount of fungus tends to be higher. Moreover, the total number of microorganisms showed an increasing trend. Combined with the growth indicators of the two kinds of plants, a decrease in bacteria and an increase in fungi allowed the plants to grow more luxuriantly. The number of microorganisms in soil plays an important role in plant growth [37]. Interplanting can significantly increase the number of soil microorganisms, thus promoting the release of soil nutrients, so that plants absorb nutrients more conducive to growth [38–40]. The IFT pattern enhanced the relative number of potentially beneficial microorganisms in the soil, which is crucial to decreasing the prevalence of soil-borne diseases and keeping soil healthy. To some level, it may be attributed to complex interactions between plants, such as allelopathy. Plant roots can release chemicals into the soil affecting the soil microbial community and can thus affect soil properties in many ways [41].

Soil enzymes are a kind of special metabolite in soil, and also an indicator of relatively stable and sensitive soil biological activity [42]. A superior understanding of the soil enzymatic activity in different planting patterns with a profundity effect can lead to a better knowledge about how interplanting patterns enhance soil microbial activity and soil fertility. Soil enzyme activity can affect soil physical and chemical properties, thus affecting plant growth. Acid phosphatase is the main enzyme in the cycle of phosphorus in soil and can catalyze the transformation of soil organic phosphorus into inorganic phosphorus available to plants [43–45]. Sucrase is an important catalytic enzyme for carbon cycling and its activity can reflect soil ripening degree and fertility level [46,47]. Protease enzyme plays a crucial role in the catalysis of nitrogen minerals and nitrogen cycling [48] while polyphenol oxidase could protect roots from bacteria and viruses as they grow [49]. Urease's primary function is to catalyze the transformation of soil amide nitrogen into ammonium nitrogen that can be absorbed directly by plants and its activity is closely related to nitrogen use efficiency [50]. It has been reported that dehydrogenase activity participates in the biological oxidation of soil organic matter, proteolytic activity, and respiration [51]. To some level, catalase activity is a critical factor that prevents the oxidation of nutrients in root protoplasts [52,53].

In our present study, a very significant improvement in seven soil enzymatic activities was observed in the interplanting pattern compared to in the monoculture pattern. *F. carica* and *T. cuspidata* in interplanting pattern could absorb more nutrients, make more direct use of ammonium nitrogen, protein, and inorganic phosphorus in the soil, and can protect from the threat of bacteria and viruses in the soil, allowing the plant to grow more smoothly. These results indicated that an interplanting pattern can improve the level of soil fertility and increase the utilization rate of nitrogen and phosphorus [54], which may be another important reason for the increase of the growth biomass of these two plants, especially *F. carica*. Nutrients, enzymes, and microorganisms work together to achieve a dynamic balance and have a positive impact on plant growth. Therefore, from the perspective of soil enzyme activity index, the IFT pattern was superior to that of MF and MT.

Psoralen is a secondary metabolite in *F. carica* and paclitaxel a secondary metabolite in *T. cuspidata*. In our study, the IFT pattern significantly increased the psoralen content in *F. carica*. This is similar to the result of where the IFT pattern could increase the paclitaxel content in *T. cuspidata*. As an agrotechnical approach, interplanting normally has a crucial factor in the production and accumulation of plant secondary metabolites [55,56]. Secondary metabolites secreted can also have a wide range of biological activities that either protect the plant against pests and pathogens or act as plant growth promotors which can be beneficial for the agricultural crops [57,58]. This study forms the theoretical basis for a sustainable way to optimize medicinal-agroforestry systems. *F. carica* provides shade for the *T. cuspidata*, reducing the cost of building shade sheds. Thus, medicinal-agroforestry systems enhance land utilization, realize a joint increase in production, and enhance economic benefits. In addition, further long-term monitoring is needed to confirm the usefulness of such a plant combination for human health.

5. Conclusions

In conclusion, the IFT pattern can enhance the ecological environment of soil and soil quality by increasing plant biomass, photosynthesis, soil enzyme activities, number of potentially beneficial microorganisms, and secondary metabolite content, hence improving the yield and quality of *F. carica* and *T. cuspidata*. The results of our study can provide a feasible theoretical basis for the large-scale establishment of *F. carica* and *T. cuspidata* mixed forests and obtain high-quality medicine sources for extracting psoralen and paclitaxel. The IFT pattern is an excellent example of a new strategy for ecological medicinal plants planting, which plays a crucial role in long-term sustainable development and production of medicinal plants. Future management should be cognizant of medicinal-agroforestry interplanting patterns, as they have multiple benefits over monoculture patterns.

Author Contributions: C.Z. and C.L. conceived the underlying ideas of the study; X.Y. and Y.L. contributed to the practical aspects of the research work; Q.L., B.Q. and S.S. analyzed the data; X.Y. and Y.L. wrote the manuscript. All authors have read and agreed to the published version of the manuscript.

Funding: This research was funded by the Fundamental Research Fund for Central Universities (2572019CZ01), the National Natural Science Foundation (31870609), Heilongjiang Touyan Innovation Team Program (Tree Genetics and Breeding Innovation Team), and the 111 Project of China (B20088).

Institutional Review Board Statement: Not applicable.

Informed Consent Statement: Not applicable.

Data Availability Statement: Data are contained within the article.

Acknowledgments: The authors thank the Rongcheng healthy group Co., Ltd. for supplying the seedlings of *F. carica* and *T. cuspidata*.

Conflicts of Interest: The authors have declared no conflict of interest.

References

1. Barolo, M.I.; Ruiz Mostacero, N.; López, S.N. *Ficus carica* L. (Moraceae): An ancient source of food and health. *Food Chem.* **2014**, *164*, 119–127. [CrossRef]
2. Li, C.; Yu, M.; Li, S.; Yang, X.; Qiao, B.; Shi, S.; Zhao, C.; Fu, Y. Valorization of Fig (*Ficus carica* L.) Waste Leaves: HPLC-QTOF-MS/MS-DPPH System for Online Screening and Identification of Antioxidant Compounds. *Plants* **2021**, *10*, 2532. [CrossRef]
3. Sun, R.; Sun, L.; Jia, M. Analysis of psoralen and mineral elements in the leaves of different fig (*Ficus carica*) cultivars. *Acta Hortic.* **2017**, *1173*, 293–296. [CrossRef]
4. Wojdyło, A.; Nowicka, P.; Carbonell-Barrachina, Á.A.; Hernández, F. Phenolic compounds, antioxidant and antidiabetic activity of different cultivars of *Ficus carica* L. fruits. *J. Funct. Foods* **2016**, *25*, 421–432. [CrossRef]
5. Abdel-Rahman, R.; Ghoneimy, E.; Abdel-Wahab, A.; Eldeeb, N.; Salem, M.; Salama, E.; Ahmed, T. The therapeutic effects of *Ficus carica* extract as antioxidant and anticancer agent. *S. Afr. J. Bot.* **2021**, *141*, 273–277. [CrossRef]
6. Jiang, P.; Zhao, Y.J.; Xiong, J.; Wang, F.; Xiao, L.J.; Bao, S.Y.; Yu, X.D. Extraction, Purification, and Biological Activities of Flavonoids from Branches and Leaves of *Taxus cuspidata* S. et Z. *Bio. Resources* **2021**, *16*, 2655–2682. [CrossRef]
7. Bajpai, V.K.; Sharma, A.; Moon, B.; Baek, K.H. Chemical Composition Analysis and Antibacterial Mode of Action of Taxus Cuspidata Leaf Essential Oil against Foodborne Pathogens. *Ital. J. Food Saf.* **2014**, *34*, 9–20. [CrossRef]
8. Rozendaal, E.V.; Lelyveld, G.P.; Beek, T.V. Screening of the needles of different yew species and cultivars for paclitaxel and related taxoids. *Phytochemistry* **2000**, *53*, 383–389. [CrossRef]
9. Sollen-Norrlin, M.; Ghaley, B.B.; Rintoul, N.L.J. Agroforestry Benefits and Challenges for Adoption in Europe and Beyond. *Sustainability* **2020**, *12*, 7001. [CrossRef]
10. Ilany, T.; Ashton, M.S.; Montagnini, F.; Martinez, C. Using agroforestry to improve soil fertility: Effects of intercropping on *Ilex paraguariensis* (yerba mate) plantations with Araucaria angustifolia. *Agrofor. Syst.* **2010**, *80*, 399–409. [CrossRef]
11. Pala, N.A. Soil Microbial Characteristics in Sub-Tropical Agro-Ecosystems of North Western Himalaya. *Curr. Sci.* **2018**, *115*, 1956–1959.
12. Yeomans, J.C.; Bremner, J.M. A rapid and precise method for routine determination of organic carbon in soil. *Commun. Soil Sci. Plant Anal.* **1988**, *19*, 1467–1476. [CrossRef]
13. Gallaher, R.N.; Weldon, C.O.; Boswell, F.C. A semiautomated procedure for total nitrogen in plant and soil samples. *Soil Sci. Soc. Am. J.* **1976**, *40*, 887–889. [CrossRef]
14. Gu, L.; Xu, B.; Liang, Q.; Yin, T. Impact and colonisation ability of on lawn soil microflora. *Acta Pratoculturae Sin.* **2013**, *22*, 321–326.
15. Tang, X.; Jiang, J.; Huang, Z.; Wu, H.; Wang, J.; He, L.; Xiong, F.; Zhong, R.; Liu, J.; Han, Z. Sugarcane/peanut intercropping system improves the soil quality and increases the abundance of beneficial microbes. *J. Basic Microbiol.* **2021**, *61*, 165–176. [CrossRef]
16. Zhang, Q.; Liu, X.; Ma, X.; Fang, J.; Fan, T.; Wu, F.; An, L.; Feng, H. Microcalorimetric study of the effects of long-term fertilization on soil microbial activity in a wheat field on the Loess Plateau. *Ecotoxicology* **2014**, *23*, 2035–2040. [CrossRef] [PubMed]
17. Tang, X.; Zhong, R.; Jiang, J.; He, L.; Huang, Z.; Shi, G.; Wu, H.; Liu, J.; Xiong, F.; Han, Z. Cassava/peanut intercropping improves soil quality via rhizospheric microbes increased available nitrogen contents. *BMC Biotechnol.* **2020**, *20*, 13. [CrossRef] [PubMed]
18. Guan, S.Y.; Zhang, D.; Zhang, Z. *Soil Enzyme and Its Research Methods*; Agriculture Press: Beijing, China, 1986.
19. Cordero, I.; Snell, H.; Bardgett, R.D. High throughput method for measuring urease activity in soil. *Soil Biol. Biochem.* **2019**, *134*, 72–77. [CrossRef]

20. Zhang, J.; Yu, X.; Xu, B.; Yagoub, A.E.G.A.; Mustapha, A.T.; Zhou, C. Effect of intensive pulsed light on the activity, structure, physico-chemical properties and surface topography of polyphenol oxidase from mushroom. *Innov. Food Sci. Emerg. Technol.* **2021**, *72*, 102741. [CrossRef]
21. Wyk, D.A.B.v.; Adeleke, R.; Rhode, O.H.J.; Bezuidenhout, C.C.; Mienie, C. Ecological guild and enzyme activities of rhizosphere soil microbial communities associated with Bt-maize cultivation under field conditions in North West Province of South Africa. *J. Basic Microbiol.* **2017**, *57*, 781–792. [PubMed]
22. Friedel, J.K.; Mölter, K.; Fischer, W.R. Comparison and improvement of methods for determining soil dehydrogenase activity by using triphenyltetrazolium chloride and iodonitrotetrazolium chloride. *Biol. Fertil. Soils* **1994**, *18*, 291–296. [CrossRef]
23. Mueller, S.; Riedel, H.D.; Stremmel, W. Determination of catalase activity at physiological hydrogen peroxide concentrations. *Anal. Biochem.* **1997**, *245*, 55–60. [CrossRef]
24. Kazemipoor, M.; Lorestani, M.A.; Ansari, M. Extraction and determination of biomarkers in *Ficus carica* L. leaves from various species and different cultivars by HPLC. *J. Liq. Chromatogr. Relat. Technol.* **2012**, *35*, 2831–2844. [CrossRef]
25. Gill, H.; Vasundhara, M. Isolation of taxol producing endophytic fungus *Alternaria brassicicola* from non-Taxus medicinal plant Terminalia arjuna. *World J. Microbiol. Biotechnol.* **2019**, *35*, 74. [CrossRef]
26. Kinoshita, T.; Yamazaki, H.; Inamoto, K. Effects of interplanting on fruit yield and dry matter production in greenhouse-grown tomato by integratingtwo different crop periods. *JARQ Jpn. Agr. Res. Q.* **2019**, *53*, 295–304. [CrossRef]
27. Wang, X.; Zhang, R.; Wang, J.; Di, L.; Sikdar, A. The Effects of Leaf Extracts of Four Tree Species on *Amygdalus pedunculata* Seedlings Growth. *Front Plant Sci.* **2021**, *11*, 587579. [CrossRef] [PubMed]
28. Pandey, M.; Singh, T. Effect of Intercropping Systems and Different Levels of Nutrients on Dry Matter Accumulation and Physiological Growth Parameters of Bed Planted Wheat (*Triricum aestivum* L.). *Indian. J. Sci. Technol.* **2015**, *8*, 11. [CrossRef]
29. Marconi, L.; Armengot, L. Complex agroforestry systems against biotic homogenization: The case of plants in the herbaceous stratum of cocoa production systems. *Agric. Ecosyst. Environ.* **2020**, *287*, 106664. [CrossRef]
30. Szw, A.; Fsc, A.; Xfh, B.; Yang, Z.A.; Xmf, A. Urbanization aggravates imbalances in the active C, N and P pools of terrestrial ecosystems—ScienceDirect. *Glob. Ecol. Conserv.* **2020**, *21*, e00831.
31. Loveland, P.; Webb, J. Is there a critical level of organic matter in the agricultural soils of temperate regions: A review. *Soil Tillage Res.* **2003**, *70*, 1–18. [CrossRef]
32. Jiang, F.; Drohan, P.J.; Cibin, R.; Preisendanz, H.E.; Veith, T.L. Reallocating crop rotation patterns improves water quality and maintains crop yield. *Agric. Syst.* **2021**, *187*, 103015. [CrossRef]
33. Jensen, E.S. Grain yield, symbiotic N_2 fixation and interspecific competition for inorganic N in pea-barley intercrops. *Plant Soil* **1996**, *182*, 25–38. [CrossRef]
34. Jensen, E.S.; Carlsson, G.; Hauggaard-Nielsen, H. Intercropping of grain legumes and cereals improves the use of soil N resources and reduces the requirement for synthetic fertilizer N: A global-scale analysis. *Agron. Sustain. Dev.* **2020**, *40*, 5. [CrossRef]
35. Zhang, F.; Shen, J.; Zhang, J.; Zuo, Y.; Chen, X. Rhizosphere Processes and Management for Improving Nutrient Use Efficiency and Crop Productivity. *Adv. Agron.* **2010**, *107*, 1–32.
36. Shen, J.; Li, C.; Mi, G.; Li, L.; Yuan, L.; Jiang, R.; Zhang, F. Maximizing root/rhizosphere efficiency to improve crop productivity and nutrient use efficiency in intensive agriculture of China. *J. Exp. Bot.* **2013**, *64*, 1181–1192. [CrossRef] [PubMed]
37. Ehrmann, J.; Ritz, K. Plant: Soil interactions in temperate multi-cropping production systems. *Plant Soil* **2014**, *376*, 1–29. [CrossRef]
38. Li, Y.; Li, Z.; Arafat, Y.; Lin, W.; Jiang, Y.; Weng, B.; Lin, W.X. Characterizing rhizosphere microbial communities in long-term monoculture tea orchards by fatty acid profiles and substrate utilization—ScienceDirect. *Eur. J. Soil Biol.* **2017**, *81*, 48–54. [CrossRef]
39. Langenberger, G.; Cadisch, G.; Martin, K.; Min, S.; Waibel, H. Rubber intercropping: A viable concept for the 21st century? *Agrofor. Syst.* **2017**, *91*, 577–596. [CrossRef]
40. Qin, X.; Wei, C.; Li, J.; Chen, Y.; Chen, H.S.; Zheng, Y.; Nong, Y.; Liao, C.; Chen, X.; Luo, Y.; et al. Changes in Soil Microbial Community Structure and Functional Diversity in the Rhizosphere Surrounding Tea and Soybean. *J. Agri. Sci. Sri Lanka* **2017**, *12*, 1. [CrossRef]
41. Mortimer, P.E.; Gui, H.; Xu, J.; Zhang, C.; Barrios, E.; Hyde, K.D. Alder trees enhance crop productivity and soil microbial biomass in tea plantations. *Appl. Soil Ecol.* **2015**, *96*, 25–32. [CrossRef]
42. Schandry, N.; Becker, C. Allelopathic Plants: Models for Studying Plant–Interkingdom Interactions. *Trends Plant Sci.* **2020**, *25*, 176–185. [CrossRef]
43. Zhao, J.; Wu, X.; Nie, C.; Wu, T.; Dai, W.; Liu, H.; Yang, R. Analysis of unculturable bacterial communities in tea orchard soils based on nested PCR-DGGE. *World J. Microbiol. Biotechnol.* **2012**, *28*, 1967–1979. [CrossRef] [PubMed]
44. Chen, L.; Rossi, F.; Deng, S.; Liu, Y.; Wang, G.; Adessi, A.; Philippis, R.D. Macromolecular and chemical features of the excreted extracellular polysaccharides in induced biological soil crusts of different ages. *Soil Biol. Biochem.* **2014**, *78*, 1–9. [CrossRef]
45. Wang, Y.; Dong, J.; Zheng, X.; Zhang, J.; Zhou, P.; Song, X.; Song, W.; Wang, S. Wheat straw and biochar effect on soil carbon fractions, enzyme activities, and nutrients in a tobacco field. *Can. J. Soil Sci.* **2021**, *101*, 1–12. [CrossRef]
46. Rashid, M.; Tigabu, M.; Chen, H.; Farooq, T.H.; Wu, P. Calcium-mediated adaptive responses to low phosphorus stress in Chinese fir. *Trees* **2020**, *34*, 825–834. [CrossRef]
47. Kwiatkowski, C.A.; Harasim, E.; Feledyn-Szewczyk, B.; Antonkiewicz, J. Enzymatic Activity of Loess Soil in Organic and Conventional Farming Systems. *Agriculture* **2020**, *10*, 135. [CrossRef]

48. Jan, M.T.; Roberts, P.; Tonheim, S.K.; Jones, D.L. Protein breakdown represents a major bottleneck in nitrogen cycling in grassland soils. *Soil Biol. Biochem.* **2009**, *41*, 2272–2282. [CrossRef]
49. Roohi, M.; Arif, M.S.; Yasmeen, T.; Riaz, M.; Bragazza, L. Effects of cropping system and fertilization regime on soil phosphorous are mediated by rhizosphere-microbial processes in a semi-arid agroecosystem. *J. Environ. Manag.* **2020**, *271*, 111033. [CrossRef] [PubMed]
50. Li, Y.; Wang, C.; Gao, S.; Wang, P.; Shang, S. Impacts of simulated nitrogen deposition on soil enzyme activity in a northern temperate forest ecosystem depend on the form and level of added nitrogen. *Eur. J. Soil Biol.* **2021**, *103*, 103287. [CrossRef]
51. Uren, N.C. Types, Amounts, and Possible Functions of Compounds Released into the Rhizosphere by Soil-Grown Plants. *Rhizosph. Biochem. Org. Subst. Soil Plant Interface* **2007**, *2*, 1–21.
52. Farhangi-Abriz, S.; Ghassemi-Golezani, K.; Torabian, S. A short-term study of soil microbial activities and soybean productivity under tillage systems with low soil organic matter. *Appl. Soil Ecol.* **2021**, *168*, 104122. [CrossRef]
53. Wen, F.; Vanetten, H.D.; Tsaprailis, G.; Hawes, M. Extracellular proteins in pea root tip and border cell exudates. *Plant Physiol.* **2007**, *143*, 773–783. [CrossRef] [PubMed]
54. Mobley, H.; Hu, L.T.; Foxal, P.A. Helicobacter pylori Urease: Properties and Role in Pathogenesis. *Scand. J. Gastroenterol.* **2009**, *26*, 39–46. [CrossRef]
55. Bainard, L.D.; Koch, A.M.; Gordon, A.M.; Klironomos, J.N. Growth response of crops to soil microbial communities from conventional monocropping and tree-based intercropping systems. *Plant Soil* **2013**, *363*, 345–356. [CrossRef]
56. Ngwene, B.; Neugart, S.; Baldermann, S.; Ravi, B.; Schreiner, M. Intercropping Induces Changes in Specific Secondary Metabolite Concentration in Ethiopian Kale (*Brassica carinata*) and African Nightshade (*Solanum scabrum*) under Controlled Conditions. *Front. Plant Sci.* **2017**, *8*, 1700. [CrossRef] [PubMed]
57. Figueiredo, A.C.; Barroso, J.G.; Pedro, L.G.; Scheffer, J.J.C. Factors affecting secondary metabolite production in plants: Volatile components and essential oils. *Flavour. Fragr. J.* **2008**, *23*, 213–226. [CrossRef]
58. Keswani, C.; Singh, H.B.; García-Estrada, C.; Caradus, J.; Sansinenea, E. Antimicrobial secondary metabolites from agriculturally important bacteria as next-generation pesticides. *Appl. Microbiol. Biotechnol.* **2020**, *104*, 1013–1034. [CrossRef] [PubMed]

Article

The Use of Compost Increases Bioactive Compounds and Fruit Yield in Calafate Grown in the Central South of Chile

Fernando Pinto-Morales [1], Jorge Retamal-Salgado [2,*], María Dolores Lopéz [1], Nelson Zapata [1], Rosa Vergara-Retamales [3] and Andrés Pinto-Poblete [3]

[1] Faculty of Agronomy, Universidad de Concepción, Vicente Méndez 595, Casilla 537, Chillán 3812120, Chile; fernandopinto@udec.cl (F.P.-M.); mlopezb@udec.cl (M.D.L.); nzapata@udec.cl (N.Z.)
[2] Instituto de Investigaciones Agropecuarias, INIA-Quilamapu, Avenida Vicente Méndez 515, Chillán 3800062, Chile
[3] Faculty of Engineering and Business, Universidad Adventista de Chile, km 12 Camino a Tanilvoro, Chillán 3780000, Chile; rosavergara@unach.cl (R.V.-R.); andrespinto@unach.cl (A.P.-P.)
* Correspondence: jorge.retamal@inia.cl; Tel.: +56-9-9153-7288

Citation: Pinto-Morales, F.; Retamal-Salgado, J.; Lopéz, M.D.; Zapata, N.; Vergara-Retamales, R.; Pinto-Poblete, A. The Use of Compost Increases Bioactive Compounds and Fruit Yield in Calafate Grown in the Central South of Chile. *Agriculture* **2022**, *12*, 98. https://doi.org/10.3390/agriculture12010098

Academic Editors: Mario Licata, Antonella Maria Maggio, Salvatore La Bella and Teresa Tuttolomondo

Received: 7 December 2021
Accepted: 8 January 2022
Published: 11 January 2022

Publisher's Note: MDPI stays neutral with regard to jurisdictional claims in published maps and institutional affiliations.

Copyright: © 2022 by the authors. Licensee MDPI, Basel, Switzerland. This article is an open access article distributed under the terms and conditions of the Creative Commons Attribution (CC BY) license (https://creativecommons.org/licenses/by/4.0/).

Abstract: Different concentrations of compost applied as organic fertilizer can modify productive, quality, and chemical parameters in several fruit tree species. The objective of this study was to determine the effect of increasing applications of compost on physiological, productive, and quality parameters in calafate fruit during the seasons of 2018–2019 and 2019–2020. The study was conducted on a commercial calafate orchard using a randomized complete block design with four treatments (CK: no compost application, T1: 5 Ton ha^{-1}, T2: 10 Ton ha^{-1}, and T3: 15 Ton ha^{-1}), each with four repetitions. The results did not show statistical significance for stomatal conductance (Gs), quantum yield of PSII, or photosynthetic active radiation (PAR) within treatments. As for fruit yield, a statistical difference was found between the control treatment and T1, which were lower than T2 and T3 in both seasons. The trees reached a higher leaf area index with T2 in both seasons. The highest antioxidant capacity was obtained with T3 and T2 for the first and second season, respectively. Polyphenols and total anthocyanin production showed statistical significance, with a higher content at the second season with T2. It is concluded that the dose under which yield, quality, and nutraceutical content of calafate fruit are optimized is the one used in T2, 10 Ton ha^{-1}.

Keywords: polyphenols; Berberis; negative fruits; organic agriculture

1. Introduction

Calafate (*Berberis microphylla* G. Forst) is a bush native of the Chilean and Argentine Patagonia. In Chile, it can be found from the metropolitan region to Punta Arenas (34° 59′0″ South to 53° 28′33″ South). However, its existence is concentrated in the Aisén and Magallanes regions. In these regions, an increasing demand for products made from calafate has been observed [1,2]. Currently, this plant is a subject of study due to its biological properties, attributed mainly to the content of polyphenols present in it [3]. The antioxidant capacity of *B. microphylla* compared with other species has shown to be up to 10 times more than apples, oranges, and pears, and more than four times higher than blueberries [4]. Different studies have detected 18 anthocyanins in the calafate fruit, with a total concentration between 14.2 and 26 µmol g^{-1} of fresh weight [5,6]. These polyphenols (PF) substantially reduce the presence of degenerative, cardiovascular, and carcinogenic diseases [7]; therefore, they are beneficial compounds for human health [8].

Most studies of *B. microphylla* have been developed in the so-called Austral Zone of Chile; all of them use wild calafate for which vegetative growth takes place by mid-spring [9]. Furthermore, a comparative study about the morphology and anatomy of mature leaves of calafate, growing under two different conditions, showed that leaves change their morphology and structure to adapt to new environmental conditions [10]. This makes it important to analyze the behavior and/or adaptation of B. microphylla focusing on both physiology and phenology mainly in agroclimatic zones different from its natural habitat, which could also lead to negative effects of the quality of the product [11,12] as well as variations in photosynthetic efficiency [13,14]. Thus, a question that the present research addresses is: could the introduction of B. microphylla into an intensive commercial environment, different from the original, generate changes on productive and quality parameters of the fruit? [15].

Currently, there is an unsatisfied demand for this fruit that wild species have not been able to meet [16], generated by educated consumers who are eager for a more natural and nutritious alimentation to have a healthier life. This has increased safe and environmental friendly food production [17].

Calafate is a species capable of growing under several environmental conditions [5]. However, no agronomic research has been conducted under intensive-commercial conditions that allow for optimized productivity and polyphenolic content in the fruit [10]. Despite the fact that compost application is a widely used technique of organic fertilization to use bio-residues [18], however, it is known that compost usage has beneficial effects on both quality and soil fertility but also on the environment [19].

Compost, as an organic fertilizer, has been shown in other species to enhance fruit quality, as pointed out by Vásquez and Maravi [20], where applications of 10 Ton ha^{-1} of compost in *Morus alba* L. significantly increased yield and quality of the fruit. Similarly, in a study on strawberries (*Fragaria* × *ananassa* Duch) cvs. Allstar and Honeoye grown in pots with an organic fertilization, the concentration of anthocyanins, phenolic content, and antioxidant capacity of the fruit increased with the increasing compost doses [21].

Furthermore, in macadamia (*Macadamia integrifolia*), compost application to the soil increased its total cationic exchange capacity, organic matter, potassium (K), calcium (Ca), and magnesium (Mg), among other micronutrients [22]. Compost also modifies physical properties in the soil, such as total porosity and water retention capacity. It has also been observed on studies on wine grapes (*Vitis vinifera* cv. Chardonnay), improving nitrogen mineralization and its availability to the studied crop [23]. Additionally, in research conducted for 21 years of organic applications to eroded soils that were nutritionally deficient and with low pH, it significantly increased soil pH, organic carbon content, total nitrogen, phosphorus, potassium, available nitrogen, and biological activity [24].

According to the stated above, the objective of this research was to determine the effect of different doses of compost on productive and physiologic parameters, including polyphenolic composition and antioxidant activity of the fruit of calafate grown under an intensive agronomic management in the central zone of Chile.

2. Materials and Methods

2.1. General Characteristics of the Site of the Study and Orchard Establishment

The study was conducted at the Universidad Adventista de Chile (UnACh), located in the Kilometer 12 route to Tanilvoro, province of Chillán, region of Ñuble (36°31′ S; 71° 54′ W), Chile. The site of the study has volcanic soil (Melanoxerand) (Stolpe, 2006), a temperate Mediterranean climate, hot and dry summers, cold and humid winters, with an annual precipitation of 815 mm concentrated in winter and early spring [25]. The study was conducted in a commercial calafate orchard established in August of 2017, using two-year-old plants with an average height of 70 cm. The plants were multiplied from seeds in 2015. The plant population density was 1 m within a row and 3 m between rows, planted on berms 1 m wide and 20 cm high. After establishing the orchard, the soil was physically and chemically characterized at a depth of 0 to 40 cm, where most parts of the roots are found [9] (Table 1). These analyses were carried out at the laboratory of chemistry and physics of soils of the Agricultural Research Institute of Chile (INIA Quilamapu, Chillán, Chile).

Table 1. Physical-chemical analysis of the soil before treatments.

Analysis [1]	Unit	Result
Organic matter	%	9.70
Water pH		6.40
Nitrogen availability	mg kg^{-1}	19.00
Phosohorus availability	mg kg^{-1}	15.30
Potassium availability	mg kg^{-1}	496.00
Sulfur availability	mg kg^{-1}	24.00
Exchangeable calcium	cmol + kg^{-1}	8.70
Exchangeable magnesium	cmol + kg^{-1}	1.60
Exchangeable potassium	cmol + kg^{-1}	1.30
Exchangeable sodium	cmol + kg^{-1}	0.01
Sum of bases	cmol + kg^{-1}	11.60
Interchangeable aluminum	cmol + kg^{-1}	0.02
CEC	cmol + kg^{-1}	11.59
Aluminum saturation	%	0.14
Boron	mg kg^{-1}	0.40
Copper	mg kg^{-1}	1.63
Zinc	mg kg^{-1}	0.90
Iron	mg kg^{-1}	44.00
Manganese	mg kg^{-1}	3.02

[1] Samples were obtained at the beginning of the study in August of 2017, 0–40 cm depth. CEC = cation exchange capacity of soil.

At establishment, a base fertilization was applied into the plantation hole of 150 g of urea (45% N), 200 g of triple superphosphate (46% P_2O_5). and 200 g of potassium sulfate (50 % K_2O) [19]. In addition, hydraulic replenishment was standardized for all treatments and estimated according to the daily potential evapotranspiration of the crop (ETCc) using the methodology suggested by Romero et al. [26]. This was performed with the objective of maintaining optimum humidity levels in the soil during the entire development of the crop. Weed control was also standardized for all treatments and consisted of the manual elimination of them when establishing the orchard, plus three times a year, equally distributed, according to the annual cycle of the crop. In parallel, the same phytosanitary management was applied to all treatments, which consisted of six annual applications during the growing season, alternating two active ingredients, which were Tebuconazole (Orius 43 SC) at a concentration of 25.8 g per hectoliters and cuprous oxide (Cuprodul WG) at a concentration of 180 g per hectoliter of the active ingredient.

The experimental design used for the study was a randomized complete block design with a total of 4 treatments of compost doses, with 4 repetitions per treatment. Each treatment and repetition consisted of 4 plants in which the 2 central plants were evaluated. Additionally, there were border rows to help diminish the border effect. Compost treatments consisted of: (1) control treatment (CK) with no compost application, (2) 5 tons per hectare (Ton ha^{-1}) of compost (T1), (3) 10 Ton ha^{-1} of compost (T2), and (4) 15 Ton ha^{-1} of compost (T3). All treatments were applied each year in August to each experimental unit, administering the first application at the plantation.

The compost used in the study was commercially produced by the composting and recycling center of the Universidad Adventista of Chile, which was created from chicken manure, produced at the same institution, and oat bales. The manure and bales were mixed at a rate of 3:1 (Vol/Vol). The creation process lasted 5 months, controlling during that time the temperature, humidity, and ventilation [27–29]. The physical–chemical characterization of the compost used for this study is detailed in Table 2.

Table 2. Physical–chemical analysis of compost used for this study.

Analysis	Units	Result
Humidity (dry basis)	%	22.20
Apparent density (dry basis <16 mm)	Kg m^{-3}	NS *
Porosity (sample <16 mm)	mg kg^{-1}	NS *
pH in water 1:5		7.41
Electric conductivity 1:5	dS/m	0.19
Organic matter	%	21.60
Organic carbon	%	12.00
Total nitrogen	%	0.87
Nitrogen–ammonia (N–NH$_4^+$)	mg kg^{-1}	0.50
Nitrogen–Nitric (N–NO$_3^-$)	mg kg^{-1}	59.64
Carbon/Nitrogen Ratio	–	13.78
Ammonium/Nitrate Ratio	–	0.008

* Undetermined.

2.2. Characterization of Physiological and Environmental Conditions of the Plant

2.2.1. PAR Radiation and Leaf Area Index

For the purposes of environmental records, the photosynthetic photon flux density (PPFD, µmol m^{-2} s^{-1}) was quantified at five times of the day: 09:00, 11:00, 13:00, 15:00, and 17:00 h, in ambient conditions of a completely sunny day. The radiation parameters correspond to the direct, diffuse, residual, and reflected photosynthetically active radiation of the soil and plant throughout the development of the crop, and with these parameters, the intercepted PAR was estimated. For this, an AccuPAR LP-80 ceptometer (Decagon Devices Inc., Washington, DC, USA) was used, which delivers the average of 80 quantum sensors. The readings of the leaf area index (LAI; m^2 m^{-2}) were made at noon and were measured in post-harvest (January) when the growth of the plant had already stopped, using the same instrument and in parallel to the PAR radiation measurements [30].

2.2.2. Chlorophyll Fluorescence and Stomatal Conductance

The maximum intensity of fluorescence (Fm) was measured as well as the minimum intensity of fluorescence (Fo) of chlorophyll. This was achieved by using a portable fluorimeter model OS-5p (Opti-Sciences, Hudson, NH, USA) during a clear day at five times of the day, respectively: 09:00, 11:00, 13:00, 15:00, and 17:00 h, on leaves exposed to the sun and in the second third of a branch of the season [31].

Both Fo and Fm were determined after a period of 30 min in which the leaves were adapted to darkness [31,32]. For this, foliar clips that included a mobile obturation plate were used. With these parameters, the maximum photochemical efficiency of photosystem II (Fv/Fm) was quantified using the following relationship proposed by Kooten and Snell and Maxwell and Johnson [33,34]: Fv/Fm = (Fm − Fo)/Fm.

At the same time, stomatal conductance measurements (Gs, mmol m^{-2} s^{-1}) were performed. For this, a portable porometer model SC-1 (Decagon Devices INC, Washington, DC, USA) was used. The Gs measurements were performed on fully illuminated leaves of the same plant, shoots, location, and frequencies used in the chlorophyll fluorescence measurements. Using the same equipment and the same frequencies as for Fv/Fm, a record of the leaf temperature (Tf; °C) was kept. In order for the data collected to be representative, these were taken on leaves exposed to the sun and in the second third of a branch of the season [14].

2.3. Yield and Chemical Parameters of the Fruit

2.3.1. Calafate Fruit Productivity

The harvest was carried out 130 days after full flower for both study seasons. The harvest was done manually in which the total weight (g) of fruits per plant was quantified.

2.3.2. Determination of Total Polyphenol Concentration

Total polyphenols were determined by colorimetry using the method of Folin Ciocalteu in the food chemistry laboratory of the Universidad de Concepcion, Chillán, Chile. To calculate the polyphenol content, a calibration curve with gallic acid was used, with concentrations between 0 to 1000 mg L^{-1} of gallic acid according to the methodology proposed by Yıldırım et al. [35]. The results are expressed in mg of gallic acid 100 g^{-1} [10].

2.3.3. Determination of Anthocyanin Content

Total anthocyanins were determined by a differential pH technique. The determination of the anthocyanin content is based on the Lambert–Beer Law (A = ε * C * L), where A corresponds to the absorbance that is measured with a spectrophotometer; ε corresponds to the molar absorbance, a constant physics for molecular species in a solvent at a given wavelength; C is the molar concentration; and L is the length of the route, expressed in cm. Molar absorbance values for purified pigments were obtained from the literature. The concentration in mg L^{-1} was determined by multiplying by the molecular weight (MW) of the pigment. To calculate the anthocyanin content, the molecular weight and molar absorbance of the anthocyanin pigment present in the highest proportion were used [5]. The calculation of the anthocyanin concentration was carried out with the equation shown below, and data was expressed as mg of cyanidin 100 g^{-1} of fresh weight:

$$\frac{A \times 1000 \times 449.2}{26,900} \times \frac{3000}{100} \times \frac{5}{1000 \times g\ sample} \times 100 \quad (1)$$

2.3.4. Determination of Antioxidant Capacity

The DDPH antioxidant capacity was determined through the decolorization of the 1.1-Diphenyl-2-picrylhydrazyl free radical, proposed by Brand-Williams et al. [36]. The DPPH radical is reduced in the presence of antioxidants, manifesting a color change in the solution over time. To quantify the inhibition, a calibration curve was elaborated using the TROLOX reagent in methanol, achieving concentrations of 25, 50, 75, 100, 150, 200, 250, 300, 350, and 400 ppm. A methanol solution was used as a blank, and all solutions were incubated in the dark for 30 min; absorbance was measured in the spectrophotometer at 515 nm after 60 min. The antioxidant capacity was expressed in µmol Trolox equivalent (TE) 100 g^{-1} fresh weight).

2.4. Statistical Analysis

The effect of the treatments was estimated by an ANOVA and the Fischer LSD test, with a level of statistical significance of 0.05; for this, the INFOSTAT software was used (Infostat, Cordoba, Argentina, 2015).

3. Results

3.1. Edaphoclimatic and Physiological Parameters of Calafate

In Figure 1a, the average of direct photosynthetically active radiation (µmol $m^{-2}\ s^{-1}$) can be observed in the 2018–2019 and 2019–2020 seasons, quantified at five times of the day: 09:00 a.m., 11:00 a.m., 1:00 p.m., 3:00 p.m., and 5:00 p.m, presenting the lowest PAR values at 09:00 for both seasons, with values close to 750 µmol $m^{-2}\ s^{-1}$. In both seasons, the same trend of increasing PAR was observed from the first hours of the day until reaching the maximum values, close to 2000 µmol $m^{-2}\ s^{-1}$, towards the end of the day, with values close to 1500 µmol $m^{-2}\ s^{-1}$, in both seasons (Figure 1a). In Figure 1b, the absorbed photosynthetically active radiation can be observed, which did not show significant differences between treatments for each of the seasons under evaluation, registering similar values ($p < 0.05$) in both seasons, which were on average 866, 878, 893, and 873 µmol $m^{-2}\ s^{-1}$ for CK, T1, T2, and T3, respectively.

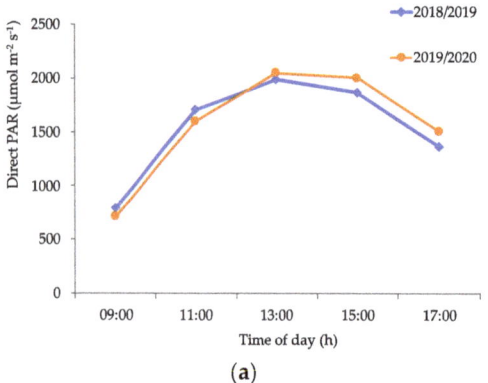
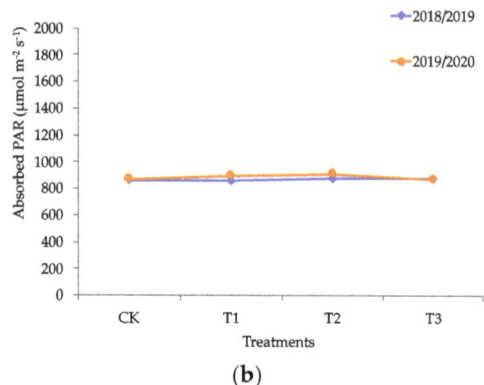

Figure 1. Average photosynthetically active radiation (PAR); (**a**) direct PAR (μmol m^{-2} s^{-1}) and (**b**) absorbed PAR (μmol m^{-2} s^{-1}) for the different treatments for the seasons 2018–2019 and 2019–2020. For (**b**): according to Fischer's LSD test ($p < 0.05$), there are no significant differences between treatments; the experimental error was very small, so the error bars were not observed.

Table 3 shows the variations in the physicochemical parameters of the soil at the end of the study for the different compost treatments. It should be noted that the percentage of organic matter increased in the different treatments as the volume of compost applied increased, being CK < T1 < T2 < T3, with values of 9.8%, 10.40%, 11.00%, and 11.90% for each treatment, respectively. For the pH parameter, no significant modifications were observed between treatments towards the end of the study, corresponding to 6.59, 6.56, 6.50, and 6.60 for CK, T1, T2, and T3, respectively. The nutritional levels of the soil tended to improve with increases in the dose of compost applied, mainly in the case of P, K, Ca, S, and Mg (Table 3). Together, the cation exchange capacity increased close to 12% in the treatments with the application of compost (Table 3). On the contrary, the concentration of iron and manganese decreased by 12 and 89%, respectively.

Table 3. Physicochemical analysis of the soil in the 2019/2020 season at the end of the study.

Analysis	Units	Treatments [1]			
		CK	T1	T2	T3
Organic matter	%	9.80	10.40	11.00	11.90
Water pH	—	6.59	6.56	6.50	6.60
Nitrogen available	mg kg^{-1}	16.00	18.00	16.00	13.00
Available phosphorus	mg kg^{-1}	10.00	14.00	9.00	13.00
Available potassium	mg kg^{-1}	342.00	332.00	372.00	359.00
Available sulfur	mg kg^{-1}	9.00	12.00	29.00	27.00
Exchangeable calcium	cmol+ kg^{-1}	9.25	10.90	10.04	10.86
Exchangeable magnesium	cmol+ kg^{-1}	2.17	2.33	1.82	2.19
Exchangeable potassium	cmol+ kg^{-1}	0.87	0.85	0.95	0.92
Exchangeable sodium	cmol+ kg^{-1}	0.30	0.27	0.29	0.29
Sum of bases	cmol+ kg^{-1}	12.59	14.36	13.10	14.26
Interch. aluminum	cmol+ kg^{-1}	0.010	0.001	0.010	0.010
CEC	cmol+ kg^{-1}	12.60	14.37	13.11	14.27
Aluminum saturation	%	0.08	0.07	0.08	0.07
Boron	mg kg^{-1}	0.45	0.48	0.50	0.45

Table 3. Cont.

Analysis	Units	Treatments [1]			
		CK	T1	T2	T3
Copper	mg kg^{-1}	2.06	2.11	2.14	1.85
Zinc	mg kg^{-1}	0.90	1.62	1.17	1.74
Iron	mg kg^{-1}	35.00	33.10	31.30	30.80
Manganese	mg kg^{-1}	3.02	3.50	3,20	6.00

[1] CK = Control treatment (without compost application); T1 = Treatment 1 (5 Ton ha^{-1}); T2 = Treatment 2 (10 Ton ha^{-1}); T3 = Treatment 3 (15 Ton ha^{-1}). CEC = cation exchange capacity of soil; all samples were obtained at the end of the study in August 2020 at a depth between 0–40 cm.

In Figure 2, the results obtained for the leaf area index parameter can be observed. In the first season, the treatment that registered the highest leaf area index (LAI) value was T2 (LAI: 2.5), which was significantly higher than CK and T3 and with no significant differences from T1 (Figure 2). For the second season, the same trend was observed, with T2 showing the highest LAI value (2.41; $p < 0.05$) compared to treatments CK, T1, and T3, which did now show statistical significance ($p > 0.05$) between the three of them; the results were 1.9, 1.9 and 2.08 LAI, respectively.

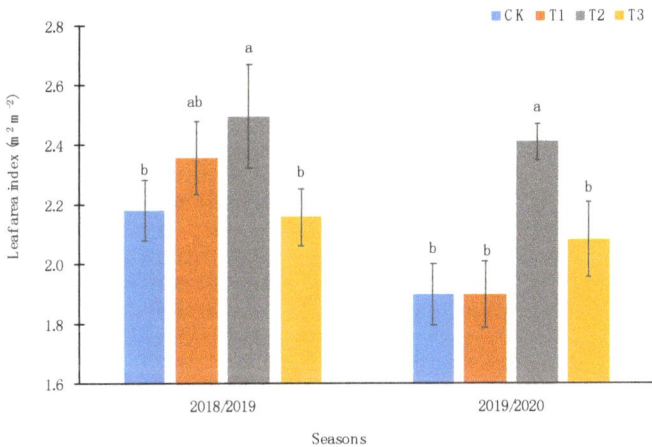

Figure 2. Leaf area index (LAI; m^2 m^{-2}) of calafate (*Berberis mycrophilla* G. Forst) for the different compost treatments for the 2018–2019 and 2019–2020 seasons. For each figure: CK = Control treatment (without compost application); T1 = Treatment 1 (5 Ton ha^{-1}); T2 = Treatment 2 (10 Ton ha^{-1}); T3 = Treatment 3 (15 Ton ha^{-1}). For each season, different lowercase letters indicate significant differences for the leaf area index between treatments according to Fischer's LSD test ($p < 0.05$). The bars correspond to the experimental error of each treatment.

The results for the variation of maximum quantum yield of photosystem II (Fv/Fm) analysis can be observed in Figure 3a,b for the 2018/2019 and 2019/2020 seasons, respectively. Emphasizing that, in the first season, all treatments presented similar values at the beginning of the day ($p < 0.05$), close to 0.8. Similar results were observed at 11:00, 13:00, 15:00, and 17:00, where the average values recorded for Fv/Fm were 0.79, 0.78, 0.76, and 0.77, respectively. In all treatment values, Fv/Fm decreased as the day passed (Figure 3a) without significant differences between the treatments ($p < 0.05$). For the second season, the same trend and values obtained in the first season were observed, on average, with only a lower average value of Fv/Fm towards the end of the day in the second season under evaluation (Figure 3b).

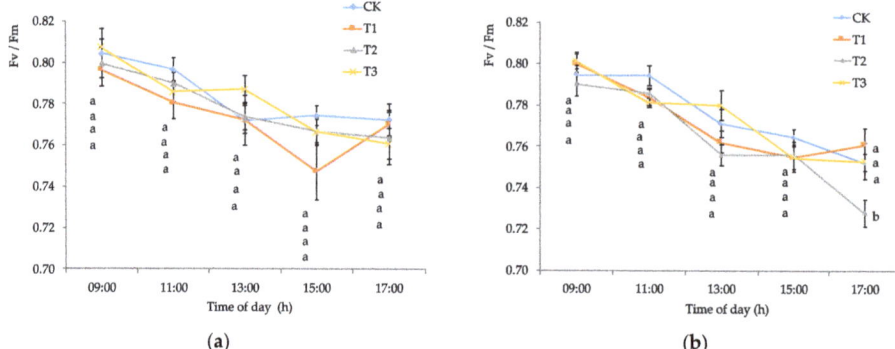

Figure 3. Variation of maximum quantum yield of photosystem II (Fv/Fm) in calafate plants (*Berberis mycrophilla* G. Forst) for the different compost treatments, evaluated at different times of the day, 09:00, 11:00, 13:00, 15:00 and 17:00, corresponding to the figure: (**a**) season 2018–2019; (**b**) season 2019–2020. For each figure: CK = Control treatment (without compost application); T1 = Treatment 1 (5 Ton ha^{-1}); T2 = Treatment 2 (10 Ton ha^{-1}); T3 = Treatment 3 (15 Ton ha^{-1}). For each hour of the day, different lowercase letters indicate significant differences between treatments according to Fischer's LSD test ($p < 0.05$). The bars correspond to the experimental error of each treatment.

In Figure 4, it is possible to observe the registered values of the stomatic conductance (mmol m^{-2} s^{-1}) of the calafate leaf at different hours of the day and for different compost dose treatments. Highlighting that, for the 2018–2019 season, no significant differences were observed among treatments, with average values of 260, 289, 308, and 291 mmol m^{-2} s^{-1} ($p > 0.05$) for CK, T1, T2, and T3, respectively. However, for the control treatment, from 11:00 until the end of the day, Gs values were always above 260 mmol m^{-2} s^{-1} unlike the rest of the treatments, which, after obtaining the maximum values of stomatal conductance, their values decreased by up to 40% (data not shown). For the 2019/2020 season, the same trend was observed as in the 2018/2019 season. However, recorded values were slightly lower than the previous season, with averages of Gs for CK, T1, T2, and T3 of 227, 214, 257, and 230 mmol m^{-2} s^{-1} ($p > 0.05$), respectively.

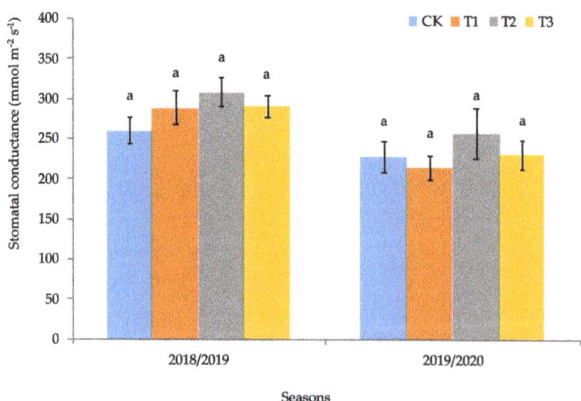

Figure 4. Effect of different doses of compost on stomatal conductance (mmol m^{-2}s^{-1}) of the calafate leaf (*Berberis mycrophilla* G. Forst) for the 2018–2019 and 2019–2020 seasons. CK = Control treatment (without compost application); T1 = Treatment 1 (5 Ton ha^{-1}); T2 = Treatment 2 (10 Ton ha^{-1}); T3 = Treatment 3 (15 Ton ha^{-1}). For each treatment, different lowercase letters indicate significant differences in both seasons according to Fischer's LSD test ($p < 0.05$). The bars correspond to the experimental error of each treatment.

3.2. Productive and Quality Parameters of the Calafate Fruit

In Figure 5, the antioxidant capacity of the calafate fruit can be observed for the 2018–2019 and 2019–2020 seasons. For the first season under study, the highest value of DDPH antioxidant capacity was recorded at the dose of 15 Ton ha^{-1} (T3; $p < 0.05$) with 4900 µmol TE/100 g fw, followed by treatments CK, T1, and T2 with 3961, 4130, and 4172 µmol TE/100 g fw, respectively, being similar to each other ($p > 0.05$). For the second season, a greater effect was observed where T2 was the one that reported the highest antioxidant capacity of the calafate fruit with 4905 µmol TE/100 g fw ($p < 0.05$), followed by treatments T1 and T3, with values of 4406 and 4435 µmol TE/100 g fw, respectively, both without significant differences. Finally, the control treatment showed the lowest DDPH antioxidant capacity of all the treatments ($p > 0.05$) with a value of 3958 µmol TE/100 g fw (Figure 5).

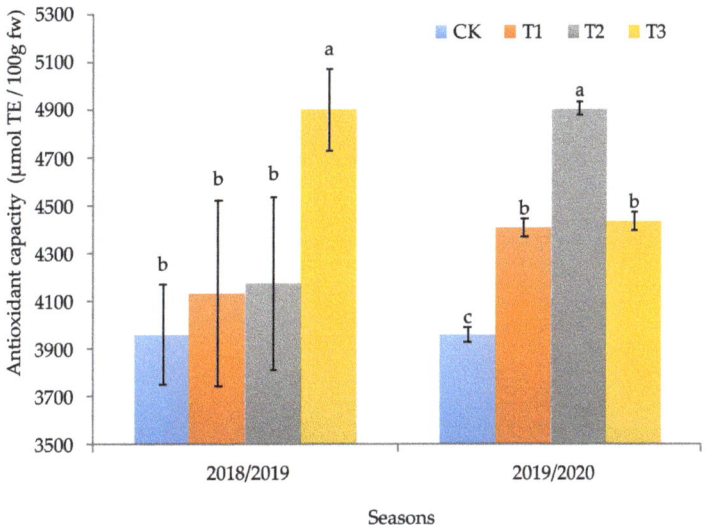

Figure 5. Effect of different doses of compost on the antioxidant capacity of the Calafate fruit (*Berberis mycrophilla* G. Forst) for the 2018–2019 and 2019–2020 seasons. CK = Control treatment (without compost application); T1 = Treatment 1 (5 Ton ha^{-1}); T2 = Treatment 2 (10 Ton ha^{-1}); T3 = Treatment 3 (15 Ton ha^{-1}). For each treatment, different lowercase letters indicate significant differences in both seasons according to Fischer's LSD test ($p < 0.05$). The bars correspond to the experimental error of each treatment.

Figure 6 shows the total content of polyphenols in the calafate fruit grown with different doses of compost for the 2018–2019 and 2019–2020 seasons. In the first season, no significant differences were observed between treatments. However, for the 2019/2020 season, the average polyphenolic content of the fruit decreased by 25% compared with CK. T2 and T3 were the ones that contributed the highest CPT values ($p < 0.05$), these being 764 and 718 mg of gallic acid/100 g of fresh weight, respectively (Figure 6). However, T3 did not show significant differences with T1 (645 mg of gallic acid/100 g of fresh weight). CK was the one that registered the lowest PFT content (543 mg of gallic acid/100 g of fresh weight) ($p < 0.05$) among all treatments.

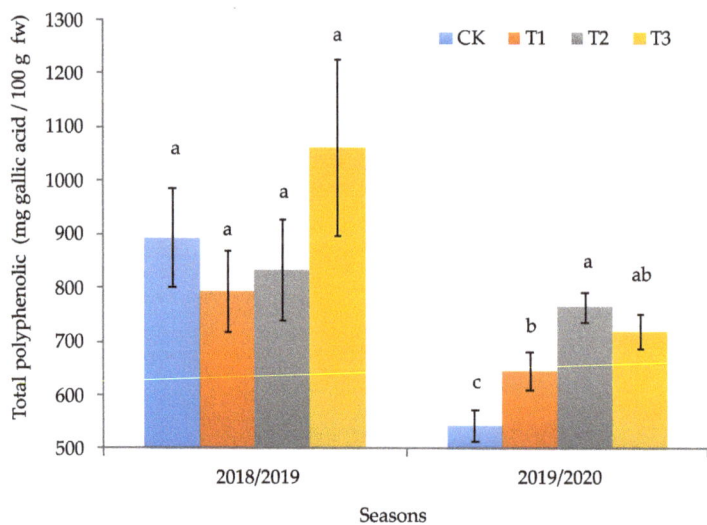

Figure 6. Effect of different doses of compost on the total polyphenolic content of calafate fruit (*Berberis mycrophilla* G. Forst) for the 2018–2019 and 2019–2020 seasons. CK = Control treatment (without compost application); T1 = Treatment 1 (5 Ton ha^{-1}); T2 = Treatment 2 (10 Ton ha^{-1}); T3 = Treatment 3 (15 Ton ha^{-1}). For each treatment, different lowercase letters indicate significant differences in both seasons according to Fischer's LSD test ($p < 0.05$). The bars correspond to the experimental error of each treatment.

Figure 7 shows the total anthocyanin content (TAC) in the calafate fruit for the different compost treatments in the 2018–2019 and 2019–2020 seasons. For the first season under evaluation, the same behavior as the CPT was observed, not registering significant differences between CK, T1, T2, and T3, whose values averaged 573 mg cyanidin-3-glucoside/100 g fw. For the second season, more noticeable effects were observed, although there was a decrease in TAC in most treatments. T2 was the one that registered the highest TAC ($p < 0.05$), with 545 mg cyanidin-3-glucoside/100g fw. This was followed by T1 and T3, without significant differences between them, with values of 445 and 431 mg cyanidin-na-3-glucoside/100 g fw, respectively. Finally, the one that registered the lowest TAC ($p < 0.05$) was CK, with a value of 363 mg cyanidin-3-glucoside/100g fw (Figure 7).

In Figure 8, the yields of fresh calafate fruit (g plant^{-1}) grown under different doses of compost in the 2018–2019 and 2019–2020 seasons are observed. For the first season, the treatments that recorded the highest fruit production were T2 and T3 ($p < 0.05$), without significant differences, with values of 629 and 561 g plant^{-1}, respectively. On the contrary, the treatment that registered a lower production was CK, with an average fruit yield of 249 g plant^{-1} ($p < 0.05$). In the second season, there was an increase in performance in most treatments, with T3 and T2 as the ones that registered the highest average fruit production of 924 and 726 g plant^{-1} ($p > 0.05$), respectively, and without significant differences between them. Finally, lower fruit production T1 and CK registered an average production of 424 and 370 g plant^{-1} ($p > 0.05$), respectively.

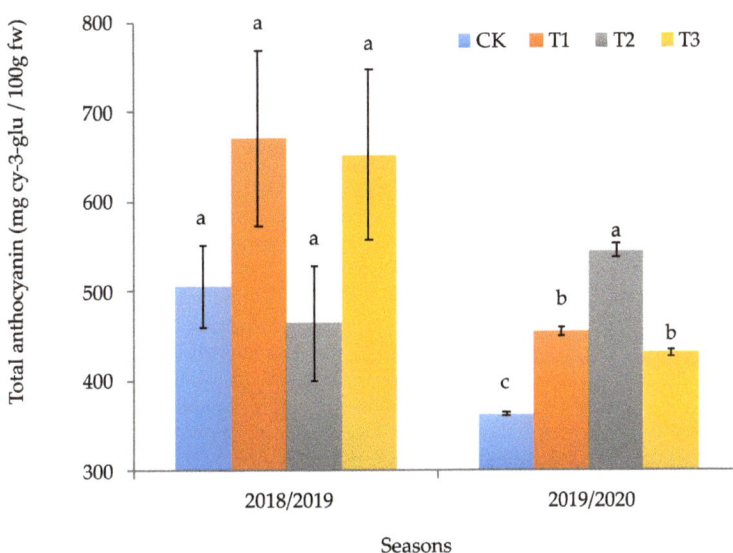

Figure 7. Effect of different doses of compost on the total anthocyanin content of the calafate fruit (*Berberis microphylla* G. Forst) for the 2018–2019 and 2019–2020 seasons. CK = Control treatment (without compost application); T1 = Treatment 1 (5 Ton ha^{-1}); T2 = Treatment 2 (10 Ton ha^{-1}); T3 = Treatment 3 (15 Ton ha^{-1}). For each treatment, different lowercase letters indicate significant differences in both seasons according to Fischer's LSD test ($p < 0.05$). The bars correspond to the experimental error of each treatment.

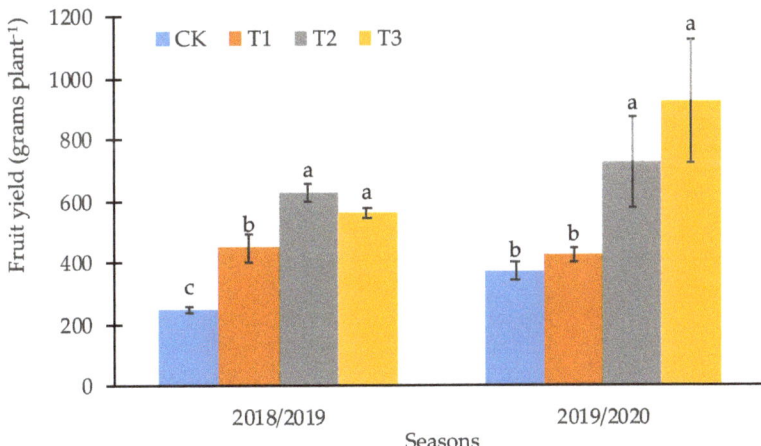

Figure 8. Effect of different doses of compost on the yield of calafate fruit (*Berberis microphylla* G. Forst) for the 2018–2019 and 2019–2020 seasons. CK = Control treatment (without application of com-post); T1 = Treatment 1 (5 Ton ha^{-1}); T2 = Treatment 2 (10 Ton ha^{-1}); T3 = Treatment 3 (15 Ton ha^{-1}). For each season, different lowercase letters indicate significant differences in the different treatments according to Fischer's LSD test ($p < 0.05$). The bars correspond to the experimental error of each treatment.

4. Discussion

Despite that photosynthetically active radiation (Figure 1a) was 26% higher at the site of study with respect to the habitat of origin of the plants (1600 µmol m^{-2} s^{-1}; Valdivia, Chile), no symptoms of excess radiation were observed. This was possibly due to the great structural and physiological plasticity that this species possesses [15,37–41], managing to adapt to higher ambient temperature conditions, as indicated by Radice and Arena [10]. The above is confirmed by Romero-Román et al. [42], who observed that productive and some physiological parameters of the calafate plant were not influenced by extreme temperatures; however, these environmental conditions could influence chemical parameters of the fruit [42] in response to higher levels of radiation, such as those observed at the study site (Figure 1a). Even though the radiation was high, no higher levels of absorbed PAR radiation between treatments and seasons were registered (Figure 1b), which could be showing that calafate, despite having morphological plasticity, has low variability of its light saturation point. The light saturation point of calafate is 800 µmol m^{-2} s^{-1} [43], and at values higher than 1000 µmol m^{-2} s^{-1}, photosynthesis rates would be constant. Therefore, higher levels of PAR radiation could be generating photo-oxidative damage due to an excess of radiation. Studies developed by Arenas et al. [15] point out that low levels of irradiation improve plant development and nutrient content in the calafate leaf. In this study, however, no improvements were observed in indirect parameters used as indicators of photosynthetic performance, such as stomatal conductance (Figure 4). Results showed that this species does not respond strongly to changes in fertilization levels in relation to Gs, with values between 200 and 300 mmol m^{-2} s^{-1} between treatments and evaluation seasons ($p > 0.05$). These results are in contrast to those found in a study carried out on blueberries, where Gs was affected by the fertilization doses together with the water regime, where a correlation was found between the fertilization dose and moisture content in the soil [44]. As the soil moisture was constant for all treatments in this study, it could be strongly influential so that no significant differences in Gs are observed between treatments [44].

On the other hand, the application of compost to the soil increased levels of organic matter (OM) in all treatments (Tables 1 and 2), which could be affecting the moisture retention capacity, aeration, porosity, and soil carbon content as reported in multiple studies [45–47]. Said OM modifications in the soil could be having an impact not only on the physical parameters of the soil, but also to the nutritional status and biomass of the plant and consequently on the foliar area of the plant [44,45]. In the present study, the LAI was higher in both seasons in Q3 (Figure 2), showing an increase in the second season of more than 20% compared to CK and T1 ($p < 0.05$) and close to 10% higher than T3 ($p < 0.05$). These results are consistent with other authors, who point out that the application of compost not only increases the vegetative development of the plant, but also the total chlorophyll content of the leaf [15,45].

In a study carried out on vine (*Vitis vinifera* cv. Chardonnay), during 9 years of compost applications, equal yields were observed with inorganic fertilization. However, fertilization with compost significantly increased levels of organic matter in the soil; in addition, there was a substantial increase in the concentrations of mineralizable nitrogen in the soil [23]. The former could have happened in this study since the applied levels of nitrogen reached 60 mg kg^{-1} of soil (Table 2), but they were not strongly affected until the third year from the implementation of the study (Table 3). Interestingly, nitrogen levels in the soil fluctuated between 13 and 16 mg kg^{-1} in all treatments; this response of low nitrogen availability in the soil could be generated by the high levels of nitrogen that the plant is extracting to satisfy the greater vegetative development [48] (Figure 2) in conjunction with the higher levels of productivity [45] (Figure 8), as those observed in T2 and T3, which were 100% higher in both seasons for the treatments of 10 and 15 ton ha^{-1} of compost, compared to the control treatment. The treatments that obtained the best productive results, both at the beginning and at the end of the study, were the treatments with the highest doses of compost, corresponding to 10 and 15 tons ha^{-1} and with productivity levels close to 1000 g per plant (Figure 8), coinciding with the results of other authors [45,48,49]. These results suggest that the long-term application of compost to the soil, in addition to improving the physical and chemical properties of the soil, as indicated above, could improve its biological activity. [24]. Although, the aforementioned is an uncertainty in this new species for commercial purposes since there are no studies related to the microbial activity in the soil and rhizosphere. Therefore, this study opens the door to future investigations that propose to understand and/or analyze the interaction of agronomic management with the activity and microbiological diversity of the soil and the response of the calafate plant [15,42].

In the present study, it can be observed that the evaluated nutraceutical parameters were positively influenced by the dose of compost application when compared with the control treatment (Figures 5–7). Regarding the content of polyphenols (Figure 6) and total anthocyanins of the calafate fruit (Figure 7), these decreased in the second evaluation season by 25% on average between treatments. The compost treatment of 15 Ton ha^{-1} was the one that contributed the greatest decrease in polyphenolic content in the second season. However, the average total polyphenol contents observed in this study (764 mg of gallic acid/100 g fw) were below the polyphenolic contents observed in wild plants in studies developed by other authors [50]. Nevertheless, the results obtained in the compost treatments were superior to the control treatment. This has been corroborated in other species, such as strawberries, where the effect of an organic fertilization based on compost increased the contents of anthocyanins, phenolic contents, and antioxidant capacity, and also in the Rhubarb (*Rheum rhabarbarum* L.) crop, where organic fertilization also improved the levels of polyphenolic content and antioxidant capacity of the fruit [48,51]. In a study conducted by Cojocaru et al., no increase in fruit yield was observed as it was observed in this study (Figure 8) and was probably associated with the low doses of compost (2.4 Ton ha^{-1}) used in their study [51], which was associated with the high levels of extraction given by the levels of fruit production.

Regarding the contents of total anthocyanins, despite the fact that these plants are being cultivated and subjected to intensive agronomic management, the total concentrations of anthocyanins observed were higher than the results obtained in wild calafate plants in different locations in Usuahia (Argentina) and Buenos Aires (Argentina) [10], with values close to 118 and 316 mg cyanidin-3-glucoside/100 g fw, respectively, as well as values lower than the results observed in Chile [16,43,52], with values over 1000 mg cyanidin-3-glucoside/100 g of fresh weight. These differences in concentrations of anthocyanins and total polyphenols could be stimulated by multiple factors, which could affect the biosynthesis of bioactive compounds, such as different light intensities, ultraviolet radiation, extreme temperatures, and availability of nutrients and water, among other factors, specific to each environment where this species grows [15,53]. However, it is suggested that, among the factors that could be influencing the differences in results between the different studies, the most influential is the great variability in the opportune moment of harvest, which is influenced by the aforementioned parameters. It is important to mention that in this study, the harvest time corresponded to 130 days after full flower, which is longer than the harvest dates observed in other polyphenolic evaluation studies of wild calafate fruits, fluctuating from 98 to 126 days after full flower [40,54].

5. Conclusions

The use of increasing doses of compost turned out to be beneficial to the physiological, productive, and quality parameters of calafate during the studied seasons. Treatment T2 at a rate of 10 Ton ha^{-1} obtained the highest index of foliar area, antioxidant capacity, total polyphenols, and total anthocyanins in the second study season. The compost application rates of 10 and 15 Ton ha^{-1} obtained the highest fruit production per plant, with a production of 3300 kilos per hectare. On the other hand, increasing compost doses generated an increase in organic matter in the soil and nutritional content of the soil. Therefore, the dose that optimizes the yield, fruit quality, and nutraceutical content of the calafate fruit is set at a rate of 10 Ton ha^{-1}. However, this study opens the doors to future research in this matter to answer questions regarding the behavior of soil microbial activity and its interaction between agronomic management and the calafate plant, which could alter the nutraceutical properties of calafate fruits.

Author Contributions: Conceptualization, J.R.-S.; methodology, J.R.-S., F.P.-M. and N.Z.; software, J.R.-S., A.P.-P. and R.V.-R.; validation, J.R.-S.; formal analysis, J.R.-S. and F.P.-M.; investigation, F.P.-M. and J.R.-S.; resources, F.P.-M. and J.R.-S.; data curation, M.D.L. and J.R.-S.; writing—original draft preparation, F.P.-M. and J.R.-S.; writing—review and editing, R.V.-R., N.Z., M.D.L. and A.P.-P.; project administration, J.R-S.; funding acquisition, J.R.-S. and F.P.-M. All authors have read and agreed to the published version of the manuscript.

Funding: The present work benefited from the project of research N° 106, granted by the Universidad Adventista de Chile, Chile. We are grateful to Fondecyt 1191141 ANID Chile for the scientific instruments provided for this research.

Institutional Review Board Statement: Not applicable.

Informed Consent Statement: Not applicable.

Data Availability Statement: Corresponding author.

Acknowledgments: We are grateful for the support of the students of the Adventist University of Chile, for actively participating in the data collection: Samuel Contreras, Juan Pablo Molina and Sebastiancamilo Ospino.

Conflicts of Interest: The authors declare no conflict of interest.

References

1. Mc Leod, C.; Pino, M.T.; Aguila, K. Calafate (Berberis microphylla): Otro Superberries Chileno. Available online: https://biblioteca.inia.cl/bitstream/handle/123456789/4637/NR40444.pdf?sequence=1 (accessed on 10 November 2021).
2. Pino, M.T.; Mc Leod, C.; Ojeda, A.; Zamora, O.; Saavedra, J. Characterization and Clonal Selection of *Berberis Microphylla* G. Forst in the Chilean Patagonia Region for Natural Colorant Purposes. In *XIV EUCARPIA Symposium on Fruit Breeding and Genetics*; ISHS: Bologna, Italy, 2017; pp. 249–254. [CrossRef]
3. Arena, M.E.; Zuleta, A.; Dyner, L.; Constenla, D.; Ceci, L.; Curvetto, N. Berberis Buxifolia Fruit Growth and Ripening: Evolution in Carbohydrate and Organic Acid Contents. *Sci. Hortic.* **2013**, *158*, 52–58. [CrossRef]
4. Rodoni, L.M.; Feuring, V.; Zaro, M.J.; Sozzi, G.O.; Vicente, A.R.; Arena, M.E. Ethylene Responses and Quality of Antioxidant-Rich Stored Barberry Fruit (*Berberis microphylla*). *Sci. Hortic.* **2014**, *179*, 233–238. [CrossRef]
5. Ruiz, A.; Hermosín-Gutiérrez, I.; Mardones, C.; Vergara, C.; Herlitz, E.; Vega, M.; Dorau, C.; Winterhalter, P.; von Baer, D. Polyphenols and Antioxidant Activity of Calafate (*Berberis microphylla*) Fruits and Other Native Berries from Southern Chile. *J. Agric. Food Chem.* **2010**, *58*, 6081–6089. [CrossRef] [PubMed]
6. Ruiz, A.; Zapata, M.; Sabando, C.; Bustamante, L.; von Baer, D.; Vergara, C.; Mardones, C. Flavonols, Alkaloids, and Antioxidant Capacity of Edible Wild Berberis Species from Patagonia. *J. Agric. Food Chem.* **2014**, *62*, 12407–12417. [CrossRef] [PubMed]
7. Manach, C.; Mazur, A.; Scalbert, A. Polyphenols and prevention of cardiovascular diseases. *Curr. Opin. Lipidol.* **2005**, *16*, 77–84. [CrossRef] [PubMed]
8. Albrecht, C.; Pellarin, G.; Rojas, M.J.; Albesa, I.; Eraso, A.J. Beneficial Effect of *Berberis Buxifolia* Lam., Zizyphus Mistol Griseb and Prosopis Alba Extracts on Oxidative Stress Induced by Chloramphenicol. *Medicina* **2010**, *70*, 65–70.
9. Arena, M.E.; Vater, G.; Peri, P. Fruit Production of *Berberis Buxifolia* Lam. in Tierra Del Fuego. *HortScience* **2003**, *38*, 200–202. [CrossRef]
10. Radice, S.; Arena, M.E. Environmental Effect on the Leaf Morphology and Anatomy of *Berberis Microphylla* G. Forst. *Int. J. Plant Biol.* **2015**, *6*. [CrossRef]
11. Sandri, M.A.; Andriolo, J.L.; Witter, M.; Dal Ross, T. Effect of Shading on Tomato Plants Grow under Greenhouse. *Hortic. Bras.* **2003**, *21*, 642–645. [CrossRef]
12. Challa, H.; Schapendonk, A.H.C.M. Quantification of Effects of Light Reduction in Greenhouses on Yield. *Acta Hortic.* **1984**, 501–510. [CrossRef]
13. Medina, C.L.; Souza, R.P.; Machado, E.C.; Ribeiro, R.V.; Silva, J.A.B. Photosynthetic Response of Citrus Grown under Reflective Aluminized Polypropylene Shading Nets. *Sci. Hortic.* **2002**, *96*, 115–125. [CrossRef]
14. Retamal-Salgado, J.; Vásquez, R.; Fischer, S.; Hirzel, J.; Zapata, N. Decrease in Artificial Radiation with Netting Reduces Stress and Improves Rabbit-Eye Blueberry (*Vaccinium Virgatum* Aiton) Cv. Ochlockonee Productivity. *Chil. J. Agric. Res.* **2017**, *77*, 226–233. [CrossRef]
15. Arena, M.E.; Pastur, G.M.; Lencinas, M.V.; Soler, R.; Bustamante, G. Changes in the Leaf Nutrient and Pigment Contents of *Berberis Microphylla* G. Forst. in Relation to Irradiance and Fertilization. *Heliyon* **2020**, *6*, e03264. [CrossRef] [PubMed]

16. Romero, M.E.; Noriega Vásquez, F.; Farías Villagra, M.; López Belchi, M.D.; Jara Zapata, P.; Vera Flores, B. Nuevas fuentes de antioxidantes naturales: Caracterización de compuestos bioactivos en cinco frutos nativos de Chile. *Perfiles* **2019**, *2*, 34–41. [CrossRef]
17. Valenzuela, C.; Pérez, P. Actualización En El Uso de Antioxidantes Naturales Derivados de Frutas y Verduras Para Prolongar La Vida Útil de La Carne y Productos Cárneos. *Rev. Chil. Nutr.* **2016**, *43*, 188–195. [CrossRef]
18. Radice, S.; Alonso, M.; Arena, M. *Berberis Microphylla*: A Species with Phenotypic Plasticity in Different Climatic Conditions. *Int. J. Agric. Biol.* **2018**, *20*, 2221–2229. [CrossRef]
19. Hirzel, J. *Diagnostico Nutricional y Principios de Fertilizacion en Frutales y Vides. Segunda Edicion Aumentada y Corregida*; Colección Libros INIA—Instituto de Investigaciones Agropecuarias Trama Impresores S.A.: Hualpén, Chile, 2014.
20. Vásquez, H.V.; Maraví, C. Efecto de fertilización orgánica (biol y compost) en el establecimiento de morera (*Morus alba* L.). *Rev. De Investig. En Cienc. Y Biotecnol. Anim.* **2017**, *1*, 33–39. [CrossRef]
21. Wang, S.Y.; Lin, H.-S. Compost as a Soil Supplement Increases the Level of Antioxidant Compounds and Oxygen Radical Absorbance Capacity in Strawberries. *J. Agric. Food Chem.* **2003**, *51*, 6844–6850. [CrossRef]
22. Bittenbender, H.C.; Hue, N.V.; Fleming, K.; Brown, H. Sustainability of organic fertilization of macadamia with macadamia husk-manure compost. *Commun. Soil Sci. Plant Anal.* **1998**, *29*, 409–419. [CrossRef]
23. Mugnai, S.; Masi, E.; Azzarello, E.; Mancuso, S. Influence of Long-Term Application of Green Waste Compost on Soil Characteristics and Growth, Yield and Quality of Grape (*Vitis Vinifera* L.). *Compost Sci. Util.* **2012**, *20*, 29–33. [CrossRef]
24. Zhong, W.; Gu, T.; Wang, W.; Zhang, B.; Lin, X.; Huang, Q.; Shen, W. The Effects of Mineral Fertilizer and Organic Manure on Soil Microbial Community and Diversity. *Plant Soil* **2010**, *326*, 511–522. [CrossRef]
25. INIA Red Agrometeorológica de INIA. Available online: http://agromet.inia.cl (accessed on 27 April 2020).
26. Romero, P.; Fernández-Fernández, J.I.; Martinez-Cutillas, A. Physiological Thresholds for Efficient Regulated Deficit-Irrigation Management in Winegrapes Grown under Semiarid Conditions. *Am. J. Enol. Vitic.* **2010**, *61*, 300–312.
27. Evanylo, G.; Sherony, C.; Spargo, J.; Starner, D.; Brosius, M.; Haering, K. Soil and Water Environmental Effects of Fertilizer-, Manure-, and Compost-Based Fertility Practices in an Organic Vegetable Cropping System. *Agric. Ecosyst. Environ.* **2008**, *127*, 50–58. [CrossRef]
28. Bohórquez, W. *El proceso de Compostaje*; Universidad de la Salle: Bogota, DC, Colombia, 2019; ISBN 978-958-54-8668-3.
29. Sayara, T.; Basheer-Salimia, R.; Hawamde, F.; Sánchez, A. Recycling of Organic Wastes through Composting: Process Performance and Compost Application in Agriculture. *Agronomy* **2020**, *10*, 1838. [CrossRef]
30. Sonnentag, O.; Talbot, J.; Chen, J.M.; Roulet, N.T. Using Direct and Indirect Measurements of Leaf Area Index to Characterize the Shrub Canopy in an Ombrotrophic Peatland. *Agric. For. Meteorol.* **2007**, *144*, 200–212. [CrossRef]
31. Retamal-Salgado, J.; Bastías, R.M.; Wilckens, R.; Paulino, L. Influence of Microclimatic Conditions under High Tunnels on the Physiological and Productive Responses in Blueberry Cv. O'Neal. *Chil. J. Agric. Res.* **2015**, *75*, 291–297. [CrossRef]
32. Reyes-Diaz, M.; Alberdi, M.; de la Luz Mora, M. Short-Term Aluminum Stress Differentially Affects the Photochemical Efficiency of Photosystem II in Highbush Blueberry Genotypes. *J. Am. Soc. Hortic. Sci.* **2009**, *134*, 14–21. [CrossRef]
33. van Kooten, O.; Snel, J.H. The Use of Chlorophyll Fluorescence Nomenclature in Plant Stress Physiology. *Photosynth. Res.* **1990**, *25*, 147–150. [CrossRef]
34. Maxwell, K.; Johnson, G.N. Chlorophyll Fluorescence—A Practical Guide. *J. Exp. Bot.* **2000**, *51*, 659–668. [CrossRef]
35. Yıldırım, A.; Mavi, A.; Kara, A.A. Determination of Antioxidant and Antimicrobial Activities of Rumex Crispus L. Extracts. *J. Agric. Food Chem.* **2001**, *49*, 4083–4089. [CrossRef]
36. Brand-Williams, W.; Cuvelier, M.E.; Berset, C. Use of a Free Radical Method to Evaluate Antioxidant Activity. *LWT—Food Sci. Technol.* **1995**, *28*, 25–30. [CrossRef]
37. Giordani, E.; Gori, M.; Arena, M.; Radice, S. Caracterización molecular de especies nativas de Berberis del NOA. *Investig. Cienc. Univ.* **2019**, *3*, 118.
38. Giordani, E.; Müller, M.; Gambineri, F.; Paffetti, D.; Arena, M.; Radice, S. Genetic and Morphological Analysis of *Berberis Microphylla* G. Forst. Accessions in Southern Tierra Del Fuego. *Plant Biosyst.—Int. J. Deal. Asp. Plant Biol.* **2017**, *151*, 715–728. [CrossRef]
39. Arena, M.E.; Lencinas, M.V.; Radice, S. Variability in Floral Traits and Reproductive Success among and within Populations of *Berberis Microphylla* G. Forst., an Underutilized Fruit Species. *Sci. Hortic.* **2018**, *241*, 65–73. [CrossRef]
40. Arena, M.E.; Postemsky, P.D.; Curvetto, N.R. Changes in the Phenolic Compounds and Antioxidant Capacity of *Berberis Microphylla* G. Forst. Berries in Relation to Light Intensity and Fertilization. *Sci. Hortic.* **2017**, *218*, 63–71. [CrossRef]
41. Arena, M.E.; Radice, S. Shoot Growth and Development of *Berberis Buxifolia* Lam. in Tierra Del Fuego (Patagonia). *Sci. Hortic.* **2014**, *165*, 5–12. [CrossRef]
42. Romero-Román, M.E.; Schoebitz, M.; Bastías, R.M.; Fernández, P.S.; García-Viguera, C.; López-Belchi, M.D. Native Species Facing Climate Changes: Response of Calafate Berries to Low Temperature and UV Radiation. *Foods* **2021**, *10*, 196. [CrossRef] [PubMed]
43. Peri, P.L.; Arena, M.; Martínez Pastur, G.; Lencinas, M.V. Photosynthetic Response to Different Light Intensities, Water Status and Leaf Age of Two Berberis Species (*Berberidaceae*) of Patagonian Steppe, Argentina. *J. Arid Environ.* **2011**, *75*, 1218–1222. [CrossRef]
44. Guo, X.; Li, S.; Wang, D.; Huang, Z.; Sarwar, N.; Mubeen, K.; Shakeel, M.; Hussain, M. Effects of water and fertilizer coupling on the physiological characteristics and growth of rabbiteye blueberry. *PLoS ONE* **2021**, *16*, e0254013. [CrossRef] [PubMed]

45. Hasnain, M.; Chen, J.; Ahmed, N.; Memon, S.; Wang, L.; Wang, Y.; Wang, P. The Effects of Fertilizer Type and Application Time on Soil Properties, Plant Traits, Yield and Quality of Tomato. *Sustainability* **2020**, *12*, 9065. [CrossRef]
46. Medina, J.; Calabi-Floody, M.; Aponte, H.; Santander, C.; Paneque, M.; Meier, S.; Panettieri, M.; Cornejo, P.; Borie, F.; Knicker, H. Utilization of Inorganic Nanoparticles and Biochar as Additives of Agricultural Waste Composting: Effects of End-Products on Plant Growth, C and Nutrient Stock in Soils from a Mediterranean Region. *Agronomy* **2021**, *11*, 767. [CrossRef]
47. Maselesele, D.; Ogola, J.B.O.; Murovhi, R.N. Macadamia Husk Compost Improved Physical and Chemical Properties of a Sandy Loam Soil. *Sustainability* **2021**, *13*, 6997. [CrossRef]
48. Mohamed, M.H.M.; Petropoulos, S.A.; Ali, M.M.E. The Application of Nitrogen Fertilization and Foliar Spraying with Calcium and Boron Affects Growth Aspects, Chemical Composition, Productivity and Fruit Quality of Strawberry Plants. *Horticulturae* **2021**, *7*, 257. [CrossRef]
49. Wen, M.; Zhang, J.; Zheng, Y.; Yi, S. Effects of Combined Potassium and Organic Fertilizer Application on Newhall Navel Orange Nutrient Uptake, Yield, and Quality. *Agronomy* **2021**, *11*, 1990. [CrossRef]
50. Arena, M.E.; Curvetto, N. Berberis Buxifolia Fruiting: Kinetic Growth Behavior and Evolution of Chemical Properties during the Fruiting Period and Different Growing Seasons. *Sci. Hortic.* **2008**, *118*, 120–127. [CrossRef]
51. Cojocaru, A.; Vlase, L.; Munteanu, N.; Stan, T.; Teliban, G.C.; Burducea, M.; Stoleru, V. Dynamic of Phenolic Compounds, Antioxidant Activity, and Yield of Rhubarb under Chemical, Organic and Biological Fertilization. *Plants* **2020**, *9*, 355. [CrossRef]
52. Fredes, C.; Parada, A.; Salinas, J.; Robert, P. Phytochemicals and Traditional Use of Two Southernmost Chilean Berry Fruits: Murta (Ugni Molinae Turcz) and Calafate (*Berberis Buxifolia* Lam.). *Foods* **2020**, *9*, 54. [CrossRef] [PubMed]
53. Giordani, E.; Radice, S.; Lencinas, M.V.; Arena, M. Variabilidad fenotípica de *Berberis microphylla* G. Forst. en poblaciones naturales de Tierra del Fuego, una especie frutal forestal no maderable con posibilidad de diversificar la producción agroforestal patagónica. *Investig. Cienc. Univ.* **2019**, *3*, 91.
54. Arena, M.E. *Estudio de Algunos Fenómenos Morfofisiológicos y Cambios Bioquímicos en Berberis microphylla G. Forst. (sinónimo B. Buxifolia Lam.) Asociados a la Formación y Maduración de Frutos en Tierra de Fuego y su Relación con la Producción de Metabolitos Útiles*; Sebelas Maret University: Surakarta City, Indonesia, 2016.

Iron Chelate Improves Rooting in Indole-3-Butyric Acid-Treated Rosemary (*Rosmarinus officinalis*) Stem Cuttings

Zeinab Izadi [1,*], Abdolhossein Rezaei Nejad [1] and Javier Abadía [2,*]

[1] Department of Horticultural Sciences, Faculty of Agriculture, Lorestan University, P.O. Box 465, Khorramabad 68151-44316, Iran; rezaeinejad.h@lu.ac.ir
[2] Department of Plant Nutrition, Estación Experimental de Aula Dei (EEAD-CSIC), Av. Montañana 1005, 50059 Zaragoza, Spain
* Correspondence: izadi.ze@fa.lu.ac.ir (Z.I.); jabadia@eead.csic.es (J.A.)

Abstract: Adventitious root formation in stem cuttings is affected by exogenous and endogenous factors. The study assessed the effect of Fe(III)-EDDHA (ethylenediamine-N,N-bis 2-hydroxyphenyl acetic acid) on the rooting of 4 indol-3-butyric acid (IBA)-treated hardwood cuttings of the aromatic and medicinal species *Rosmarinus officinalis*. Cuttings treated with 0, 1000, 2000 or 3000 mg L^{-1} IBA were placed in pots filled with sand:perlite mixture and irrigated daily with nutrient solution pH 5.8, containing 0, 5, 10 or 20 µM Fe(III)-EDDHA. Ten days later, the number of new emerging roots were recorded. After 20 days, leaf photosynthetic pigments and morphological traits, including root number, fresh (FW) and dry weight (DW), shoot FW and DW, mean length of the longest roots, number of new shoots and new growth in old shoots, were measured. Finally, plants were transplanted to pots filled with a sand:soil mixture and survival was measured after 10 days. Results indicate that Fe application promotes root emergence and improves root and shoot biomass, leaf photosynthetic pigment concentrations and survival percentage. This indicates that using low concentrations of Fe(III)-EDDHA (5–20 µM) in the growth medium could be a good management strategy to facilitate the production of vigorous *R. officinalis* plants from hardwood cuttings.

Keywords: rooting; hardwood cuttings; iron chelates

1. Introduction

Adventitious root formation in plant cuttings is influenced by a large set of exogenous and endogenous factors [1]. Root initiation involves de-differentiation of specific cells leading to the formation of the root meristems [1]. Endogenous factors that could act as rooting co-factors and auxin transport modulators are transferred from the stock plants to the propagules [2]. These include auxin and carbohydrates [3], mineral nutrients [4] and other metabolites, including phenolic compounds [5].

Among the exogenous rooting factors, the auxin IBA (indole-3-butyric acid) is widely used to stimulate rooting processes in cuttings, because of its high ability to promote root initiation. This effect of IBA is thought to be due to its conversion in the plant tissue to indole-3-acetic acid (IAA), which is needed for the rooting process. Endogenous IAA can be readily oxidized in plants by peroxidase, but IBA is quite stable and is only slowly transported from the site of application at the base of the cuttings, resulting in a localized IAA production [6]. Exogenous IBA application has been shown to have positive rooting effects in many woody plant species, including *Citrus medica* [7], *R. damascena* [8], *Hibiscus rosa-sinensis* [9], *Olea europaea* [10,11], *Zizyphus jujuba* [12], *Tilia rubra* [13], *Eucaliptus* spp. [14,15], *Sterculia foetida* [16], *Castanea* spp. [17] and *Populus* [18]. For instance, in *Cinnamomum bodinieri*, exogenous IBA was shown to modify the auxin signaling pathway and carbohydrate metabolism, improve the formation of lateral root initiation site and root cell elongation, and enhance d-glucose synthesis as well as sucrose and starch utilization [19].

Other exogenous factors involved in rooting are mineral nutrients, which are involved in many metabolic processes associated with differentiation and root meristem formation, which is essential for root initiation [1]. Transcriptome analysis of adventitious root formation in *Petunia × hybrida* revealed an increase, starting from the initiation phase, in the expression of 18 genes involved in the uptake and assimilation of N, P, K, S, Fe and Zn [4]. For instance, within this period a high transcript abundance was observed for a plasma membrane H-ATPase, which may energize nutrient uptake [4]. The mineral nutrient composition of the cuttings, especially regarding micronutrients such as Fe, Zn, Mn and B, plays a key role in controlling root morphogenesis. Iron and Mn are cofactors and structural components of peroxidase and can therefore directly affect IAA catabolism [20]. Iron is an essential micronutrient for plants, which plays vital roles in many metabolic processes in plants, including photosynthesis, respiration and N_2 fixation [21]. Additionally, auxin is involved in the root responses to Fe deficiency [21–23]. A boosting effect of mineral nutrients (including Fe) on propagation of plant cuttings has also been reported in different studies [24,25]. In hardwood cuttings from Fe-deficient peach trees, the application of Fe compounds significantly reduced chlorotic symptoms and improved rooting [26].

Rosemary (*Rosmarinus officinalis*) is a xerophytic, aromatic, evergreen shrub widely used for food and as an ornamental species in gardens. Because of its hardiness under environmental stress, it is also used to protect against soil erosion and is planted in fire-damaged areas. Rosemary is a medicinal species that contains polyphenols, resulting in a number of pharmacological effects, including antioxidant, antitumor, antidiabetic and antibacterial ones [27–29]. Rosemary plants can be propagated by seed and stem cuttings, but propagation from seeds is rarely used, because of the long times needed for blooming and germination and the low germination rates (10–20%) [30,31]. Rooting of *R. officinalis* cuttings is facilitated by using hormone treatments [32,33], and different approaches are being used to further improve the rooting ability, with the aim to reduce costs and allow for mass production. For instance, it has been recently shown that using blue light induces an upregulation of auxin signaling and leads to better root formation [30].

The aim of this work was to assess the hypothesis that Fe supplementation in the form of Fe(III)-chelate can promote rooting in stem cuttings of the medicinal and aromatic plant *R. officinalis* treated with IBA. Four concentrations each of Fe and IBA were used and shoot and root biomass as well as rooting parameters were studied.

2. Materials and Methods

2.1. Greenhouse and Propagation Conditions

This study was carried out in March 2018, in an experimental greenhouse of the Lorestan University, Khorramabad, Lorestan Province, in the western part of Iran (33°45′ N 48°26′ E). The greenhouse was north–south oriented, the mean temperature and relative humidity were maintained at 22/28 °C (night/day) and 55–75%, respectively, and the light intensity was approximately 500 ± 100 µmol quanta $m^{-2} s^{-1}$ (photosynthetically active photon flux density).

2.2. Plant Material and Growth Conditions

Cuttings were taken from vigorous and healthy bushes of *R. officinalis*, 10 years old, growing in the Faculty of Agriculture, Lorestan University (originated from cuttings obtained at the National Botanical Garden of Iran, Tehran). Cuttings consisted of hardwood branches (including the apical meristem), excised approximately 1 cm below a leaf node, at least 15 cm in length and with 7–8 nodes. In each cutting, leaves in the lower 5 cm were removed, and the basal 1 cm was dipped for 5 s [34] in a solution containing different IBA (CAS number 133-32-4) concentrations (0, 1000, 2000 or 3000 mg IBA L^{-1} –0, 4.9, 9.8 or 14.8 mM, respectively; these concentrations were thereafter called 0, 1000, 2000 and 3000 IBA). Cuttings were then placed in pots (28.0 cm height and 25.5 cm in diameter; ten cuttings per pot) filled with a sand:perlite (1:1, w:w) mixture, and irrigated daily with a nutrient solution containing (in mM) 0.1 KH_2PO_4, 0.1 $MgSO_4$, 0.25 $CaCl_2$ and 2 NH_4NO_3, and (in µM) 50 H_3BO_3 and 5 $MnSO_4$; the solution also contained 1 mM MES (2-(N-morpholino)ethanesulfonic acid), and the pH was 5.8. The nutrient solution was supplemented with 0, 5, 10 or 20 µM Fe(III)-EDDHA (ethylenediamine-N,N-bis 2-hydroxyphenyl acetic acid; Sequestrene 138 Fe, 6% chelated Fe, Syngenta, Basel, Switzerland). These Fe concentrations are thereafter called 0, 5, 10 and 20 Fe. Pictures of the plants are shown in Figure S1 in the Supplementary File. Pots were covered with a clear polyethylene sheet to keep the medium moist. Pots were irrigated daily with the same nutrient solutions (1 L per pot), with the excess being drained. After 10 days new roots had emerged, and after 20 days new shoot tissue and leaves had developed. At that date, three plants per treatment were transferred to pots filled with a sand:soil mixture (1:2, w:w) and grown for 10 more days. Four replications (pots) were used for all treatments.

2.3. Plant Morphological Traits

Ten days after the root induction treatment, three cuttings were taken from each pot (a total of 12 cuttings per treatment) and used to count the number of newly developed roots. Twenty days after rooting induction, four plants from each pot were used to determine morphological traits, including root number, fresh (FW) and dry weight (DW), shoot FW and DW, mean length of the longest roots, number of new shoots and new growth in old shoots (a total of 16 cuttings were considered per treatment). The three remaining plants from each pot were transferred to the sand:soil mixture, and plant survival was recorded after 10 days.

2.4. Pigment Analysis

Chlorophyll a (Chl a), Chl b, total Chl (Chl a + b) and carotenoid (Car) concentrations were determined 20 days after the rooting induction treatment, in leaves of the same plants considered for analysis of morphological parameters. Leaf tissue (0.1 g FW) was collected from young, fully expanded leaves pooled from four plants in each pot, ground in liquid N_2 with mortar and pestle, homogenized with 10 mL 100% acetone, centrifuged for 15 min at 4000 rpm, and the supernatant collected. The absorbance of the extracts was measured as 470, 662 and 645 nm using a spectrophotometer (Mapada UV-1800, Shanghai, P.R. China), and the leaf pigment concentrations were calculated as follows: Chl a = $11.24 \times A_{662} - 2.04 \times A_{645}$; Chl b = $20.13 \times A_{645} - 4.19 \times A_{662}$; Chl a + b = Chl a + Chl b; Car = $(1000 \times A_{470} - 1.90$ Chl a $- 63.14$ Chl b$)/214$ [35]. Leaf pigments were expressed as (mg g FW^{-1}).

2.5. Statistical Analysis

The experiment was carried out with a completely randomized design with four replications per IBA × Fe treatment (16 treatments in total, four IBA and four Fe concentrations; 64 pots in total). All data were subjected to analysis of variance (SAS 9.1.3, SAS Institute Inc., Cary, NC, USA), and normality and homogeneity tested. Post hoc multiple comparison of means corresponding to the different treatments was carried out (at $p \leq 0.05$) using a LSD test; comparisons were carried out for the Fe treatments in a given IBA treatment, and also for the IBA treatments in a given Fe treatment (significance letters are shown in all Figures in lower case and capitals, respectively). Values shown are means of four replications—pots—each averaging values from three and four plants per pot in the cases of root number and the rest of parameters, respectively.

3. Results

Both factors that were used, IBA and Fe doses, had statistically significant effects on all parameters analyzed, with the only exception of Car content for Fe, and the interaction IBA × Fe was also significant (Table S1 in the Supplementary File).

3.1. Application of Fe Enhance the Rooting Performance in IBA-Treated Cuttings

Ten days after the IBA/Fe treatments, the percentage of rooting was between 0 and 100% (Figure 1). At that time, the control 0 IBA/0 Fe cuttings did not show any rooting signs, whereas cuttings under the 2000–3000 IBA/5 Fe, 1000–3000 IBA/10 Fe and 1000–3000 IBA/20 Fe treatments showed 100% rooting. The 1000–3000 IBA/0 Fe, 0–1000 IBA/5 Fe, 0 IBA/10 Fe and 0 IBA/20 Fe treatments led to intermediate percentages of rooting. In the absence of Fe, treatments with 1000–3000 IBA increased this parameter.

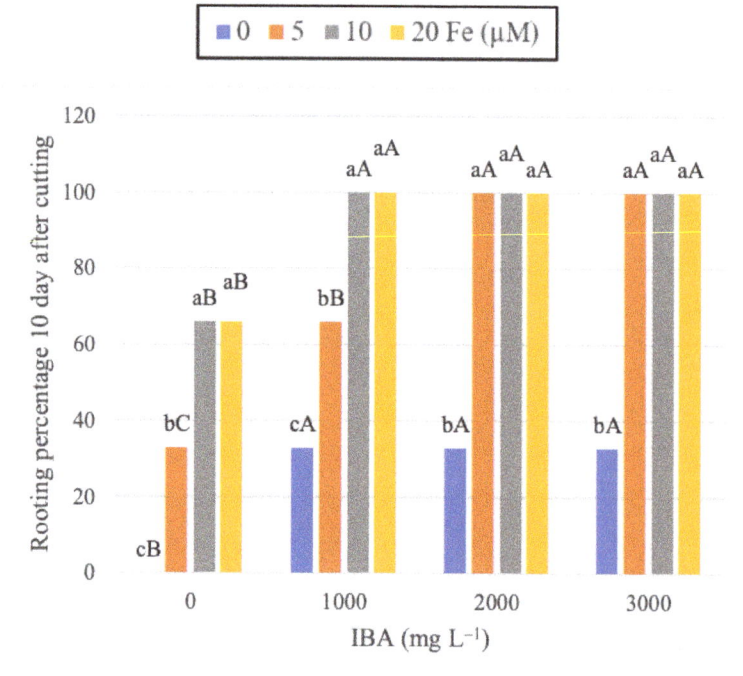

Figure 1. Percentages of rooting 10 days after the rooting induction treatments in *Rosmarinus officinalis* stem cuttings. Cuttings were treated with different concentrations of IBA (0, 1000, 2000 and 3000 mg L^{-1}) at the start of the experiment and then grown with different concentrations of Fe (0, 5, 10 and 20 μM). Values shown are means ± SE (n = 4 pots). Letters above the columns indicate significant differences at $p \leq 0.05$ for the Fe treatments in a given IBA treatment (in lower case) and for the IBA treatments in a given Fe treatment (in capitals).

3.2. Application of Fe Increase Biomass in IBA-Treated Root Cuttings

Twenty days after the start of the experiment, the root FW was between 0.19 and 0.66 g per plant, depending on the treatment (Figure 2A). The highest values were observed in some treatments including IBA and Fe (3000 IBA/5 Fe, 2000 IBA/10 Fe and 1000 IBA/20 Fe), and the lowest in the 0 IBA/0 Fe and the 0 IBA/5 Fe treatments, whereas other treatments led to intermediate values. The 10–20 Fe treatments increased root FW in cuttings not treated with IBA. On the other hand, in the absence of Fe treatments with 1000–3000 IBA increased this parameter.

The root DW was between 0.013 and 0.077 g per plant depending on the treatments (Figure 2B). The highest values were observed in some treatments including IBA and Fe (1000–3000 IBA/5 Fe, 1000–2000 IBA/10 Fe and 1000–3000 IBA/20 Fe), and the lowest in the 0 IBA/0 Fe control, whereas other treatments led to intermediate values. All treatments with Fe increased root DW in cuttings not treated with IBA. In the absence of Fe, treatments with 1000–3000 IBA increased this parameter.

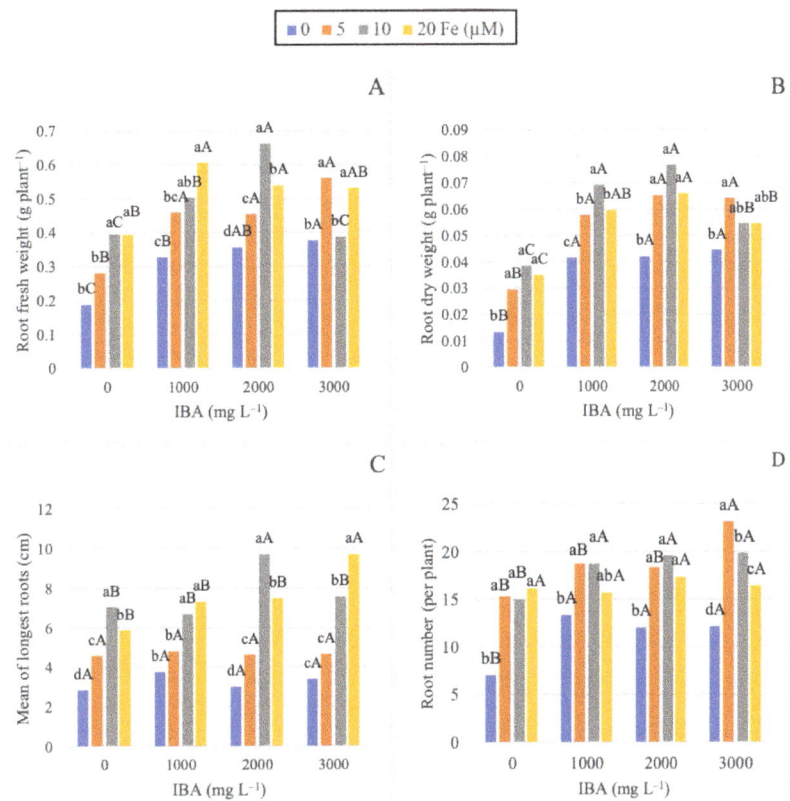

Figure 2. Root parameters 20 days after the rooting induction treatments in *Rosmarinus officinalis* stem cuttings. Cuttings were treated with different concentrations of IBA (0, 1000, 2000 and 3000 mg L^{-1}) at the start of the experiment and then grown with different concentrations of Fe (0, 5, 10 and 20 µM). Root fresh weight ((**A**), in g plant^{-1}), root dry weight ((**B**), in g plant^{-1}), mean of longest roots ((**C**), in cm) and root number per plant (**D**). Values shown are means ± SE (n = 4 pots). Letters above the columns indicate significant differences at $p \leq 0.05$ for the Fe treatments in a given IBA treatment (in lower case) and for the IBA treatments in a given Fe treatment (in capitals).

3.3. Application of Fe Increase Root Length and Number in IBA-Treated Root Cuttings

The mean length of the longest roots was in the range 2.9–9.7 cm, and it was markedly affected by the IBA/Fe regimes (Figure 2C). The highest value was observed in the 2000 IBA/10 Fe and 3000 IBA/20 Fe, and the lowest in the 0 Fe treatments. All treatments with Fe increased this parameter in cuttings not treated with IBA, whereas in the absence of Fe treatments with IBA did not have any effect.

The number of roots per plant was in the range 7–23, and it was markedly affected by the IBA/Fe regimes (Figure 2D). The highest value was observed in the 3000 IBA/5 Fe treatment, and the lowest in the 0 IBA/0 Fe. All treatments with Fe increased root number in cuttings not treated with IBA. In the absence of Fe, treatments with 1000–3000 IBA increased root number.

3.4. Application of Fe-Chelate Causes Positive Changes in Shoot Parameters by IBA

Shoot FW was between 2.1 and 4.7 g per plant depending to the treatment (Figure 3A). The highest values were observed in the 1000–3000 IBA/5 Fe, 1000–2000 IBA/10 Fe and 1000 IBA/20 Fe treatments, and the lowest in the 0 IBA/0 Fe control one. All treatments with Fe increased shoot FW in cuttings not treated with IBA. Treatments with 1000–3000 IBA increased this parameter in the absence of Fe.

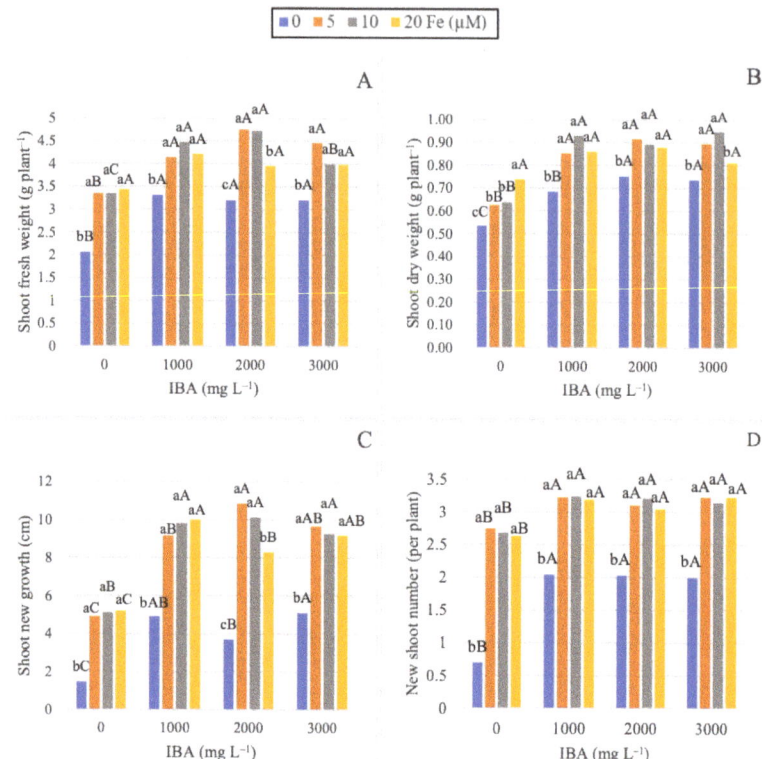

Figure 3. Shoot parameters 20 days after the rooting induction treatments in *Rosmarinus officinalis* stem cuttings. Cuttings were treated with different concentrations of IBA (0, 1000, 2000 and 3000 mg L^{-1}) at the start of the experiment and then grown with different concentrations of Fe (0, 5, 10 and 20 μM). Shoot fresh weight ((**A**), in g plant^{-1}), shoot dry weight ((**B**), in g plant^{-1}), shoot new growth ((**C**), in cm), and new shoot number per plant (**D**). Values shown are means ± SE (*n* = 4 pots). Letters above the columns indicate significant differences at $p \leq 0.05$ for the Fe treatments in a given IBA treatment (in lower case) and for the IBA treatments in a given Fe treatment (in capitals).

Total shoot DW was between 0.54 and 0.95 g per plant, and values were markedly affected by the Fe regime (Figure 3B). The highest values were observed in the 1000–3000 IBA/5–10 Fe and 1000–2000 IBA/20 Fe treatments and the lowest in the 0 IBA/0 Fe control. Treatments with Fe increased shoot DW in cuttings not treated with IBA, and in the absence of Fe treatments with 1000–3000 IBA increased this parameter.

Shoot new growth was between 1.5 and 10.9 cm depending on the treatment (Figure 3C). The highest values were observed in the 1000–3000 IBA/5–10 Fe and 1000 and 3000 IBA/20 Fe treatments, and the lowest in the 0 IBA/0 Fe control one. Treatments with Fe increased shoot new growth in cuttings not treated with IBA, whereas in the absence of Fe treatments with 1000–3000 IBA increased this parameter.

The number of new shoots per plant was in the range 0.7–3.2, and it was affected by the IBA/Fe regimes (Figure 3D). The highest values were observed in the 1000–3000 IBA/5–20 Fe and the lowest in the 0 IBA/0 Fe control one. All treatments with Fe increased the number of new shoots in cuttings not treated with IBA, and treatment with 1000–3000 IBA increased this parameter in the absence of Fe.

3.5. Fe-Chelate Increase Leaf Photosynthetic Pigment Concentration of Cuttings

The concentrations of Chl a, Chl b, total Chl and Car were in the ranges 4.0–9.7, 1.1–6.0, 5.2–15.2 and 0.7–2.5 mg g FW^{-1}, respectively, and values were markedly affected by the IBA/Fe regimes (Figure 4). The highest total Chl was observed in the 2000–3000 IBA/10 Fe and 1000–3000 IBA/20 Fe treatments, the highest Car value was in the 1000 IBA/20 Fe treatment, and the minimum value for all pigments were observed in the 0 Fe treatments. In cuttings not treated with IBA, all treatments with Fe increased the level of Chls, and treatments with 10–20 Fe increased total Car. On the other hand, in the absence of Fe treatments with IBA did not cause any change in the leaf concentration of photosynthetic pigments.

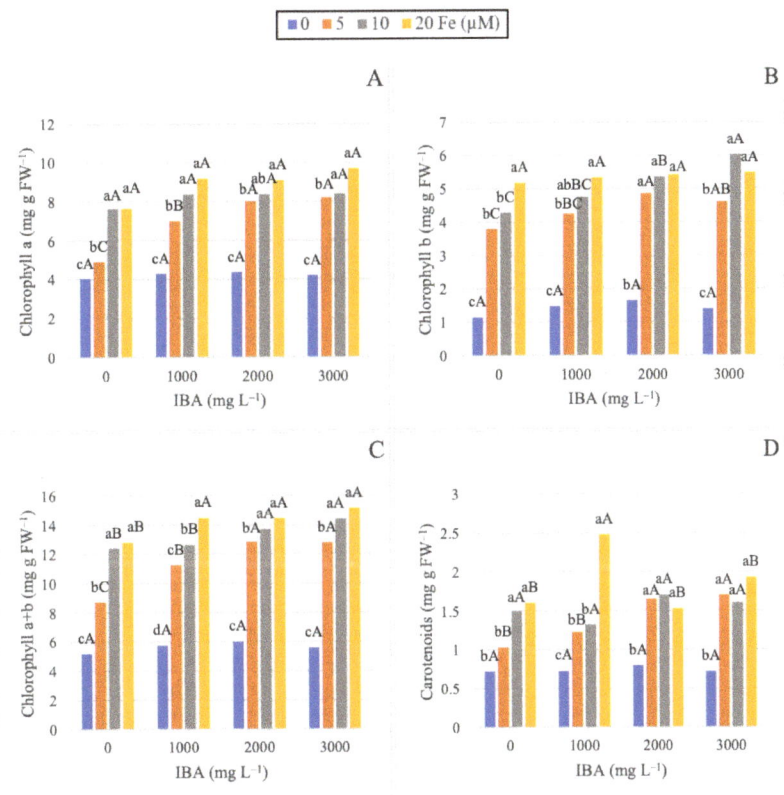

Figure 4. Leaf photosynthetic pigment concentrations 20 days after the rooting induction treatments in *Rosmarinus officinalis* stem cuttings. Cuttings were treated with different concentrations of IBA (0, 1000, 2000 and 3000 mg L^{-1}) at the start of the experiment and then grown with different concentrations of Fe (0, 5, 10 and 20 µM). Chlorophyll a (**A**), chlorophyll b (**B**), total chlorophyll (**C**), and carotenoid (**D**) concentrations. Values shown are means ± SE (n = 4 pots). Letters above the columns indicate significant differences at $p \leq 0.05$ for the Fe treatments in a given IBA treatment (in lower case) and for the IBA treatments in a given Fe treatment (in capitals).

3.6. Cutting Survival Percentage Increased by Applying IBA and Fe-Chelate Simultaneously

Ten days after transfer to the sand:soil substrate, cutting survival was between 35.3 and 89.3%, and values were markedly affected by the IBA/Fe regimes (Figure 5). The highest values were observed in the 1000–3000 IBA/5–20 Fe and the lowest in the 0 IBA/0 Fe control one. All treatments with Fe increased survival significantly in cuttings not treated with IBA. In the absence of Fe, treatments including 1000–3000 IBA increased survival.

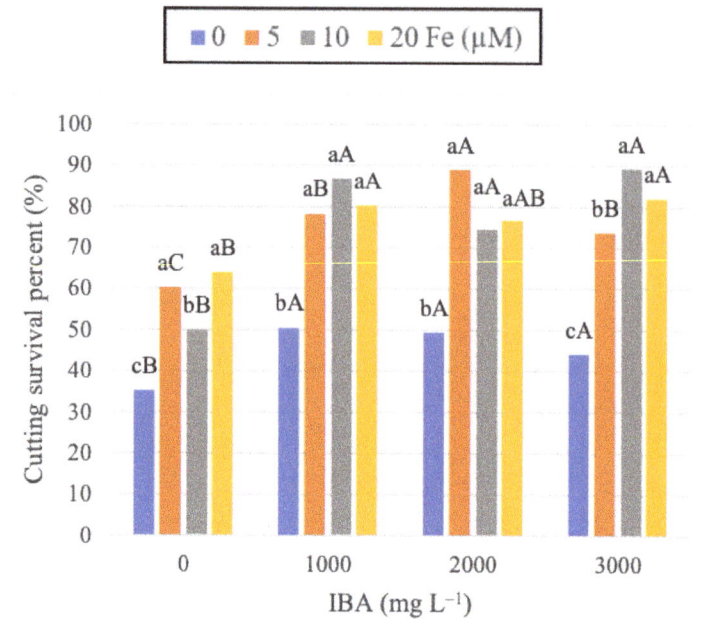

Figure 5. Survival (in %) of *Rosmarinus officinalis* stem cuttings 10 days after transfer to a sand:soil mixture. Cuttings were treated with different concentrations of IBA (0, 1000, 2000 and 3000 mg L^{-1}) at the start of the experiment, then grown for 20 days with different concentrations of Fe (0, 5, 10 and 20 µM) and finally transplanted to the sand:soil mixture. Values shown are means ± SE (n = 4 pots). Letters above the columns indicate significant differences at $p \leq 0.05$ for the Fe treatments in a given IBA treatment (in lower case) and for the IBA treatments in a given Fe treatment (in capitals).

4. Discussion

Results confirm that the application of IBA improves rooting in *R. officinalis* cuttings, in line with previous results obtained in this plant species [30–33], as well as in many other woody plants [7–18]. Data shown here indicate that when Fe supplementation is not used, an IBA concentration of 1000 mg L^{-1} appears to be adequate for *R. officinalis*, since higher IBA concentrations (2000–3000 mg L^{-1}) do not provide any supplementary advantage. Each plant species needs an appropriate concentration of IBA to promote cell proliferation and expansion [19,36], and excessive IBA concentrations may impair development [11,15,37]. For instance, *O. europaea* needs a 3500 mg L^{-1} IBA concentration [11], and 2000 mg L^{-1} IBA induced a higher percentage of adventitious rooting in *Eucalyptus benthamii* [15].

Even in the absence of IBA, Fe(III)-EDDHA supplementation improves to some extent rooting (at 10 days), root biomass and number, shoot biomass, new growth and number of new shoots, and leaf photosynthetic pigment levels (at 20 days), as well as cutting survival (10 days after transplant to sand:soil mixture). The low root biomass in cuttings grown with 0 µM Fe may be related to the auxin increases known to occur in Fe-deficient plants, which usually exhibit in roots morphological changes such as inhibition of elongation and swollen root tips [21,22,38]. The reason behind the positive effects of Fe(III)-EDDHA in the absence of IBA are not known at the current stage, although Fe is a co-factor of peroxidase, which is known to mediate the catabolism of auxin in the rooting process [38]. Evidence for a role of mineral nutrients (including Fe) in the basal part of the cutting during rooting has been shown in *Petunia × hybrida* [4] and *Euphorbia pulcherrima* [39], as well as in the woody species *Eucalyptus globulus* [14], *Prunus persica* [26] and *Pinus taeda* [40]. In *Petunia × hybrida* leaf cuttings, it has been shown that adventitious root formation depends on the local provision of Fe, since shoot-to-root translocation of Fe from the aerial part of the cuttings is ineffective [4]. Stimulation of adventitious root development by Fe requires auxin and involves auxin polar transport, but both the fact that spatial distribution and activity of the auxin-reporter GFP-GUS are not affected by Fe supply and the additive effect of Fe and 1-naphthaleneacetic acid suggest that Fe and auxin may have parallel mechanisms of stimulation of adventitious root formation [4].

The Fe-mediated improvements in all parameters studied are generally more marked in the presence than in the absence of IBA. These results confirm the hypothesis that supplementing IBA-treated *R. officinalis* cuttings with 5–20 µM Fe(III)-EDDHA improves rooting (at 10 days), and root biomass and number, shoot biomass and number of new shoots and the leaf levels of photosynthetic pigments (at 20 days), as well as cutting survival (10 days after transplant to sand:soil mixture). Generally speaking, the treatments including 10 µM Fe and 2000 mg L^{-1} IBA appear to give adequate values for most parameters measured.

Application of Fe increased leaf photosynthetic pigment levels, in line with previous studies with other plant species [41,42], including woody ones such as *Pyrus communis* [43] and *P. persica* [44]. The increase with Fe was more marked for Chl b than for Chl a and Car, also in agreement with previous studies [43]. Iron plays roles in chlorophyll [45] and carotenoid biosynthesis [46] and is also part of many components in the chloroplast thylakoid membrane, which can be assembled only when all of them are present [45,47]. For instance, *Calendula officinalis* grown under low Fe showed decreases in Chl and Car concentrations under low Fe in the growth media [48,49]. An increase in photosynthetic pigment levels leads to higher photosynthetic rates, and therefore increases the resources for the formation and development of the root system. This would better facilitate water and nutrient uptake, therefore favouring plant survival [42,50].

In the present study, shoot new growth, number of new shoots, and shoot FW and DW were positively correlated with the root number (R^2 values of 0.70, 0.76, 0.72 and 0.64, respectively, data not shown). This is in line with the finding that in IBA-treated *Hibiscus rosa-sinensis* rootstock, propagated using stenting, there was a positive correlation between shoot and root number [51]. This may be caused by a higher cytokinin generation in cuttings with a higher root number. Cytokinins are mainly synthesized in roots and transported to the shoot in the xylem transpiration stream, and they affect many aspects of plant development, including morphogenesis and shoot initiation [52,53].

5. Conclusions

Results indicate that the application of 5–20 µM Fe(III)-EDDHA and 1000–3000 mg L^{-1} IBA can improve rooting, root and shoot biomass, photosynthetic pigment levels and plant survival in cuttings of the aromatic and medicinal species *R. officinalis*. These results show that the application of Fe(III)-chelate during rooting can lead to the production of vigorous new plants in a shorter time. The application of this type of treatment for the propagation of other rare and valuable aromatic and medicinal plant species via cuttings would deserve further studies.

Supplementary Materials: The following supporting information can be downloaded at: https://www.mdpi.com/article/10.3390/agriculture12020210/s1, Figure S1: Pictures of the cuttings a few days after placing them in pots with the sand:perlite mixture; Table S1: Analysis of variance (ANOVA) of morphological and biochemical traits in *R. officinalis* treated with different concentrations of IBA (0, 1000, 2000 and 3000 mg L^{-1}) at the start of the experiment and then grown with different concentrations of Fe (0, 5, 10 and 20 µM).

Author Contributions: Design of the experiment and methodology, Z.I. and A.R.N.; data analysis, Z.I., A.R.N. and J.A.; writing—original draft preparation, Z.I., A.R.N. and J.A.; writing—review and editing,

Z.I., A.R.N. and J.A.; supervision and project administration, A.R.N. and J.A.; and funding acquisition, A.R.N. and J.A. All authors have read and agreed to the published version of the manuscript.

Funding: The study was funded by Lorestan University, Iran. J.A. was supported by the Spanish Ministry of Science and Innovation (Grant PID2020-115856RB-100 funded by MCIN/AEI/10.13039/501100011033) and the Aragón Government (group A09-20R).

Conflicts of Interest: The authors declare no conflict of interest.

References

1. Hartmann, H.T.; Kester, D.E. *Plant Propagation: Principles and Practices*, 3rd ed.; Prentice Hall: Upper Saddle River, NJ, USA, 1975.
2. Bassuk, N.L.; Hunter, L.D.; Howard, B.H. The apparent involvement of polyphenol oxidase and phloridzin in the production of apple rooting cofactors. *J. Hortic. Sci.* **1981**, *56*, 313–322. [CrossRef]
3. Ling, W.X.; Zhong, Z. Seasonal variation in rooting of the cuttings from tetraploid locust in relation to nutrients and endogenous plant hormones of the shoot. *Turk. J. Agric. For.* **2010**, *36*, 257–266.
4. Hilo, A.; Shahinnia, F.; Druege, U.; Franken, P.; Melzer, M.; Rutten, T.; von Wirén, N.; Hajirezaei, M.-R. A specific role of iron in promoting meristematic cell division during adventitious root formation. *J. Exp. Bot.* **2017**, *68*, 4233–4247. [CrossRef] [PubMed]
5. Wu, H.C.; du Toit, E.S.; Reinhardt, C.F.; Rimando, A.M.; van der Kooy, F.; Meyer, J.J.M. The phenolic, 3,4-dihydroxybenzoic acid, is an endogenous regulator of rooting in *Protea cynaroides*. *Plant Growth Regul.* **2007**, *52*, 207–215. [CrossRef]
6. Epstein, E.; Lavee, S. Conversion of indole-3-butyric acid to indole-3-acetic acid by cuttings of grapevine (*Vitis vinifera*) and olive (*Olea europaea*). *Plant Cell Physiol.* **1984**, *25*, 697–703.
7. Al-Zebari, S.M.K.; Al-Brifkany, A.-A.A.M. Effect of cutting type and IBA on rooting and growth of citron (*Citrus medica* L.). *J. Exp. Agric. Int.* **2015**, *5*, 134–138. [CrossRef]
8. Nasri, F.; Fadakar, A.; Saba, M.K.; Yousefi, B. Study of indole butyric acid (IBA) effects on cutting rooting improving some of wild genotypes of damask roses (*Rosa damascena* Mill.). *J. Agric. Sci.* **2015**, *60*, 263–275. [CrossRef]
9. Izadi, Z.; Zarei, H. Evaluation of propagation of chinese hibiscus (*Hibiscus rosa-sinensis*) through stenting method in response to different IBA concentrations and rootstocks. *Am. J. Plant Sci.* **2014**, *5*, 1836–1841. [CrossRef]
10. Peixe, A.; Raposo, A.; Lourenço, R.; Cardoso, H.; Macedo, E. Coconut water and BAP successfully replaced zeatin in olive (*Olea europaea* L.) micropropagation. *Sci. Hortic.* **2007**, *113*, 1–7. [CrossRef]
11. Hussain, K.; Qadri, R.; Akram, M.T.; Nisar, N.; Iqbal, A.; Yang, Y.; Khan, M.M.; Haq, I.U.; Khan, R.I.; Iqbal, M.A. Clonal propagation of olive (*Olea europaea*) through semi-hardwood cuttings using IBA under shaded polyethylene tunnels (SPTS). *Fresenius Environ. Bull.* **2020**, *29*, 8131–8137.
12. Shao, F.; Wang, S.; Huang, W.; Liu, Z. Effects of IBA on the rooting of branch cuttings of Chinese jujube (*Zizyphus jujuba* Mill.) and changes to nutrients and endogenous hormones. *J. For. Res.* **2017**, *29*, 1557–1567. [CrossRef]
13. Amini, A.; Tabari Kouchaksaraei, M.; Hosseini, S.M.; Yousefzadeh, H. Influence of Hormones Of IAA, IBA, And NAA On Improvement of rooting and early growth gf *Tilia Rubra* Subsp. Caucasica Form Angulata (Rupr.) V. Engler. *Ecopersia* **2019**, *3*, 169–174.
14. Schwambach, J.; Fadanelli, C.; Fett-Neto, A.G. Mineral nutrition and adventitious rooting in microcuttings of *Eucalyptus globulus*. *Tree Physiol.* **2005**, *25*, 487–494. [CrossRef]
15. Brondani, G.E.; Baccarin, F.J.B.; de Wit Ondas, H.W. Low temperature, IBA concentrations and optimal time for adventitious rooting of *Eucalyptus benthamii* mini-cutting. *J. For. Res.* **2012**, *23*, 583–592. [CrossRef]
16. Azad, M.S.; Alam, M.J.; Mollick, A.S.; Khan, M.N.I. Rooting of cuttings of the Indian almond (*Sterculia foetida*) enhanced by the application of the indole-3-butyric acid (IBA) under leafy and non-leafy conditions. *Rhizosphere* **2017**, *5*, 8–15. [CrossRef]
17. Vielba, J.M.; Vidal, N.; José, M.C.S.; Rico, S.; Sánchez, C. Recent advances in adventitious root formation in chestnut. *Plants* **2020**, *9*, 1543. [CrossRef]
18. Bannoud, F.; Bellini, C. Adventitious rooting in *Populus* species: Update and perspectives. *Front. Plant Sci.* **2021**, *12*, 918. [CrossRef]
19. Xiao, Z.; Jin, Z.; Zhang, B.; Li, F.; Yu, F.; Zhang, H.; Lu, X.; Zhang, J. Effects of IBA on rooting ability of *Cinnamomum bodinieri* citral type micro-shoots from transcriptomics analysis. *Plant Biotechnol. Rep.* **2020**, *14*, 467–477. [CrossRef]
20. Otiende, M.A.; Nyabundi, J.O.; Ngamau, K.; Opala, P. Effects of cutting position of rose rootstock cultivars on rooting and its relationship with mineral nutrient content and endogenous carbohydrates. *Sci. Hortic.* **2017**, *225*, 204–212. [CrossRef]
21. Marschner, H. *Marschner's Mineral Nutrition of Higher Plants*, 3rd ed.; Marschner, P., Ed.; Academic Press: Boston, MA, USA, 2012.
22. Landsberg, E.-C. Hormonal regulation of iron-stress response in sunflower roots: A morphological and cytological investigation. *Protoplasma* **1996**, *194*, 69–80. [CrossRef]
23. Zhu, X.F.; Wang, B.; Song, W.F.; Zheng, S.J.; Shen, R.F. Putrescine alleviates iron deficiency via NO-dependent reutilization of root cell-wall Fe in Arabidopsis. *Plant Physiol.* **2016**, *170*, 558–567. [CrossRef]
24. Johnson, C.R.; Hamilton, D.F. Effects of media and controlled-release fertilizers on rooting and leaf nutrient composition of *Juniperus conferta* and *Ligustrum japonicum* cuttings. *J. Am. Soc. Hortic. Sci.* **1977**, *102*, 320–322.
25. Ward, J.D.; Whitcomb, C.E. Nutrition of Japanese holly during propagation and production. *J. Am. Soc. Hortic. Sci.* **1979**, *104*, 523–526.

26. Tsipouridis, C.; Thomidis, T.; Zakinthinos, Z. Iron deficiency and adventitious rooting in peach hardwood cuttings (cv. Early Crest). *Austr. J. Exp. Agric.* **2006**, *46*, 1629–1632. [CrossRef]
27. De Pasquale, C.; La Bella, S.; Cammalleri, I.; Gennaro, M.C.; Licata, M.; Leto, C.; Tuttolomondo, T. Agronomical and postharvest evaluation of the essential oils of Sicilian rosemary (*Rosmarinus officinalis* L.) biotypes. *Acta Hortic.* **2019**, *1255*, 139–144. [CrossRef]
28. Tuttolomondo, T.; Dugo, G.; Ruberto, G.; Leto, C.; Napoli, E.M.; Cicero, N.; Gervasi, T.; Virga, G.; Leone, R.; Licata, M.; et al. Study of quantitative and qualitative variations in essential oils of Sicilian *Rosmarinus officinalis* L. *Nat. Prod. Res.* **2015**, *29*, 1928–1934. [CrossRef]
29. La Bella, S.; Virga, G.; Iacuzzi, N.; Licata, M.; Sabatino, L.; Consentino, B.B.; Leto, C.; Tuttolomondo, T. Effects of irrigation, peat-alternative substrate and plant habitus on the morphological and production characteristics of Sicilian rosemary (*Rosmarinus officinalis* L.) biotypes grown in pot. *Agriculture* **2021**, *11*, 13. [CrossRef]
30. Gil, C.S.; Kwon, S.J.; Jeong, H.Y.; Lee, C.; Lee, O.J.; Eom, S.H. Blue light upregulates auxin signaling and stimulates root formation in irregular rooting of rosemary cuttings. *Agronomy* **2021**, *11*, 1725. [CrossRef]
31. Kiuru, P.; Muriuki, S.; Wepukhulu, S.; Muriuki, S. Influence of growth media and regulators on vegetative propagation of rosemary (*Rosmarinus officinalis* L.). *East Afr. Agric. For. J.* **2015**, *81*, 105–111. [CrossRef]
32. Koleva Gudeva, L.; Trajkova, F.; Mihajlov, L.; Troiciki, J. Influence of different auxins on rooting of rosemary, sage and elderberry. *Ann. Res. Rev. Biol.* **2017**, *12*, 1–8. [CrossRef]
33. Poornima, K.; Chandregowda, M.; Pushpa, T.; Srikantaprasad, D. Studies on effect of growth regulators on rooting of two rosemary types and estimation of biochemical changes associated with rooting. *Crop Res.* **2012**, *43*, 245–248.
34. Hartmann, H.T.; Kester, D.E.; Davies, F.T., Jr.; Geneve, R.L. *Plant Propagation: Principles and Practice*, 7th ed.; Prentice Hall: Englewood Cliffs, NJ, USA, 2002.
35. Lichtenthaler, H.K. Chlorophylls and carotenoids: Pigments of photosynthetic biomembranes. *Meth. Enzymol.* **1987**, *148*, 350–382.
36. Björkman, T. Effect of *Trichoderma* colonization on auxin-mediated regulation of root elongation. *Plant Growth Regul.* **2004**, *43*, 89–92. [CrossRef]
37. Sun, W.-Q.; Bassuk, N.L. Effects of banding and IBA on rooting and budbreak in cuttings of apple rootstock 'MM. 106' and Franklinia. *J. Environ. Hortic.* **1991**, *9*, 40–43. [CrossRef]
38. Trejgell, A.; Libront, I.; Tretyn, A. The effect of Fe-EDDHA on shoot multiplication and in vitro rooting of *Carlina onopordifolia* Besser. *Acta Physiol. Plant.* **2012**, *34*, 2051–2055. [CrossRef]
39. Svenson, S.E.; Davies, F.T. Change in tissue mineral elemental concentration during root initiation and development of poinsettia cuttings. *HortScience* **1995**, *30*, 617–619. [CrossRef]
40. Rowe, D.B.; Blazich, F.A.; Weir, R.J. Mineral nutrient and carbohydrate status of loblolly pine during mist propagation as influenced by stock plant nitrogen fertility. *HortScience* **1999**, *34*, 1279–1285. [CrossRef]
41. Abadía, J.; Vázquez, S.; Rellán-Álvarez, R.; El-Jendoubi, H.; Abadía, A.; Álvarez-Fernández, A.; López-Millán, A.F. Towards a knowledge-based correction of iron chlorosis. *Plant Physiol. Biochem.* **2011**, *49*, 471–482. [CrossRef]
42. Briat, J.F.; Dubos, C.; Gaymard, F. Iron nutrition, biomass production, and plant product quality. *Trend. Plant Sci.* **2015**, *20*, 33–40. [CrossRef]
43. Morales, F.; Abadía, A.; Belkhodja, R.; Abadía, J. Iron-deficiency-induced changes on the photosynthetic pigment composition of field-grown pear (*Pyrus communis* L.) leaves. *Plant Cell Environ.* **1994**, *17*, 1153–1160. [CrossRef]
44. Abadía, J.; Tagliavini, M.; Grasa, R.; Belkhodja, R.; Abadía, A.; Sanz, M.; Faria, E.A.; Tsipouridis, C.; Marangoni, B. Using the flower Fe concentration for estimating chlorosis status in fruit tree orchards: A summary report. *J. Plant Nutr.* **2000**, *23*, 2024–2033. [CrossRef]
45. Terry, N.; Abadía, J. Biochemistry and physiology of iron. *J. Plant Nutr.* **1986**, *9*, 609–646. [CrossRef]
46. Kim, S.H.; Ahn, Y.O.; Ahn, M.-J.; Lee, H.-S.; Kwak, S.-S. Down-regulation of β-carotene hydroxylase increases β-carotene and total carotenoids enhancing salt stress tolerance in transgenic cultured cells of sweetpotato. *Phytochemistry* **2012**, *74*, 69–78. [CrossRef]
47. Vigani, G.; Zocchi, G.; Bashir, K.; Philippar, K.; Briat, J.-F. Signals from chloroplasts and mitochondria for iron homeostasis regulation. *Trends Plant Sci.* **2013**, *18*, 305–311. [CrossRef]
48. Izadi, Z.; Rezaei Nejad, A.; Abadía, J. Physio-morphological and biochemical responses of pot marigold (*Calendula officinalis* L.) to split iron nutrition. *Acta Physiol. Plant.* **2020**, *42*, 6. [CrossRef]
49. Izadi, Z.; Rezaei Nejad, A.; Abadía, J. Foliar applications of thidiazuron and putrescine increase leaf iron and chlorophyll concentrations in iron-deficient pot marigold (*Calendula officinalis* L.). *Acta Physiol. Plant.* **2021**, *43*, 122. [CrossRef]
50. El-Jendoubi, H.; Melgar, J.C.; Álvarez-Fernández, A.; Sanz, M.; Abadía, A.; Abadía, J. Setting good practices to assess the efficiency of iron fertilizers. *Plant Physiol. Biochem.* **2011**, *49*, 483–488. [CrossRef]
51. Izadi, Z.; Zarei, H.; Alizadeh, M. Effect of time, cultivar and rootstock on success of rose propagation through stenting technique. *Am. J. Plant Sci.* **2014**, *5*, 1644–1650. [CrossRef]
52. Müller, D.; Leyser, O. Auxin, cytokinin and the control of shoot branching. *Ann. Bot.* **2011**, *107*, 1203–1212. [CrossRef]
53. Skalák, J.; Vercruyssen, L.; Claeys, H.; Hradilová, J.; Černý, M.; Novák, O.; Plačková, L.; Saiz-Fernández, I.; Skaláková, P.; Coppens, F. Multifaceted activity of cytokinin in leaf development shapes its size and structure in Arabidopsis. *Plant J.* **2019**, *97*, 805–824. [CrossRef]

MDPI
St. Alban-Anlage 66
4052 Basel
Switzerland
Tel. +41 61 683 77 34
Fax +41 61 302 89 18
www.mdpi.com

Agriculture Editorial Office
E-mail: agriculture@mdpi.com
www.mdpi.com/journal/agriculture

www.ingramcontent.com/pod-product-compliance
Lightning Source LLC
LaVergne TN
LVHW070408100526
838202LV00014B/1410